Prehistoric Copper Mining
in Michigan

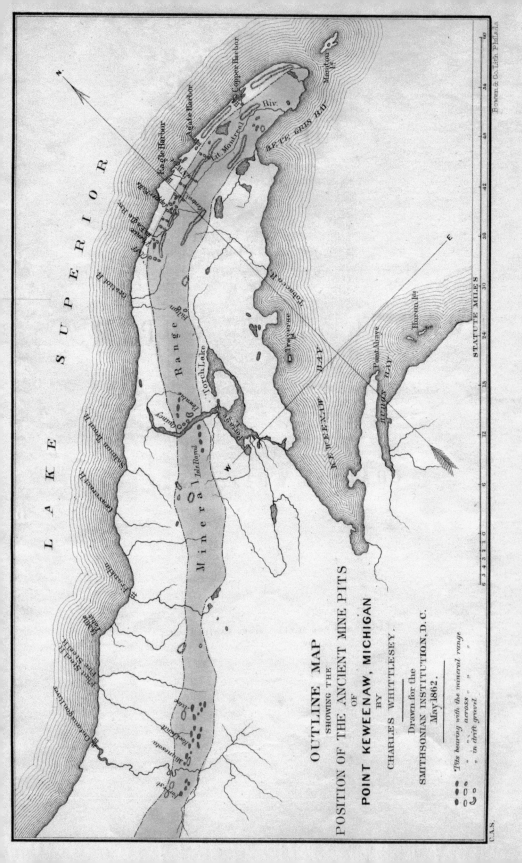

OUTLINE MAP
SHOWING THE
POSITION OF THE ANCIENT MINE PITS
OF
POINT KEEWENAW, MICHIGAN
BY
CHARLES WHITTLESEY.

Drawn for the
SMITHSONIAN INSTITUTION, D.C.
May 1862.

Pits bearing with the mineral range
" " across "
" " in drift gravel

L A K E   S U P E R I O R

Bowen & Co. lith. Philada.

STATUTE MILES

Anthropological Papers
Museum of Anthropology, University of Michigan
Number 99

Prehistoric Copper Mining
in Michigan

The Nineteenth-Century Discovery of "Ancient Diggings" in the Keweenaw Peninsula and Isle Royale

by

John R. Halsey

Ann Arbor, Michigan
2018

Printed in the United States of America
ISBN 978-0-915703-89-0

Cover design by John Klausmeyer and Kay Clahassey

The Museum currently publishes two monograph series: Anthropological Papers and Memoirs. Contact the Museum by email at umma-pubs@umich.edu; by phone at 734-764-0485; or through our website at www.lsa.umich.edu/ummaa/publications.

Library of Congress Cataloging-in-Publication Data

Names: Halsey, John R., author.
Title: Prehistoric copper mining in Michigan : the nineteenth-century discovery of ancient diggings in the Keweenaw Peninsula and Isle Royale / by John R. Halsey.
Description: Ann Arbor, Michigan : University of Michigan, 2018. | Series: Anthropological papers museum of anthropology, University of Michigan ; number 99. | Includes bibliographical references. |
Identifiers: LCCN 2018009873 (print) | LCCN 2018015765 (ebook) | ISBN 9780915703906 (e-book) | ISBN 9780915703890 (pbk. : alk. paper)
Subjects: LCSH: Indians of North America--Michigan--Upper Peninsula--Antiquities. | Copper mines and mining, Prehistoric--Michigan--Upper Peninsula. | Archaeology--Michigan--Upper Peninsula--19th century.
Classification: LCC E78.M6 (ebook) | LCC E78.M6 H347 2018 (print) | DDC 622/.34309774995--dc23
LC record available at https://lccn.loc.gov/2018009873

The paper used in this publication meets the requirements of the ANSI Standard Z39.48-1984 (Permanence of Paper).

Frontispiece: Hand-tinted map showing prehistoric copper mining pits in Michigan's Keweenaw Peninsula. (Reprinted from Whittlesey 1863: frontispiece.)

Dedication

To all those people who have documented the incredible efforts
of the first miners of the Keweenaw Peninsula and Isle Royale

Table of Contents

List of Illustrations

Foreword

The deposits of native copper in the Lake Superior region are among the largest and richest on earth. The casual observer driving on the Canadian side of Lake Superior can view the twisted veins of copper in the road cuts. In other instances, massive ingots of pure copper can be found sitting in glacial isolation on the lake bottom. Euro-American prospectors took note of these incredibly pure deposits, and Lake Superior copper was soon filling the holds of vessels bound for industries on the lower lakes.

Yet as these deposits began to be mapped and exploited, it became clear that the American and Canadian miners were not the first to discover or work the rich deposits. While Euro-Americans observed no active Native American mines during the contact period, they found numerous open pits, many filled with rubble, crude mining implements of stone and wood, and intentionally fashioned copper tools. These discoveries broadly co-occurred with the discovery of elaborately worked copper tools and implements associated with the burial mounds and massive earthworks in the Midwest. Metalworking was well known in prehistoric Europe, but the natives of North America were thought incapable of such technology. This, along with the monumental constructions themselves, led to a belief in a prior Mound Builder culture that predated the native inhabitants of the region. Modern research has clarified the situation. The elaborate earthwork structures of the American Midwest and the copper tools were indeed the product of not one but multiple Native American cultures over time, dating back to the Late Archaic times and continuing into the contact period. Yet even today, one can find modern claims and videos attributing the ancient mining along Lake Superior to the work of Vikings or Phoenicians or even ancient astronauts.

The native copper in the Midwest was not smelted and cast but instead was worked via the repeated process of cold hammering and annealing to produce the final ornament or implement shape. The mines themselves have much in common with the earliest copper mining in Europe, and many of the same techniques and tools were used for the job. In both regions, too, we see the preservation of wooden tools and other organic materials due to their contact with copper salts.

The copper used by Native Americans derived from sources in the Lake Superior Basin, as well in glacial drift deposits further south. And like other exotic materials such as obsidian and mica, copper was traded over vast distances. The fact that there are multiple potential sources of copper raises issues when archaeologists attempt to assess the distances over which copper was being moved and traded. It is similarly difficult to assess how much copper may have been mined or collected during prehistoric times.

It is to address these questions that John Halsey has returned to the earliest historic records and accounts of copper deposits and mining in Michigan's Upper Peninsula and Isle Royale. What did these early prospectors and miners see? What do their observations tell us about the sources and methods of copper mining that had been practiced? What was the scale of extraction? The early historic observations are particularly crucial because many of the deposits worked by Native Americans were subsequently exploited by industrial mining interests, which completely obliterated the earlier mining sites. As such, we must also ask if the traces of mining that do survive to the present are at all representative of the practices used in prehistory.

John Halsey has devoted a major portion of his career to understanding the native copper industries of the Midwest in all their varied forms and cultural contexts, from copper sourcing to the use and distribution of the finished copper artifacts. To do this he has called upon his skills as an archaeologist and as a historian in equal parts. This volume represents an exhaustive compilation of the early written and published accounts of mines and mining in Michigan's Upper Peninsula. It will prove a valuable resource to current and future scholars. Through these early historic accounts of prospectors and miners, Halsey provides a vivid picture of what once could be seen.

It is an irony of archaeological research that as our methods improve and our questions become more nuanced, the material record of the past becomes sparser and dimmer, as sites are progressively and irrevocably destroyed by modern development. It is for this reason that early historic accounts of Native American mounds, earthworks, mines, and trails are so critical. They provide a view of a material record that simply no longer exists.

JOHN M. O'SHEA
CURATOR OF GREAT LAKES ARCHAEOLOGY
UNIVERSITY OF MICHIGAN MUSEUM OF ANTHROPOLOGICAL ARCHAEOLOGY

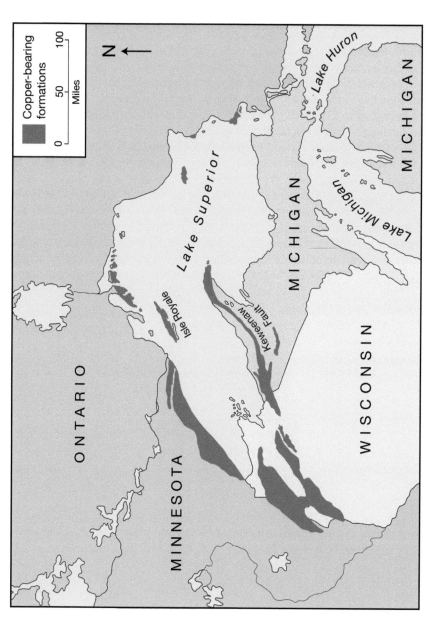

Figure 0.1. Copper-bearing formations near Lake Superior. (Redrawn by John Klausmeyer from Map 3 in Griffin 1961d:34.)

Preface

In this book I summarize the results of my effort to locate early references related to the discovery of Michigan's prehistoric copper mines. In the early 1980s, as Michigan's state archaeologist, I began to write articles on Native American use of Michigan's native copper. Over the years, realizing the overwhelming importance of nineteenth-century industrial and historical activities in the Upper Peninsula and Isle Royale, I narrowed my focus to the exploitation of the bedrock sources of Michigan copper, the largest and richest copper deposits in the world. That research culminated in this book.

My sources of information were pioneer newspapers, mining-company annual reports, and company prospectuses from Michigan's Copper Country, such as were accessible from the holdings of the Library of Michigan of Lansing, Michigan, and willing interlibrary loan lenders across the country.

This book was inspired by and should be considered an update and expansion of James B. Griffin's treatment of the topic in *Lake Superior Copper and the Indians: Miscellaneous Studies of Great Lakes Prehistory* (1961) and my own earlier efforts on the same topic (Halsey 2008, 2009).

No study of this nature is truly comprehensive because of the lack of complete records for most of the era's newspapers (Jamison 1946) and for the numerous mining-company annual reports and mine prospectuses. All copies of some newspapers, such as the *Upper Peninsular Advocate* (cited in Tenney 1857f:475), appear to have disappeared entirely.

In addition, little survives of the written correspondence between the managers at the Michigan mine sites and the home offices "back east." There must have been an astounding volume of such communication, but how much survives today and where it is located is unknown. Many of these letters undoubtedly were lost in the decades of sales and acquisitions among mining companies.

In short, the written record of day-to-day mine functions and ancient mining discoveries is patchy at best. Yet what is left to us forms the basis of our knowledge of what was going on as the prehistoric mining pits were stripped of their protective vegetation and fill in the 1840s and 1850s.

With the exception of Isle Royale, I have not included twentieth-century survey and excavation activities, as these represent a totally different milieu (Bastian 1963a, 1963b; Clark 1995; Drier 1961; Ferguson 1923, 1924; Fox 1911, 1929; Holmes 1901; West 1929). My interest is in how nineteenth-century men understood the evidence of prehistoric work. (I make an exception of Isle Royale because of the fact that many of the pits there survived into the twentieth century.)

Many people assisted in the creation of this document. I had invaluable help from the staff of the Library of Michigan, especially those with responsibility for interlibrary loans and the Martha W. Griffiths Michigan Rare Book Room.

Former colleagues at the Michigan Historical Center were also essential for their encouragement and support over many years, notably LeRoy Barnett, Sandra Clark, and Dr. Henry Wright.

I must single out the following American and Canadian authors (many of whom have passed on), because they made much of this research possible: Oliver Anttila, Charles E. Brown, James A. Brown, Claude Chapdelaine, Jacques Cinq-Mars, Norman Clermont, K. C. A. Dawson, Kathleen Ehrhardt, Kenneth B. Farnsworth, James B. Griffin, Mark A. Hill, James T. Hodge, Lewis H. Larson, Jr., Gregory D. Lattanzi, Jonathan M. Leader, Mary Ann Levine, Susan R. Martin, Ronald J. Mason and Carol I. Mason, W. C. McKern, William C. Mills, Clarence B. Moore, Warren K. Moorehead, E. J. Neiburger, Rowland B. Orr, John T. Penman, Gregory H. Perino, Thomas C. Pleger, David P. Pompeani, Mary H. Pulford, F. W. Putnam, George I. Quimby, Robert Ritzenthaler, Michael W. Spence, Jack Steinbring, and Warren L. Wittry.

This study has a unique focus, but it was not written in a vacuum. Pertinent books the reader may find of interest are: *This Ontonagon Country: The Story of an American Frontier* by James K. Jamison; *The Making of a Mining District: Keweenaw Native Copper, 1500–1870* by David J. Krause; *Beyond the Boundaries: Life and Landscape at the Lake Superior Copper Mines, 1840–1875* by Larry Lankton; and *Wonderful Power: The Story of Ancient Copper Working in the Lake Superior Basin* by Susan R. Martin. Full references will be found in the bibliography.

JOHN R. HALSEY
EAST LANSING, MICHIGAN

Introduction

A brief notice of this man who, no doubt, was respectable in his day and generation, may not be improper in this place. This man who wrought in the copper mines probably belonged to the Stone Age of the race. If born at a later period he was—although very sagacious in the matter of finding copper bearing lodes—a rude miner using rounded stone or boulders for hammers, and mining by means of fire and water, calcining the vein rock with a fire made of wood, and rending the vein by throwing water upon the heated rock. He managed to mine from 10 to 20 feet in depth, or down to the water line. Further he did not go on account, probably, of the difficulty he had in keeping his pits and trenches clear of water. He had no means of cutting up or removing any considerable masses of copper. There have been found in ancient pits masses of copper from one to three tons in weight, which show evidences of his work. The stone hammer marks can be distinctly seen; the angular pieces of the mass have been removed; in some cases the mass has been partly braised, and the blocking which shored it up remains; broken hammers, rude copper tools and plenty of charcoal are found in the pits. What a grief to the enterprising, yet unskillful miners to be obliged to abandon so fair a treasure! These ancient pits are found by our explorers by observing on the surface of the ground circular, or trench-like depressions, in line for a distance of several hundred feet. They are filled with sand, debris of rocks and vegetable mould. Large trees have been found growing in these pits. These works are common along the Trap Range.

John H. Forster
1879

Introduction

Beginning in the late 1840s, Euro-American copper miners in the western Lake Superior Basin became aware that someone had been there before them. With this realization came curiosity about who had done the mining, how long ago, and how much copper had been removed.

Eventually, evidence of prehistoric copper mining would be found over a swath 150 miles long, varying in width from four to seven miles in Michigan's Upper Peninsula counties of Keweenaw, Houghton, and Ontonagon and also on Isle Royale in Lake Superior (Houghton 1876a:79). To this day, Keweenaw County and most of Houghton, Baraga, and Ontonagon counties together make up what's commonly called "Copper Country."

As the historic miners cleared out the prehistoric pits preparatory to establishing their own operations, a significant number of them recorded their observations and ideas.

Many mine operators were ruthless in their exploratory zeal. Edwin J. Hulbert, a member of that explorer/miner fraternity, noted this trait in his colleagues with undisguised displeasure:

> It is most unfortunate for the now later students of the 'Houghton School of Mining,' that these traces have not been recorded; the reason was, that the Capitalists as a general thing, were of that class of monied-men who sought only to increase their incomes, and gave no place to a thought for those who had preceded them in the work of mining within the rocks of Lake Superior; the explorers *often and continually* related to their employers, these strange evidences of the labor of an *ancient-race*, but the information was passed bye, as not being worthy of consideration. (1893:119)

The italicized words are Hulbert's. In 1893 Hulbert prepared a list of 72 "Lake-Superior Explorers from 1819 to 1874" (Hulbert 1893:13–14; and see Appendix 2). He stated that he had prepared this list "not from *hearsay*, but from memories of a personal acquaintance with each and everyone named" and "If I shall have omitted worthy names which should have place here, such omission has not been with intention to occult them, but from temporary forgetfulness" (1893:12–13). By cross-referencing his names with my own list of those who played a part in the discovery of prehistoric mines (see Appendix 1), we see that only 22 names are on both lists. That is, only about one in four miners and administrators are known to have expressed any interest in the discovery of prehistoric mining activities. The others had to have known about the discoveries, but apparently had little or no interest in them.

This suggests to me that there is truth to Hulbert's statement quoted above—that many of the early explorers, and certainly the majority of those in charge, did not care much about the value or interest of the artifacts or mining techniques revealed in the ancient pits.

However, despite Hulbert's negative statements, it is clear that some high-ranking officials thought the evidence of the ancient mines and their contents was of exceptional importance for economic reasons, if not for their archaeological significance. Perhaps

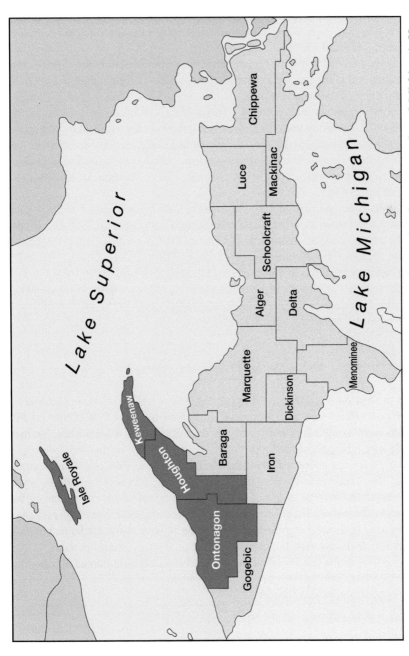

Figure 0.2. Prehistoric copper mines—and thus, historic mines—were clustered at the western edge of Michigan's Upper Peninsula, primarily on Isle Royale and in the counties of Ontonagon, Houghton, and Keweenaw. (Map by Kay Clahassey and John Klausmeyer.)

the most dramatic example of this is the prospectus of the Minong Mining Company (1875), in which eleven pages are devoted to the history and firsthand descriptions of "ancient diggings." The most valuable of these references for the mainland mines date between 1848 and 1865.

We certainly see the result of this attitude in the diminished number of prehistoric sites. Visitors in the late nineteenth and early twentieth centuries commented on the swift destruction of the ancient mines.

Pioneering American archaeologist William Henry Holmes made a visit to Isle Royale in 1892. "Desiring to examine for myself the existing traces of a great native industry, I resolved to undertake a trip to Isle Royale, since there modern mining had not so completely destroyed traces of the ancient work as on the southern shore of Lake Superior, where extensive mining operations have been carried on for many years" (1901:685–686).

George A. West, one of the founding members of the Wisconsin Archeological Society and the author of articles and monographs on copper mining and copper artifacts, lamented in 1929, "As early as 1875, when the writer conducted investigations along this copper belt [i.e. Keweenaw, Houghton, and Ontonagon counties], practically every pit had disappeared or become unrecognizable" (1929:43).

At least one new discovery of a prehistoric mine (Baum 1903) was made at the dawn of the twentieth century. The number of intact prehistoric mining pits (and there are some still surviving today) is unknown.

Who Were The Discoverers?

The first professional scientists in Copper Country were botanists, mineralogists, and geologists—the "men with hammers" described by Mentor Williams (1946). These men accompanied the various early surveys of state and federal lands, but they apparently discovered no ancient pits.

Ultimately, the most dependable information would come from the mine "agents," "captains," and "superintendents"—the highest-ranking company men on the scene and those most likely to be aware of unusual or important discoveries made by the men down in the mines. They were educated (by contemporary standards) and had a broad base of real-world experience. Some seem to have had remarkable physical properties (or incredible luck):

> The North West company have recommenced operations with commendable spirit. Their new acting Agent, Mr. Alexander, who has traversed the route from Green Bay, on snow-shoes, and was for seven days decriped by rheumatism, obliged to lie in the woods, four of which without company and in a helpless state... (Roberts 1849a).

Over the span of their careers, most of these individuals worked for several mining companies, sometimes offering assessments of the economic potential of new mining properties while employed by another company. Newspapermen such as George D. Emerson and William J. Tenney offered immediate commentary on discoveries being made in their localities.

Although most of these early mining pioneers have been forgotten, the contributions of some live on today.

Charles Whittlesey—a soldier, attorney, newspaperman, geologist, historian, and archaeologist—is one whose descriptions and maps of the region prove critical to our understanding, even two centuries later (Baldwin 1887, Tribble 2008).

Probably the most controversial and colorful of the early mining figures was Dr. Charles T. Jackson. Trained in both geology and medicine, he was appointed United States Geologist for the Lake Superior land district. His leadership at this post was such a disaster that he was dismissed. His assistants, John Wells Foster and Josiah Dwight Whitney, completed the survey. Jackson had severe mental problems. He had a propensity for claiming discovery of guncotton (nitrocellulose—an explosive), the telegraph, the anesthetic effects of ether and the digestive action of the stomach, among other items. He spent the last seven years of his life in a mental asylum and died in 1880.

Foster and Whitney fared much better. They finished the Jackson survey, and their final document of the project, *Report on the Geology and Topography of a Portion of the Lake Superior Land District in the State of Michigan* (1850), is still cited today.

Samuel W. Hill—a surveyor, geologist, mine administrator, engineer, and politician—accomplished much in his day, but his greatest claim to fame was his legendary cursing ability. It's possible that his language inspired the now-passé expletive, "What the Sam Hill!"—his name being used as a euphemism for the string of curses that he typically used in his stories (Barnett 2004; Everts & Co. 1877).

The principal problem facing these pioneers was that no miners anywhere had ever had to deal with native copper deposits of the size, richness, and irregularity of those in the western Lake Superior Basin. The miners and the mining equipment of the time could not readily deal with the enormous size of the copper masses, some of which were hundreds of tons. These large masses required the removal of much rock just to get at them. The copper masses then had to be cut into pieces small enough to be removed. Copper is malleable and not readily broken into smaller pieces.

There would also be extended periods when little or no copper was found. The inadequacies of the pioneers and the challenges they faced were concisely summed up by one of them, John H. Forster.

> As mining was to be his chief business, a knowledge of the rocks and vein phenomena had to be acquired. As a rule, this man was not familiar with the science of geology and mineralogy, or with the arts of mining and exploring; of these he scarcely understood the first principles. But this tyro, who was in time to become the skillful explorer and successful manager of mine affairs, was a bold man and full of expedients. Everything was to be learned by practical experiment, by diligent application, and at the expense of toil, hardship and sufferings untold. (1879:135) (See also Forster 1891.)

Figure 0.3. Charles Whittlesey (1808–1886) was one of the most important figures in the discovery of prehistoric copper mines in Michigan. Much of what we know of the ancient mines is due to his maps and reports. (Reprinted with permission of the Wisconsin Historical Society: WHS-36627.)

Why Did The Discoverers Care?

Today the prehistoric mining sites are precious to us for their archaeological content. But the nineteenth-century miners' observations were made in a different context: they supported requests for money. This is exemplified in a series of letters written in 1850 by Charles Whittlesey, then superintendent of the Piscataqua Mining Co. (20ON272) in Ontonagon County, to his superiors in Philadelphia. One wonders how long it took for a letter to travel from the wilds of Ontonagon to 74 Walnut St., Philadelphia, PA, home office of the Piscataqua Mining Co. They were almost certainly sent by ship.

> August 15: "We have undoubtedly one of the best locations in the country; we have cleaned up some Indian diggings, and they show copper in abundance. I trust soon to see the Piscataqua[,] one of the companies with the Cliff and Minesota."

> August 16: "Enclosed I send you [a] list of supplies needed for thirteen hands eight months. Our Indian diggings are opening beautifully, and justify a large outlay—they show abundance of copper and regular walls."

> August 21: "The vein which we are at present opening presents copper almost continually, and appears to be large and well defined, with much Indian work upon it."

> September 6: "Our hands are clearing up and stripping the vein excavating on the Indian diggings. Wherever we have yet exposed the vein, it looks well, is from two to four feet wide, and produces excellent stamp work. The Indian diggings appear to have been very extensive, though we have as yet but partially developed them. There is no doubt that a vast amount of labor has been expended on these mines by the ancient miners upon this tract; the same vein appears to extend the whole length of our location, and appears of a regularity and character that I have not met with elsewhere." (Piscataqua Mining Company 1850:17–18)

Also in regard to money, Dr. L. W. Clarke noted in a September 18, 1850, letter to A. W. Marks, in regard to the Aztec Mine (20ON271), "Besides being a correct guide to follow the veins by, they [the ancient pits] will save a large amount of money in sinking shafts" (Aztec Mining Company 1851:11).

Five years later, Thomas Lord was no less optimistic about the utility of ancient pits present at the Mass Mine, avowing:

> The gangue of the vein is composed of those minerals usually found in the adjacent mines in that Minnesota district, and its course and regularity is made apparent by a line of ancient Indian pits, which can be traced all through to the eastern boundary line.
>
> Irrespective of the above described champion lode, the existence of other veins, which traverse the entire property, is also well known to those conversant with the tract. The same infallible ancient diggings which led to the discovery and mark the course of the first vein, are to be seen at other points, and these indications will direct the miner to the richest veins, rendering the expense of exploring unnecessary, and thereby saving any outlay on this head. (Lord 1856:2)

1   Prehistoric Copper Mines

In working the surface of the vein, or of the copper-bearing bed, the ancient operators must have wrought open to the day. They no doubt commenced as low down the slope of the range as the copper appeared to them worth being taken out, and worked upwards … in order to keep their drainage. From their rude and tedious method it was of the highest consequence to cause the water to flow away behind them, without the necessity of baling [sic].

The "attle," or broken rock, was generally thrown back into the vacant space whence it had been taken: but little of it was cast out to right and left along the margin of the vein, which explains why the pits are so shallow at the present time.

Charles Whittlesey
1863

Chapter 1

To understand the nature of prehistoric mining activities, it is necessary to understand something about the nature of native copper deposits. First, native copper does occur in areas other than the western Lake Superior Basin. It is also found in the eastern Lake Superior Basin. Accessible deposits also were known in Oklahoma, the Appalachians, New Jersey, and even Nova Scotia. These are curiously devoid of any reported signs of mining pits, with the exception of one notice by Peter (Pehr) Kalm (1772:300–302).

Lake Superior native copper deposits occur principally in the form of lodes or fissures. According to the findings of archaeologist Tyler Bastian (1963b:17), lode deposits occur in conglomerate and amygdaloid bedrock and are usually a few feet thick. Fissure deposits of copper lie along ancient fractures in the rock and range in thickness from a fraction of an inch to several inches or even a foot. The copper and associated minerals, such as calcite, tend to concentrate at the intersections of fissures. These associated minerals are generally less resistant to erosion than the surrounding rock and could leave distinctive trenches. There was also more finely divided and disseminated copper, but it was of no use to aboriginal people, as it would have required the ability to smelt the ore, something they did not have.

Bastian's work on Isle Royale led him to define two general forms of mine depending on the kind of native copper deposit in which they were dug (1963b:20). He describes fissure mines as relatively deep conico-ovoid shafts sited on fissure deposits. Lode mines were relatively shallow, irregular, quarrylike excavations. Aboriginal mining always remained at this "pit mine" level, following lodes and veins only a limited distance into the ground and never approaching the deep, extensive galleries that would characterize historic American mining efforts. Working a specific vein often took the ancient miners to depths of 20 feet and more. Some ancient mines followed veins horizontally into vertical rock faces.

Charles Whittlesey, who saw many pits before they were cleaned out during historic times, left us this brief description of ancient mining at the Central Mine (20KE46) in Keweenaw County. He believed this description applied to "almost all other" pits.

> In working the surface of the vein, or of the copper-bearing bed, the ancient operators must have wrought open to the day. They no doubt commenced as low down the slope of the range as the copper appeared to them worth being taken out, and worked upwards towards…, in order to keep their drainage. From their rude and tedious method it was of the highest consequence to cause the water to flow away behind them, without the necessity of baling [sic].
>
> The "attle," or broken rock, was generally thrown back into the vacant space whence it had been taken: but little of it was cast out to right and left along the margin of the vein, which explains why the pits are so shallow at the present time. (1863:11)

The mining of enormous lodes of mass copper (those containing more than 100 tons) was a difficult task that taxed the expertise of nineteenth-century mining engineers (Blake 1875) and was clearly beyond the technological capabilities of the prehistoric miners (Steve 1953; Townsend 1864a, 1864b). It was also extremely dangerous,

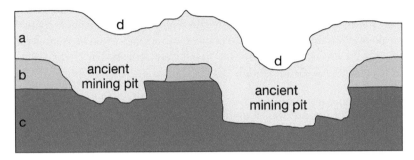

Figure 1.1. This early sketch shows the state of prehistoric mines when they were discovered in the nineteenth century: many had been filled in over the centuries until there was only a shallow depression on the surface. The original caption reads: "a, chips and moved earth in which the stone hammers are found; b, the modified drift-clay; c, the cupiferous rock; d, depressions in the surface indicating the location of the ancient pits, sometimes refilled by the old miners." (Redrawn by John Klausmeyer from Winchell 1881a:Fig. 7.)

and fatalities were relatively common (e.g. *Detroit Free Press* 1855b, 1855c). Large masses had to be cut with chisels into pieces small enough to be removed from a mine, a process that generated tons of marketable copper chips (Wilson 1856:226–227).

However, smaller exposures and masses were well within the capabilities of the ancient miners. Clearly, considerable ingenuity was expended in mining. It seems that the thousands of pits on Isle Royale and the Keweenaw Peninsula were not attempts at obtaining the maximum amount of copper, but rather at securing pieces of copper that were actually amenable to the metalworking techniques available at the time: cold-hammering and annealing (see also Barrett 1926:23). There were pure multiton copper boulders lying on the surface virtually everywhere at the western end of Lake Superior, but they were just too big or smooth to be effectively exploited with stone hammers and flint knives.

Confirmation of the ancient miners' inability to deal with large masses of copper was found at Houghton County's South Pewabic Mine (20HO302). (This mine was later consolidated with the Adams Mine to create the Atlantic Mine.) H. McKenzie, editor of the *Portage Lake Mining Gazette*, observed that:

> At the junction of the transverse and South Pewabic veins, an ancient pit had been opened to the depth of 22 feet. The ancients evidently stopped sinking at that depth on account of their not being able to either take out or avoid a large mass [of copper] lying across the bottom of the pit. (McKenzie 1865a)

Large masses of copper were of little utility to the prehistoric miners. Small thin pieces of copper were the most desirable. Even very small pieces could be flattened and combined (through folding) with other pieces to make a larger piece. When confronted by a large mass of copper with no easily removed projections, their response was to pound along a line and raise up a ridge or comb, which then could be worked off. This process and the distinctive "fingerprints" it left were described by Henry Gillman, an archaeologist, ethnologist, librarian, and U. S. government surveyor.

> Late last autumn, there was brought to our city [Detroit] a mass of copper which, in the latter part of the summer, was taken from an ancient pit on the property of the Minong Mining Company, at McCargo's Cove, Isle Royale. On cleaning out the pit of the accumulated debris, this mass was found at the bottom, at the depth of sixteen and one-half feet. It is of a crescent-like shape, and weighs nearly three tons, or exactly 5,720 pounds. Such a huge mass it was evidently beyond the ability of those ancient men to remove. They could only deal with it as they best they knew how [sic]. And as to their mode of procedure, the surroundings in the pit and the corrugated surface of the mass itself bear ample testimony. The large quantities of ash and charcoal lying around it, show that the action of fire had been brought to bear on it. A great number of the stone hammers or mauls were also found near by, many of them fractured from use. With these the surface of the mass had evidently been beaten up into projecting ridges, and then broken off. The entire upper face and sides of the relic present repeated instances of this; the depressions, several inches deep, and the intervening elevations with their fractured summits, covering every foot of the exposed superficies. (1876:330–331)

The prehistoric miners were probably most likely to seek sheetlike vein deposits—those that varied from one-quarter of an inch to an inch in thickness. They valued this copper, it seems, because it was thin and therefore readily malleable—and therefore worth the long hours of labor involved in removing cubic yards of trap rock.

U. S. geologist Charles T. Jackson, another important observer on the scene when prehistoric mining discoveries were first being made, came to a similar conclusion. On the property of the North West Company (20KE34) near Eagle Harbor, Keweenaw County, he noted "…there was a depression of six feet in the metalliferous lode, where the Indians had mined out the sheets of copper" (1850:284).

Fifty-seven years later, miner Alvinus B. Wood echoed Jackson's comments, noting, "Extensive ancient mining was done on the easterly part of Isle Royale, on narrow veins yielding sheet-copper. This form being, as already observed, suited to the needs of the old miners, they did much work on these veins" (1907: 292). Others also made similar observations (Pittsburgh and Boston Mining Company 1855:17).

Gillman noticed the emphasis on small, workable pieces of copper, and stated that while seeking sheet copper, the ancient miners were

> … rejecting as unmanageable the fragments of rock which contained even large-sized nuggets of the metal. The latter are found in large quantities in the débris forming the tumuli at the mouths of pits, as well as in the excavations themselves, where, mingled with considerable amounts of charcoal, they seemingly had been pushed behind those miners as they advanced in the exploration of the vein, the walls of which were generally left unbroken. (1873:174)

Canadian archaeologist and author Sir Daniel Wilson made a similar observation about a site in Michigan:

> ... towards the close of 1874 Mr. [A. C.] Davis, an experienced old miner of Lake Superior, recovered from another ancient trench, in the same region, a solid mass of nearly pure copper, heart-shaped, and weighing between two and three tons. It lay at a depth of seventeen feet from the surface, as when originally detached from its bed by the ancient miners. Alongside it were a number of smaller pieces, from a single ounce to seventeen pounds in weight, evidently broken off the large mass by the ancient miners. Numerous stone mauls and hammers also, weighing from ten to thirty pounds, lay scattered through the lower debris with which the trench was refilled. (1876:205)

Although ancient hard-rock mining is the theme of this book, there is abundant evidence that prehistoric peoples also conducted extensive excavations in the sand and boulder drift on the shore of Portage Lake (Whittlesey 1863:14–15) and near the site of the Ontonagon Boulder (Whittlesey 1852c:327). Other authors have supported the idea that prehistoric prospectors sought out "drift" or "float" copper—that is, copper ripped from the bedrock by glaciers and spread far and wide across the midwestern United States (e.g. Bastian 1963a:41; Clark and Martin 2005:213–214; Dustin 1957:28; Ferguson 1923, 1924; Ferone 1999:111; Halsey 2010; Holmes 1901:685; Martin 2004:9; West 1929:26–28).

The four maps that follow (Figures 1.2 to 1.5) show where nineteenth-century copper mines were established in Michigan's Upper Peninsula.

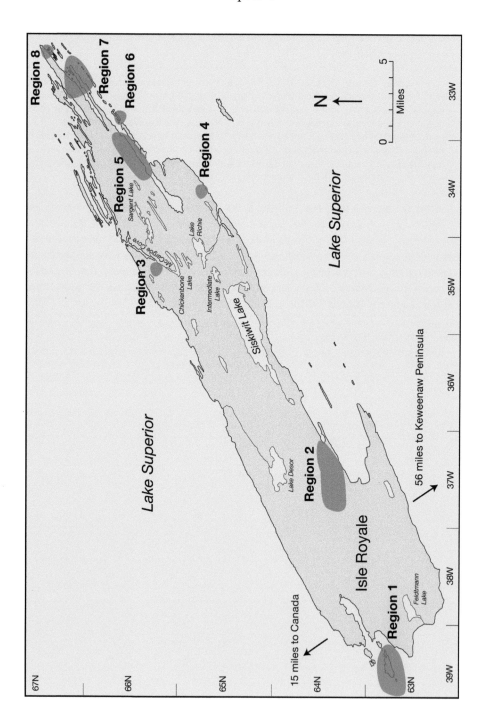

Figure 1.2. Approximate locations on Isle Royale of nineteenth-century copper mines, which were established at sites showing evidence of prehistoric mining. (Map by John R. Halsey, Kay Clahassey, and John Klausmeyer.)

Region 1: Phelps (20IR81), Singer (20IR80), and Triangle Island (20IR106)

Region 2: Dustin (20IR10), Ferguson-Dustin (20IR9), Island Mine (20IR11), and Little Siskiwit (also called Old Fort Diggings) (20IR6)

Region 3: Minong Mining Company (20IR24)

Region 4: Epidote Mine (20IR51)

Region 5: Ransom Mine (20IR43) and Siskowit (20IR13)

Region 6: Outer Hill Island (20IR37)

Region 7: Lookout Mine (20IR30), Scoville Mine (20IR32), and Stoll (20IR26)

Region 8: Third Island (20IR84)

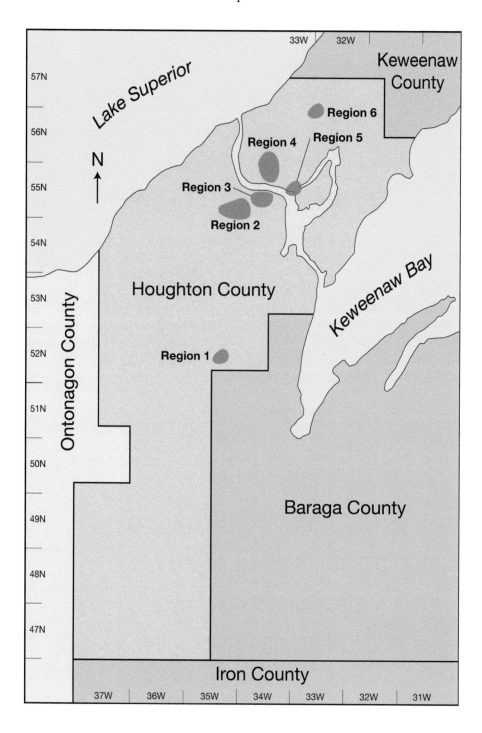

Figure 1.3. Approximate locations in Houghton County of nineteenth-century copper mines, which were established at sites showing evidence of prehistoric mining. (Map by John R. Halsey, Kay Clahassey, and John Klausmeyer.)

 Region 1: Winona Mining Company (20HO303)

 Region 2: Adams Mining Company (20HO289), Atlantic Mine (a consolidation of the Adams and South Pewabic mines), Huron Copper Company (20HO275), and South Pewabic Copper Company (20HO302)

 Region 3: Columbian Mining Company (20HO291), Isle Royale Mining Company (20HO298), Portage Mining Company (20HO300), Quincy Mining Company (20HO128), and Shelden and Columbian (20HO291)

 Region 4: Arcadian Mining Company (20HO290), Mesnard Mining Company (20HO276), Pewabic Mining Company (20HO130), and Pontiac Mining Company (20HO299)

 Region 5: Ripley Mining Company (20HO301)

 Region 6: Hulbert Mine (20HO4)

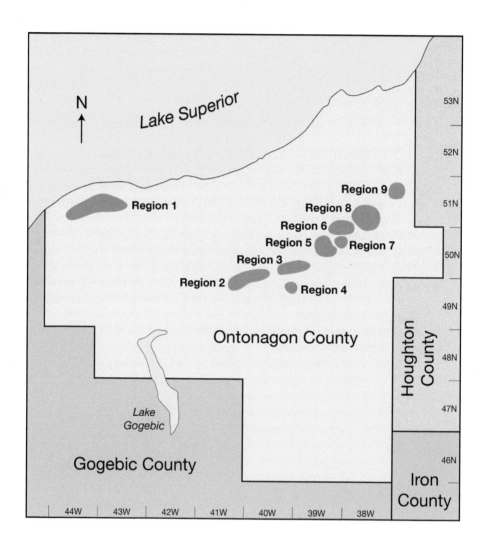

Figure 1.4. Approximate locations in Ontonagon County of nineteenth-century copper mines, which were established at sites showing evidence of prehistoric mining. (Map by John R. Halsey, Kay Clahassey, and John Klausmeyer.)

> Region 1: Carp Lake Mining Company (20ON189/190), Lafayette Mining Company (20ON182), and Piscataqua Mining Company (20ON272)
>
> Region 2: Hartford Mining Company (20ON482) and Hudson Mining Company/ Eureka Mining Company (20ON264)
>
> Region 3: Arctic Mining Company (20ON219), Devon Mining Company (20ON262), Forest Mining Company (20ON290), and Nebraska Mining Company (20ON281)
>
> Region 4: Minesota Mining Company/Michigan Copper Mining Company (20ON1), National Mining Company (20ON279), Ohio Trap Rock Company (20ON40), Ontonagon Copper Company (20ON11/290), and Ridge Copper Company (20ON429)
>
> Region 5: Caledonia Mining Company (20ON255), Flint Steel River Mining Company (20ON425), Knowlton Mining Company (20ON270), Ogima Mining Company (20ON283), Peninsula Mining Company (20ON149), and Superior Mining Company (20ON288)
>
> Region 6: Lake Superior Mining Company/Rockland Mining Company (20ON427)
>
> Region 7: Adventure Mining Company (20ON251), Aztec Mining Company (20ON271), Bohemian Mining Company (20ON272), Evergreen Bluff Mining Company (20ON265), Great Western Copper Mining Company (20ON272), Hilton Mining Company (20ON426), Mass Mining Company (20ON274-275), Ohio Mining Company (20ON426), Penn Mining Company (20ON428), Republic Mining Company (20ON430)

Author's note: Some mines share site numbers because they are essentially the same property, or because the mine was renamed. This is true for the Piscataqua, Bohemian, and Great Western mines (20ON272); Forest Mining and Ontonagon Copper (20ON290); and the Hilton and Ohio mines (20ON426).

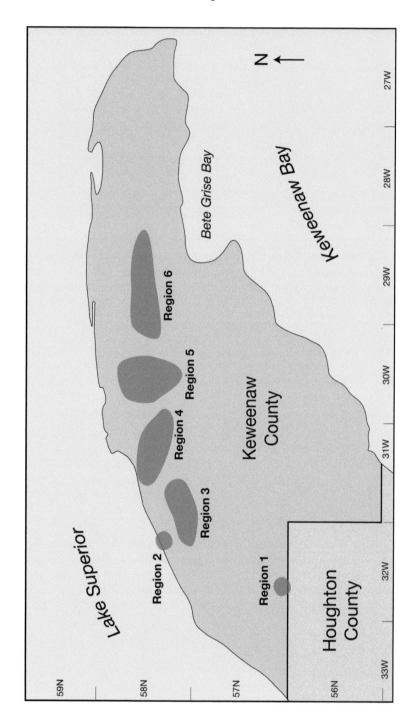

Figure 1.5. Approximate locations in Keweenaw County of nineteenth-century copper mines, which were established at sites showing evidence of prehistoric mining. (Map by John R. Halsey, Kay Clahassey, and John Klausmeyer.)

Region 1: Fulton Mining Company (20KE65)

Region 2: North Cliff Mining Company (20KE50)

Region 3: Cliff Mine (aka Pittsburgh and Boston) (20KE53), Eagle River Mining Company (20KE81), Garden City Mining Company (20KE66), Lake Superior Copper Company (20KE68), and Phoenix Mining Company (20KE68)

Region 4: Central Mining Company (20KE46), Copper Falls Mining Company (20KE63), "Humboldt Location" (20KE67), Meadow Mining Company (20KE69), and Northwestern Mining Company (20KE82)

Region 5: Agate Harbor Mining Company (20KE61), Eagle Harbor Mining Company (20KE45), Natick Copper Company (20KE70), Native Copper Company (20KE72), and Waterbury Mine (20KE73)

Region 6: Iron City Mining Company (20KE79), Mandan Mining Company (20KE78), and North West Mining Company (20KE34)

2 Early Reports

Yet I erred in saying that Silence has hitherto reigned here unbroken save by Nature's convulsions since Adam. On this very spot have been found, in this very cabin deposited, the elaborately fashioned stone-hammers wherewith some long forgotten race was accustomed to beat off scraps or fragments of the Native Copper for use ages upon ages ago—perhaps before the Deluge. Truly said the Wise King, 'There is nothing new under the sun.'

Horace Greeley
1847

It is important to remember that by the late 1840s, European explorers and missionaries had been resident in the western Lake Superior region for nearly two hundred years. The presence of large native copper boulders was well known, but the bedrock sources of these boulders were less so. (See Sagard 1939:135 and Boucher 1883:82–83) for early mentions of native copper on Isle Royale and the Keweenaw Peninsula.)

It seems likely that historic miners and other frontier types did recognize the hammerstone as a tool that was involved with prehistoric mining, even before reports on the mines appeared in the written record. For example, in 1847, Horace Greeley included the passage that opens this chapter in a report from his cabin (Clarke 1975b:19; see also Williams 1950:123) at the Northwest Copper Mining Association (later the Pennsylvania Mining Company) location.

And there are claims that others had identified the copper mines as prehistoric. For example, George R. Fox reported:

> At what date the first aboriginal pit was opened is not known and possibly never will be. The rediscovery of the pits must date before or about 1844, for at that time Dr. Jacobson, who spent several years on Lake Superior during the copper boom in the forties, told of the ancient works. (1952:238)

It is unknown who this "Dr. Jacobson" was. It was probably Charles T. Jackson, a prominent and controversial geologist in the early days of Michigan's copper boom (Woodworth 1897). Jackson was well known for his bogus claims of discovery priority in a number of high-profile cases (Krause 1992:182–187). The following quote apparently shows Jackson's efforts to claim his priority in identifying prehistoric mining efforts.

> The Jesuits nowhere describe veins of copper, or any metal in place, and do not mention any proper mining operations as having been performed by the Indians; but it was discovered, in the diggings at Eagle River, in 1844, that the aborigines had extracted the metal from the veins, and had made knives and spear-heads of the sheet copper which they obtained. This observation was very interesting to me, and, on searching at the other mines I invariably found Indian stone hammers, and proofs of superficial mining, by the native tribes. On calling the attention of the directors of the mines to these curiosities, they readily entered into the work of searching for them… (1850:284)

Unfortunately for Jackson's case, he appears to have made no published mention of ancient mining activities prior to 1849 (Jackson 1849:373–375, 445, 448; Woodworth 1897). This does not mean that Jackson had not seen and understood the nature of "ancient diggings," but rather that he hadn't put the observations into print early enough. Then there is this statement, made in 1846.

> Consolidated Mining Company. This company have five locations, four upon Isle Royal and one on Passage Island. Mr. D. B. Talbott, the agent, while here a few days since, exhibited some very excellent specimens from one of their locations on Isle Royal, of native copper, and also exceedingly rich lumps of silver. It is supposed that the vein

from which these specimens were taken is the same discovered by Dr. Houghton, and from which he took these of rich native copper now deposited in the cabinet of the State University at Ann Arbor, as there were evidences of some previous workings on the vein when discovered by the company. (Ingersoll 1846)

This is potentially a very important article, for it appears to be among the first, and quite possibly *the* first, to mention what would later be identified as prehistoric mining activities. Unfortunately, the precise location where these "workings" were discovered cannot be identified at this time. For the purposes of this study, Ingersoll certainly appears to have been the right man in the right place at the right time, with the right interests (Bayliss 1946; Jamison 1946). Although Ingersoll suggests those mentioned here are the work of Douglass Houghton's brief exploratory activities in 1840, Houghton and his party were in the vicinity of Isle Royale for less than a week, probably too short a time to have left visible traces of their presence. Given the numerous prehistoric pits on Isle Royale, the chances of it being Houghton's work are vanishingly small.

The following unpublished diary entry, written in 1847 by pioneering Isle Royale miner and explorer Cornelius G. Shaw, also points to recognition of prior ancient mining. (Spelling and punctuation here are original.)

Tuesday 21st Septr
fine weather 4 men on Shaft 1 Exploring traced sec lines & found on sec 33 a place where someone had dug for a vein to appearance more than 20 years since as timber was growing on the dirt thrown out 6 inches in diameter I intend to examine said place more (1847:24).

Although Shaw apparently did not personally write more about this discovery, he did talk to John W. Foster and Josiah D. Whitney.

Mr. C. G. Shaw pointed out to us similar evidences of mining on Isle Royale. They occur on what is known as the Middle Finger, and can be traced lengthwise for the distance of a mile. Mr. Shaw remarks that, on opening one of these pits, which had become filled up with the surrounding earth, he found the mine had been worked through the solid rock to the depth of nine feet, the wall being perfectly smooth. At the bottom he found a vein of native copper eighteen inches thick, including a sheet of pure copper lying against the foot-wall.

The workings appear to have been effected simply by stone hammers and wedges, specimens of which were found in great abundance at the bottom of the pits. He found no metallic implements of any description, and is convinced, from the appearance of the wall-rocks, the substances removed, and the multitude of hammers found, that the labor of excavating the rock must have been performed only with the instruments above named, with the aid perhaps of fire. From the appearance of the vein and the extent of the workings, he conjectures that an immense amount of labor had been expended. He endeavored to find some evidences of the antiquity of these workings, but could discover nothing very satisfactory to his own mind, except that they were made at a remote epoch. The vegetable matter had accumulated and filled up the entire opening to a level with the surrounding surface; and, in a region where it accumulates as slowly as it does on the barren and rocky parts of Isle Royale, this filling up would have been the work of centuries. Upon this vegetable accumulation he found trees growing equal in size to any in the vicinity. (1850:162)

This location later became known as the Lookout Mine (20IR30) (Clark 1995:66). Several prehistoric pits at this location were excavated by Tyler Bastian of the University of Michigan in 1960 and 1961 (Bastian 1963a:49–53). Charcoal from the fill was dated by early radiocarbon techniques to about 3360–2880 BC and later. (See this volume's Afterword for more discussion on dates.)

Samuel O. Knapp and
3 the Minesota Mine

Mr. Knapp ... has lately made some very singular discoveries here in working one of the veins which he lately found. He worked into an old cave which had been excavated centuries ago. This led them to look for other works of the same sort, and they have found a number of sinks in the earth which they have traced a long distance. By digging into those sinks they have, they find them to have been made by the hand of man.

Lake Superior News and Mining Journal
1848

The Minesota mine is two miles from the river. Upon this location was discovered the ancient diggings which are now found to exist over a vast extent of the country, and which excite the curiosity and wonder of every one.

S. V. R. Trowbridge
Assistant Agent
United States Mineral Lands
1850b

First of all, why "Minesota"? The obvious misspelling of the name, which should have been Minnesota, was apparently a clerical error made permanent at the time of the filing of the Articles of Association. (See Clarke 1978:7; Minesota Mining Company 1858a.) The following history of how the Minesota Mine came to be is derived largely from Don Clarke's self-published book, *Minesota Mining Company* (1978), Number 11 in the Copper Mines of Keweenaw series.

The land that would eventually become the Minesota Mine started out as Permit #98, issued by the United States Mineral Agency to Joseph L. Hempstead on August 5, 1844. This permit covered nine square miles of Ontonagon County in Township 50 North, Range 39 West, made up of the south halves of Sections 10, 11, 12; the entirety of Sections 13, 14, 15, 22, 23, and 24; and the north halves of Sections 25, 26, and 27. A year later, Permit #98 came into the possession of the Ontonagon Mining Company. This company built cabins for the workers on Section 15 of Permit #98 and dug exploratory shafts in Sections 11, 15, and 22. Apparently no copper was found.

On November 21, 1846, Jonas H. Titus, president of the Baltimore Mining Company, met with financier William Hickok of New York. The Baltimore Mining Company owned two mineral leases in the Ontonagon area and needed money to explore these properties.

Hickock proposed the formation of an entirely new entity to be called the Vulcan Mining Company. Vulcan was to pay $4500 in cash, and 800 shares were to be issued to the stockholders of the Baltimore Mining Company. The 800 shares were to be nonassessable—that is, the issuing company would not be allowed to impose levies on its shareholders for additional funds for further investment. This accomplished, Vulcan obtained the two permits—267 and 269—which were for land in the Porcupine Mountains.

With these deals accomplished, the Vulcan Mining Company was organized with 4000 shares of stock. Of these shares, 1000 were nonassessable; of these 1000, 800 went to the Baltimore Mining Company and 200 to Vulcan. The remaining 3000 shares were sold for $150 per share. Hickok purchased 1600 shares, which gave him control of the company. Several associates of Hickok purchased 1300 shares among them. The afore-mentioned Titus purchased 100 shares.

The 800 nonassessable shares were distributed to the stockholders of the Baltimore Mining Company on November 23, 1846. Seven days later, the Articles of Association of the Vulcan Miming Company were filed. Hickok and Knapp began to purchase shares in the Baltimore Mining Company.

Samuel Knapp, who had been named as the agent for the Vulcan Mining Company, took six men to Michigan's Upper Peninsula and examined two locations (Permits 267 and 269) in the summer of 1847. He also examined the Algonquin Mine, which showed only slight promise of an ore deposit and was nowhere near 267 and 269.

It was at this time Knapp learned of another location, Permit 98, owned by the Ontonagon Mining Company. He visited the 98 location after he had discussed

Samuel O. Knapp and the Minesota Mine

a possible purchase of land from Agent George C. Bates in Detroit. (Knapp and Hickock apparently did an amazing amount of traveling in the 1840s.) Bates agreed to sell either the north or south half of the 98 location.

Knapp returned to the location in the fall of 1847 and spent the winter exploring the location. He made a detailed study of the property. He excavated numerous already-existing pits that he believed to be mining sites of some ancient race. While cleaning out one of these pits, Knapp found a mass of native copper that was 10 feet long, 3½ feet wide, and 22 inches thick. The mass had been raised several feet above the bottom of the pit and was hammered smooth. It was eventually estimated to weigh 7½ tons.

Knapp paid the bills of several miners then occupying the property on behalf of the Ontonagon Mining Company. At this time Knapp took up George Bates' offer and selected the north half of the 98 location in April 1848.

Several of the minority stockholders of Ontonagon Mining Company objected to this purchase. Hickok and Knapp decided to form a new company to manage this new property. They called their new company the Minesota Mining Company. This gave Hickok control of the property and essentially eliminated future interference from stockholders of the Vulcan Mining Company and the Baltimore Mining Company.

On September 13, 1848, the ship bringing winter provisions to the miners—the wooden propeller-driven steamer *Goliath*—caught fire and exploded near Saginaw Bay. *Goliath*'s cargo included 200 kegs of gunpowder, 40 tons of hay, 20,000 bricks, 30,000 feet of lumber, and 2,000 barrels of provisions and merchandise. According to the *Buffalo Commercial Advertiser* of September 27, 1848, part of the cargo belonged to "S. A. Knapp, of Ontonagon."

On top of these travails, no copper had been produced.

Meanwhile, the legality of the Hickok/Knapp maneuver was questioned. Suits were brought and litigation continued for many years—although without significantly deterring operations—before finally being settled in favor of the Minnesota Mining Company.

The long battle was worth it. On March 7, 1857, workers discovered a mass of native copper 46 feet long, 18½ feet wide, and 8 to 9 feet thick. It took 15 months for 20 men to cut the mass into pieces that could be removed from the mine. There were 27 tons of "chisel chips"—debris from cutting the mass into pieces. The value of the chips alone exceeded the total cost of removing and cutting the mass into pieces. The total weight of the mass was estimated to be over 527 tons (1,054,000 pounds). This is thought to be the largest single piece of native copper ever discovered.

Unfortunately, few mines across the entire mining district ever were able to break even financially. (For those who are really interested in the origins of the Minesota Mine, Copper Country historian James K. Jamison (1950:69–83) presents a lively, likely, but largely undocumented tale of the flim-flammery perpetrated by Knapp and Hickok.)

Chapter 3

Who actually made the discoveries at the location that eventually became the Minesota Mine is a question with a murky answer. It is generally accepted that the first person to recognize the true nature of the various pits and depressions on copper veins was Samuel O. Knapp, and that the first significant written report on the ancient mines—which named Knapp as the discoverer—appeared in the winter of 1847–48.

Who Was Samuel O. Knapp?

Knapp, described by Jamison (1950:70) as "simply the smartest operator that appears on the scene of early copper mining," was born in Royalton, Vermont, on April 21, 1816, the sixth of Nathan and Mary Knapp's twelve children. The following information is based largely on an extensive retrospective of Knapp that appeared in *Portrait and Biographical Album of Jackson County, Michigan* (Chapman Brothers 1890:201–203) and Halsey's 1998 *Michigan History Magazine* article.

At the age of 10, Knapp was apprenticed to Charles Paine (later the governor of Vermont) to learn woolen manufacturing. Two years later he was put in charge of the carding department, and he was made superintendent of both carding and spinning when he was 18. In 1838, at the age of 22, he married 20-year-old Sarah Balch in Northfield, Vermont. By the time he was 28, his health had been severely affected by all those years in the mill. He then worked for a year and a half as manager of a "public house" (tavern) in Northfield, Vermont, no doubt obtaining experience that would be useful in dealing with miners later on. He then moved to Jackson, Michigan, where he set up and started machinery for woolen manufacture at the State Prison in there. For a year and a half in 1845 and 1846, he superintended the wool operation in the prison. Also in 1846, he purchased an extensive piece of property in Jackson.

With the discovery of copper riches in the Upper Peninsula, he went north. He had a pivotal role in the development of the native copper mining industry and the discovery of prehistoric mining.

In addition to mining, Knapp soon became involved in the copper-smelting process in Detroit and then sold out in 1851 (at the age of 35) and turned his protean energies to the plant nursery business, fruit growing, and real estate. He built a substantial home in Jackson and began a remarkable process of reshaping his landscape, planting exotic trees, and building a grapery and a steam-heated greenhouse in which he grew a wide variety of tropical plants. Scattered around the grounds were beds of rich tropical plants and flowers. According to a contemporary visitor, "Fountains, vine-clad rocks, rock pyramids and other curiosities add to the beauty of the place and renders it even more charming within than it seems from without" (Anonymous 1883b).

In his later years, Knapp served on several committees for local and state organizations. In Jackson he was a member of the committee that framed the first city charter, president of the school board for several years, and a member of the

Samuel O. Knapp and the Minesota Mine

Figure 3.1. Samuel O. Knapp (1816–1883), shown here with his wife Sarah in 1910, was founder and owner of the Upper Peninsula's Minesota Mine and, more significantly for this book, was the first person to recognize Michigan's prehistoric copper mining pits for what they were. (Reprinted from The Bay View Magazine 17(8):541, May 1910. HathiTrust, https://hdl.handle.net/2027/mdp.39015071503240, accessed March 20, 2018.)

Board of Public Works. At the state level, he was a member of the State Board of Agriculture, chairman of the Building Committee at Michigan Agricultural College, and long-time member and frequent vice president of the State Pomological Society.

We know little about Knapp's personality or his home life. He and Sarah had no children. His niece described him as "rather slow of speech and somewhat reserved." Despite his membership and apparently adequate performance in the organizations mentioned above, there is little written evidence of his role in their deliberations. In his politics, he was Republican.

The church played a big role in Knapp's life. From the age of sixteen, he was a devoted member of the Methodist Episcopal Church. The Reverend Eri H. Day, a Methodist missionary in the Copper Country during the 1840s, recalled Knapp's donation of a five-dollar gold piece during a time of financial distress as one of the most memorable events his mission experienced.

In the mid-1870s Michigan Methodists sought permanent grounds for summer meetings. They wanted a site in the northern part of the state, on a railroad line and adjacent to one of the Great Lakes or a large interior lake such as Walloon Lake.

During the summer of 1874, Sam and Sarah took one of the first trains to Petoskey in search of a climate that would help Sarah's respiratory problems. Their report was so positive that the location was quickly selected as the site for the meeting ground, later to be known as Bay View. Knapp not only discovered the site, he also laid out the streets and the system of blocks and lots, built the first cottage and was involved with such basic issues as development of the water supply system.

Knapp died on January 6, 1883. Sarah lived on, respiratory problems and all, until December 12, 1899. (See Anonymous 1896 for a more elaborate version of how the Bay View location was discovered.)

As far as is known, Knapp did not write a first-person account of his discovery of the mines. Despite being a very accomplished man, he apparently wrote very little—or it may be that the location of his writings is unknown. Partly because of the lack of an autobiographical report, there has been some dispute over Knapp's status in this historical tale.

There is no shortage, however, of men who did write tales of discovery in a bid for recognition.

William Spalding

In a memoir written 53 years after the fact, William Spalding claimed that he was the man who first recognized the mines.

> ...I immediately took a contract to do some mining for the Vulcan Mining company on location ninety-eight, afterward the famous Minnesota Mining company. I began work early in January, 1848.
>
> In April of that year I examined a cave that had been occupied by porcupines for many generations, and discovered that it had been worked by the hand of man. I had it cleaned out and found masses of pure copper standing up from the bottom some eighteen or twenty inches above the rock in which they were embedded. Around them were ashes, burned pieces of bark and boulders of rock weighing five to ten pounds. Around the center of these stones creases or rings had been cut, about one inch in width and an eighth to a quarter of an inch deep. It was evident that these stones had been used as hammers or mallets. Around the creases a withe had been bound for a handle. The ends of many of them were battered, showing hard use. All the indications went to show that the ancient miners, whoever they were, first built fires on the rock and then poured on water to soften it. Then they worked with these stone hammers to beat it away.
>
> This was the first discovery of the work of the ancient miners on Lake Superior. Afterwards, on the same vein on which I made this discovery, a basin like depression was found in which stood a large hemlock tree. When it was cut down it was found by its rings that it was over 400 years old. A shaft was found which had been worked by the ancients. On cleaning this out to a depth of twenty feet a drift was found leading out on the vein for about fifteen feet. In this drift, laying on oak skids—there was no oak growing in this country at the time of which I write—there was a solid mass of pure copper weighing four tons. The ancient miners, after immense labor for many years, had succeeded in detaching it from its bed in the rock and tipping it over on the skids found

they were unable to move it. The skids being covered with water preserved their shape and grain, though they were of the consistency of cheese. The shaft was continued down another twenty feet, making forty feet in all. How many years or centuries it had taken these ancients, with their crude methods, to do this work no man knows. Other works of similar description were found subsequently all over the copper country of Upper Michigan. Isle Royale was found honey-combed with these ancient pits...May 6, 1848, I finished my work, netting $1,000 profit for my four months work, and went back to the store in Ontonagon..." (1901:689–691) (See also Bryant 1881.)

It may be of significance that Spalding (Anonymous 1897a, 1897b) also claimed to have been the first to discover traces of ancient copper mining in Cook County, Minnesota.

This is significant in that it shows Spalding's willingness to take credit for discoveries he did not make. This puts much of what he has said in question. Take, for example, the following statement from Newton H. Winchell:

Some years ago Mr. William P. Spalding discovered, as he supposed, some "ancient diggings," and organized a mining company for the purpose of developing the "mines" which he thought had existed there in former times. This location was on the south side of lake Miranda, sec. 5, T. 64-2E. He worked for several years, in a feeble way, but so far as known, without success. The writer noted these so-called diggings in Seventh Annual report of the geological survey (for 1878), page 18, and gave reasons pro and con as to their being artificial, from which he was inclined to consider the 'depressions' which first attracted attention as due to natural causes, and has since not seen any reason to think otherwise. The vein which Mr. Spalding exploited was said to afford gold and silver, the latter probably in the form of argentiferous galena like that taken from the Animikie at Thunder bay. The ancients knew nothing of the smelting of such ore. No stone hammers were found in the neighborhood.
*In the *St. Paul Globe*, Feb. 8, 1897, is some account of the efforts of Mr. Spalding to revive the mining industry which he supposed to have existed formerly at lake Miranda. The statements made imply a rather unusual combination of gold, silver and copper, all in a metallic state in a quartz vein. A mound is mentioned, on sec. 3, T. 64-2E, near these "ancient diggings," and it is suggested that if explored, it might reveal some of the implements and perhaps skeletons of the ancient miners. (1911:379)

Albert Hughes

According to educator J. H. Lathrop, "the earliest record in detail of the work of the ancient miners was the discovery in 1846 by the prospector, Albert Hughes, on the Minnesota mine location in Ontonagon County and thus described by Samuel O. Knapp, then agent of the Minnesota mine" (1901b:249–250). However, there was no entity known as the Minesota Mining Company until after the 1847–48 discoveries, so Knapp could not have been the company agent in 1846. Most damning, the quotation presented—supposedly directly quoting Knapp describing Hughes' efforts—is lifted (unattributed) from Foster and Whitney (1850:159) and actually describes Knapp's discoveries.

Bruce Johanson (1985:43–44), citing multiple unidentified "early survey maps" and other sources, states that the claim that eventually became the Minesota was previously worked by the Ontonagon Mining Company. Johanson spins a tale of Hughes actually being an employee of the Ontonagon and discovering the "ancient diggings," with Knapp just happening along, seeing what had been found and, in effect, hijacking the discovery. Nevertheless, Johanson credits Knapp with being "the enterprising soul who recognized the importance of the find and capitalized on it" (1985:54).

Did Knapp Act Alone?

Albert Hagar (1865:308) asserts that it was Knapp and J. B. Townsend who made the discovery in March 1848, while Hobart and Wright (1883:512) declare the co-discoverers to have been Knapp, Townsend, and Spalding.

Perhaps the key to sorting out this confusion lies in this statement by Foster and Whitney (1850:159) about Knapp: "He saw numerous evidences to convince him that this was an artificial excavation, and at a subsequent day, with the assistance of two or three men, proceeded to explore it."

It does seem highly unlikely that Knapp alone accomplished the excavation. This probability, and the statement above, seems to indicate there was participation by Hughes, Spalding, Townsend and even more.

The Report That Changed Everything

Regardless of who actually was on the scene, an extensive and detailed account appears in 1848 that references only Mr. Knapp. The origin of this account is mysterious, but its impact was remarkable. It set a standard for detail and useful information that would be seldom met in later years. It is the "Ur-document" for our modern understanding of what historic miners thought about the "ancient diggings."

In its most common rendition, the account reads as follows:

The Copper Region.—Singular Discovery

A correspondent of the *Buffalo Express*, writing under the date of June 14th, 1848, from Ontonagon, Lake Superior, says:--

Mr. Knapp, of the Vulcan Mining Company, has lately made some very singular discoveries here in working one of the veins which he lately found. He worked into an old cave which had been excavated centuries ago. This led them to look for other works of the same sort, and they have found a number of sinks in the earth which they have traced a long distance. By digging into those sinks they have, they find them to have been made by the hand of man. It appears that the ancient miners went on a different principle from what they do at the present time. The greatest depth yet found in these holes is thirty feet. After getting down to a certain depth, they drifted along the vein, making an open cut.

Samuel O. Knapp and the Minesota Mine

These cuts have been filled nearly to a level by the accumulation of soil, and we find trees of the largest growth standing in this gutter; and also find that trees of a very large growth have grown up and died, and decayed many years since; in the same place there are now standing trees of over three hundred years' growth. Last week they dug down into a new place, and about twelve feet below the surface found a mass of copper that will weigh from eight to ten tons. This mass was buried in ashes, and it appears that they could not handle it, and had no means of cutting it, and probably built fire to melt or separate the rock from it, which might be done by heating, and then dashing on cold water. This piece of copper is as pure and clean as a new cent; the upper surface has been pounded clear and smooth.

It appears that this mass of copper was taken from the bottom of a shaft, at the depth of about thirty feet. In sinking this shaft from where the mass now lies, they followed the course of the vein, which pitches considerably. This enabled them to raise it as far as the hole came up with a slant. At the bottom of the shaft they found skids of black oak, from eight to twelve inches in diameter. These sticks were charred through, as if burnt. They found large wooden wedges in the same situation. In this shaft they found a miner's gad and a narrow chisel made of copper. I do not know whether these copper tools are tempered or not, but their make displays good workmanship.

They have taken out more than a ton of cobble stones, which have been used as mallets. These stones were nearly round, with a score cut around the centre, and look as if this score was cut for the purpose of putting a withe round for a handle. The Chippewa Indians all say that this work was never done by Indians. This discovery will lead to a new method of finding veins in this country, and may be a great benefit to some. I suppose they will keep finding new wonders for some time yet, as it is but a short time since they first found the old mine. There is copper here in abundance, and I think people will begin to dig it in a few years. Mr. Knapp has found considerable silver during this past winter. (*Lake Superior News and Mining Journal* 1848)

It is all here: what the pits looked like, why they were frequently overlooked, the presence of mining tools (hammerstones, gad, chisel), a heavily worked-over mass of copper left in the pit, a "facility" (the skids or logs placed under the copper mass), the probable use of fire in the mining process, ethnographic testimony as to a lack of native tradition regarding copper mining, an estimate of age based on an early effort at dendrochronology, and, perhaps the most important of all, the statement that, "This discovery will lead to a new method of finding veins in this country, and may be a great benefit to some." Miners quickly took this suggestion to heart and it was the death warrant for most prehistoric pits as a source of latter-day archaeological information. (See also Lawton 1883:39–40 for confirmation.)

The source of the original Knapp article is uncertain. On its face, it was written by someone in Ontonagon, identified as a "correspondent" of the *Buffalo Express*, on June 14, 1848. But that does not necessarily mean that it was actually published on June 14th or even in the *Buffalo Express*. I have examined the microfilm copy of the *Buffalo Morning Express* for that date and every issue after that, and it does not contain the article in question.

Wherever it was initially published, the article (with the author's name given only as "B.") was quickly noted. It was reprinted extensively over the summer and

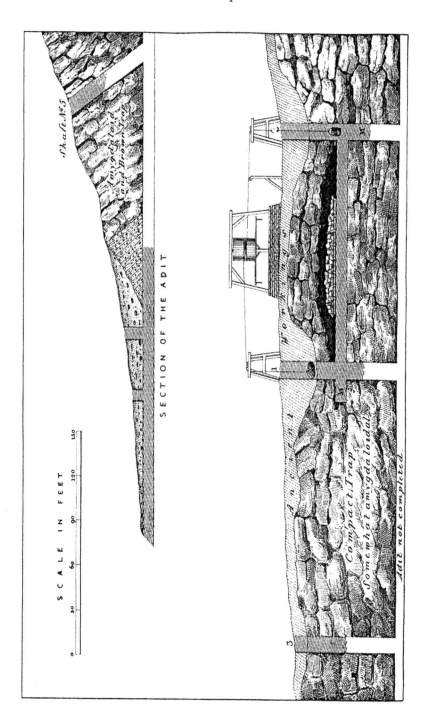

SCALE IN FEET

SECTION OF THE ADIT

Figure 3.2. (opposite page). This plan of the Minesota Mine, published in 1850, shows the ground surface along the entire length as "Ancient Workings." From the original caption: "According to the statement of Mr. Knapp, the agent of this mine . . . there were taken from this mine, last year, eleven tons of copper, seven and a half of which were included in the mass raised by the ancient miners, the position of which is indicated in the plan." Note that the mapmaker shows the famous mass of copper in the process of being lifted up through Shaft 1. (Reprinted from Foster and Whitney 1850: Plate 11. HathiTrust, http://hdl.handle.net/2027/aeu.ark:/13960/t01z52z3g, accessed March 20, 2018.)

autumn of 1848 in the US, Canada, and England. A short list of publications where it appeared includes *Massachusetts Eagle* [Lenox, Massachusetts], "The Copper Region. Singular Discovery," August 11, 1848; *Latter Day Saints' Millennial Star*, "The Copper Region. Singular Discovery," 10(22):351, 1848; *Lake Superior News and Mining Journal* [Sault Ste. Marie, Michigan], "The Copper Region. Singular Discovery," p.2, August 18, 1848; *Littell's Living Age*, "The Copper Region," 18(223):375, 1848; *Hunt's Merchants' Magazine and Commercial Review*, "Singular Discovery in the Copper Region," 19(3):340, 1848; *Niles' National Register*, "The Copper Region," 74(1910):159–160, 1848; *The Times* [London], September 15, 1848; and *Latter Day Saints' Millennial Star*, November 15, 1848. Nine months later, the news had even made its way to the other side of the planet, appearing in Australia's *Moreton Bay Courier* ("B." 1849) under a column heading "British Extracts," suggesting that a copy of the September 15 issue of *The Times* had finally made it "down under."

This listing must be taken as an absolute minimum of places where the initial telling of Knapp's discovery was published. Some sources (e.g. Beatty 1849) published distilled versions. It is unlikely that any other North American archaeological discovery of the nineteenth century excited such national and even international interest, as expressed in newspaper coverage.

Leading up to the Knapp announcement, there had been little to excite the general public's interest about ancient use of native copper. Daniel Drake, a pioneering medical doctor and resident of Cincinnati, wrote of the "excavation" and leveling of a mound in that city:

> Several copper articles, each consisting of two sets of circular concavo-convex plates: the interior one of each set connected with the other by a hollow axis around which had been wound a quantity of lint: the whole encompassed with the bones of a man's hand. Several other articles resembling this have been dug up in other parts of town. They all appear to consist of pure copper, covered with the green carbonate of that metal. After removing this incrustation of rust from two pieces, their specific gravities were found to be 7.545 and 7.857. Their hardness is about that of the sheet copper of commerce. They are not engraven or embellished with characters of any kind. (Drake 1815: 207)

In 1820, Caleb Atwater of Circleville, Ohio, (named after the circular earthwork in which the town was established) published an early compendium of archaeological sites in Ohio, assisted by a variety of lawyers, doctors, surveyors, and an Indian agent. He largely ignores the presence of copper or copper artifacts, instead publishing extensive letters by Dr. Samuel P. Hildreth, another pioneering doctor, on Hopewellian burials containing copper artifacts at Marietta, Ohio, and other locations in the vicinity (1820:168–176).

> Copper, has been found in more than twenty mounds, but generally not very well wrought. It is in all cases, like that described by Dr. Drake, already quoted. The copper, belonging to the sword found at Marietta, is wrought with the most art of which I have seen.

Samuel O. Knapp and the Minesota Mine

Pipe bowls of copper, hammered out, and not welded together, but lapped over, have been found in many tumuli. General Tupper [Benjamin Tupper, an important figure in the Revolutionary War and the founding of Marietta] described such a one to me, found by him on the elevated square at Marietta, or rather a few feet below the surface of that work. Similar ones have been discovered in other places. A bracelet of copper was found in a stone mound near Chillicothe, and forwarded to the museum in Cincinnati by the Hon. Jessup N. Couch, Esq. some time since. This was a rude ornament, and resembled somewhat the link of a common log chain; the ends passed by each other, but were not welded together. I have seen several arrow heads of this metal, some of which were five or six inches in length, and must have been used as heads of spears. Circular medals of this metal several inches in diameter, very thin and much injured by time, have often been found in the tumuli. They had no inscriptions that I could discover. Some of them were large enough to have answered for breast plates. (1820:224)

Clearly, in the first quarter of the nineteenth century, there was only minor interest in copper artifacts or the source of the native copper from which they were made. That attitude would change.

The main conveyor of that change was the publication of *Ancient Monuments of the Mississippi Valley* by Ephraim G. Squier and Edwin H. Davis. This volume, issued as Vol. I in the Smithsonian Institution's Smithsonian Contributions to Knowledge (accepted for publication in June 1847 and published in 1848), continues to be recognized as one of the most important documents ever published in the history of North American archaeology. Containing 48 plates and 207 wood engravings of artifacts, mounds, and earthworks, it opened a new world of images and thoughts concerning the ancient inhabitants of the eastern United States. Although not believing that the people who built these amazing structures were the ancestors of the local Native Americans still present in the region, their documentation set high standards for the time.

Below is what Squier and Davis had to say about native copper and its use.

From what has been presented, it appears that the mound builders were very well acquainted with the use of copper. They do not, however, seem to have possessed the secret of giving it any extraordinary degree of hardness. The axes above described were found, upon analysis, to be pure copper,--unalloyed, to any perceptible extent, by other metals. The hardness which they seem to possess, beyond the copper of commerce, is no doubt due to the hammering to which they were subjected in their manufacture. As already observed, the metal appears to have been worked, in all cases, in a cold state. This is somewhat remarkable, as the fires upon the altars were sufficiently strong, in some instances, to melt down the copper implements and ornaments deposited upon them, and the fact that the metal is fusible could hardly have escaped notice.

It has been suggested, upon the strength of the fact that some of the specimens of copper obtained from the mounds have crystals of silver attached to them, that a part of the supply of the ancient people was obtained from the shores of Lake Superior, where alone this peculiar combination is known to exist. The circumstance that the mound axes are made of unalloyed copper, does not affect this conclusion; for a large proportion of the native metal found at this locality is pure. The conclusion is further sustained by the amount of metal extracted from the mounds, implying a large original supply. Besides numerous small pieces, some large fragments are occasionally discovered. One of these weighing twenty-three pounds, and from which portions had evidently been cut, was found a few years since near Chillicothe. [Here they were talking about "drift" or "float" copper.] Still, it does not appear that copper was sufficiently abundant to entirely supersede the use of bone and stone implements. (1848:202–203)

What an incredible circumstance it was that while *Ancient Monuments* was going to press, Sam Knapp was opening the secrets of the strange trenches on good old 98. And in June of the next year, the public could learn about both *Ancient Monuments* and the Knapp report in the *Buffalo Express*.

4 Who Wrote the Knapp Report?

When he first came to the conclusion, about eighteen months ago, that the pits and trenches visible on the range were artificial, he caused one of them to be cleaned out. He found, at about eighteen feet in depth, measuring along the inclined face or floor of the vein, a mass of native copper, supported on a cobwork of timber, principally the black oak of these mountains, but which the ancient miners had not been able to raise out of the pit.

Henry Schoolcraft
1851

Where did the account of Samuel Knapp's discoveries come from? (See Chapter 3 for Knapp's report.) Did Knapp himself write it? And where was it first published? Many questions still remain.

The version published by the *Lake Superior Journal and Mining News* and *Littell's Living Age* ends with the initial "B.," which is presumably an initial of the "correspondent" (i.e., the author). This was common in nineteenth-century communications, especially newspapers: contributors often used only initials or nicknames. (See "Miskwabikokewin" 1863.)

Who was "B."? There were no newspapers in the frontier town of Ontonagon at this early date—the first printing press would not arrive for another seven years. "B." appears to have been a local citizen with an interest in the mines.

Daniel Beaser

Now more than 160 years removed from the event, it is very difficult to be certain, but one candidate stands out. The following information is derived from Hobart and Wright (1883:538–539), who presumably used information obtained directly from the subject.

Daniel Beaser was born near Buffalo, New York, on July 24, 1825. As a boy he was an avid reader of travel and adventure books describing the adventures of such notables as Captain Cook. At the age of 15, without his father's consent, he shipped on a New Bedford whaler and over a voyage of four years saw many of the places he had read about. He returned to Buffalo in 1845 and "engaged in sailing." For unknown reasons, he went to Ontonagon, arriving on April 6, 1848. He became captain of a small schooner that carried mail between Ontonagon and Eagle River. In the fall of 1864, he gave up sailing and hired on with the Lafayette Mining Company, where he was in charge of explorations in the Porcupine Mountains for two years. He was among the many who commented on "ancient diggings" (Appendix 1).

In short, 1) Beaser had close ties to the city of Buffalo; 2) his surname begins with B; 3) he was in Ontonagon at the right time; 4) he was engaged in an occupation (captain of a ship carrying mail) that could have facilitated his communication with the *Buffalo Express*; and 5) he had an interest in the discoveries in the local copper mines. Beaser also donated to at least one historical society hammerstones from Michigan's Mass and Aztec mines (Draper 1882a:19). This, of course, is not proof, but he certainly seems a likely suspect.

Charles Whittlesey

Another possibility is the omnipresent Charles Whittlesey, but he would seem to be eliminated by his third person statement concerning the Knapp discovery—"The first public announcement, so far as we are aware, of the remains of ancient mines in the copper region is that by Mr. S. O. Knapp, agent of the Minnesota Mining Company, in 1848" (Whittlesey 1863:4). Nevertheless, it is disappointing that someone as intimately involved with the 1840s copper-mining scene as Whittlesey attributes the "announcement" directly to Knapp rather than to "A correspondent of the *Buffalo Express*" aka "B." Surely Whittlesey would have taken credit for this announcement had he made it.

William Hickok

Two months after the original announcement, John N. Ingersoll, editor of the *Lake Superior News and Mining Journal*, found himself traveling with one of the officers of the company mining at the Minesota location. Ingersoll related the following.

...Having, on our way home, fallen in company with Mr. [William] HICKOK, the gentlemanly agent and part proprietor of the company (the modern not the ancient one, we mean) engaged upon the spot referred to, he exhibited to us specimens of the copper with the implements found, and furnished us with the following facts which are more in detail than any account that we have yet seen published.

The indication which led to the discovery is a sunken trench, upon the line of the vein, which being drifted into disclosed a mass of native copper, lying in the vein, estimated to weigh seven tons. The remains of large timbers were found, by which this had evidently been propped, and beneath it were several cart-loads of ashes and cinders, shewing that the miners had endeavored to reduce the mass by fire. Several of the implements used in the mining operations were found, consisting of stone hammers, a chisel and a gad of copper. The perfect state of the point of the latter would seem to indicate that a process for hardening the metal was known, for the hammer end is most battered. With the copper of this tool, were some large particles of silver. The chisel is ingeneously constructed, so as to admit a handle, probably of wood, and the edges of the part which enclosed the latter are closely united either by welding or hammering. No iron implement was discovered.

That the mining operations were conducted to a much greater extent than is practiced by any existing tribe of Indians is apparent from the fact that the trench sunk upon the vein extends for a mile in length. The accumulation of earth in the trench concealed the depth of the workings, except in the small part re-opened; but here the depth was found to be twenty-four feet, and the width of eight feet. Similar trenches exist in the neighborhood, which were traced for several mines.

Not the least interesting part of the discovery, is the evidence of the great antiquity of the workings. Large trees were growing upon the earth that had accumulated in the diggings—one which, directly over the large mass of copper, proved to be four hundred years old!—Beneath it were trunks of trees that had previously decayed or fallen in, and the whole depth of soil that by the process of time had accumulated upon this antique furnace, was eighteen feet!

This mine is about four miles east of the large mass of copper which was removed from its place, some years since, and is now in the National Cabinet at Washington.

These mementos of ancient aboriginal industry are deserving of more than a mere passing notice. They may be considered as adding to the proof that, long before the discovery of America, a race existed on this Continent among whom the arts had reached a higher grade than with the wandering tribes that have succeeded. The Indians now living in this region know nothing of the people by whom, or the time when, these operations were undertaken. They evince a concerted effort which does not characterise the present feeble efforts in the arts. It is somewhat singular that among a people so observant and persevering, the use of iron remained so wholly unknown, since some of the ores, which exist in vast abundance, and upon the surface, in the Carp river region, are found to be easily reduced to a valuable steel by the heat of a common forge. A knowledge of the use of iron might have changed the destiny of that people, as it may be said to have done that of the race who now triumph, in the pride of art and power, over their almost perished memorials.

We trust this discovery will be followed up, and that we may soon learn more on this interesting subject. (Ingersoll 1848)

This article by Ingersoll appeared a few months later in a lightly paraphrased version in Hunt 1848. William Hickok was the director of the Minesota Mining Company for many years. Hickok talked freely and knowledgeably about the Knapp discovery on at least two occasions (Ingersoll 1848; Trowbridge 1881).

Later in the summer of 1848, apparently Knapp visited Detroit and met with the publishers of the *Detroit Free Press*. On August 2 of that year, the following article appeared (Bagg and Harmon 1848a). The impression is given that it was based on an interview or conversation, but it clearly is closely based on the original June 14 report.

We were yesterday shown some curiosities taken from the Minesota Mines on Lake Superior, better known as location No. 98 on the Ontonagon. Last winter while the snow was on the ground, Mr. Knapp, the agent of the company, traveled the sinking of the ground for a great many miles. After the snow had disappeared, by digging into those sinks, they found them to have been made by the hand of man. It appears that the ancient miners went on a different principle from what they do at the present time. The greatest depth yet found in these holes is thirty feet—after getting down to a certain depth, they drifted along the vein, making an open cut. These cuts have been filled nearly to a level by the accumulation of soil, and trees of the largest growth standing in this gutter; and also trees of a very large growth have grown up and died, and decayed many years since; in the same places there are now standing trees of over four hundred years growth.

In digging down on this river, they have lately discovered a mass of pure copper estimated at seven tons weight, mined out and laying in the vein some 18 feet below the surface, resting on skids six feet from the lowest depth of the original excavation. In the bottom of this excavation there was several cart loads of ashes and charcoal, showing that whoever worked it before, had attempted to separate the mass of copper by heating and probably dealing on cold water. There were a large number of hammers found made of stone, weighing from one to twenty pounds with a groove encircling them to secure a withe to use them with greater advantage, and also copper chisels which had probably been used in detaching the rock from the native copper.

This piece of copper is as pure and as clean as a new cent, the upper surface has been pounded clear and smooth. It appears that this mass of copper was taken from the bottom of a shaft, at the depth of thirty feet. In sinking this shaft from where the mass now lies, they followed the course of the vein, which pitches considerably; this enabled them to raise it as far as the hole came up with a slant. The excavations are on an [area] of some three square miles and with all the modern improvements at the present day for mining could not be made for seventy five thousand dollars. The old Indians say that the work was never done by their race. To contemplate the great length of time that must necessarily elapse before vegetation could start in these excavations, and trees of hundreds of years growth come up, blow down and decay is a theme full of interest, and leads the imagination thousands of years back without a response to show what race of beings were the miners or when the mines were worked. The hammers and tools found are of the same workmanship of those found in the mounds throughout the western country, and it requires no great stretch of the imagination to suppose that the same race that left these monuments to commemorate their existence, worked the mines on Lake Superior.

Mr. Knapp has shown us some rich specimens of copper and silver taken from the vein and also the tools, wedges, and pieces of the skids used by these ancient miners. This is but the commencement of discoveries that may lead to matters of great interest and importance.

This account gives the impression that Knapp (and Hickok) had mounted some kind of marketing campaign—hauling around copper and silver samples and prehistoric artifacts, giving interviews, and distributing accounts of the big discovery. With regards to the discovery, it seems they worked as a team, with Knapp taking care of the ownership situation in the field and Hickock managing the legal end in New York (and occasionally visiting at the mine). It's highly likely that their goal in publicizing their discovery was selling more shares in the mine.

Schoolcraft published a near-contemporaneous description (1851:96–97). He attributes the description only to "an observer" in September 1849.

It is along the edges or out-crop of these veins that the ancients dug copper in great quantities, leaving, as external evidence of their industry, large trenches, now partly filled with rubbish, but well defined, with a breadth of ten to fifteen feet, and a variable depth of five to twenty feet. In one place the inclined roof, or upper wall work is supported by a natural pillar, which was left standing, being wrought around, but no marks of tools are visible. In another place, east of the recent works, is a cave where they have wrought along the vein a few feet without taking away the top or outside vein stone. The rubbish has been cleared away in one spot to the depth of twenty feet, to the bottom of the trench, but the Agent [Samuel Knapp] is of the opinion that deeper cuts than this will be hereafter found. When he first came to the conclusion, about eighteen months ago, that the pits and trenches visible on the range were artificial, he caused one of them to be cleaned out. He found, at about eighteen feet in depth, measuring along the inclined face or floor of the vein, a mass of native copper, supported on a cobwork of timber, principally the black oak of these mountains, but which the ancient miners had not been able to raise out of the pit.

The sticks on which it rested were not rotten, but very soft and brittle, having been covered for centuries by standing water, of which the pit was full at all times. They were from five to six inches in diameter, and had the marks of a narrow axe or hatchet about one and three quarter inches in width.

They had raised it two or three feet by means of wedges, and then abandoned it on account of its great weight, which was eleven thousand five hundred and eighty-eight pounds, (11,588,) or near six tons.

The upper surface had been pounded smooth by the 'stone hammers' and mauls, of which thousands are scattered around the diggings. These are hard, tough, water-worn pebbles, weighing from five to fifteen pounds, or even twenty pounds, around which in the middle is a groove, as though a with had been placed around it for a handle, and most of them are fractured and broken by use. Besides these mauls there has been found a copper wedge, such as miners call a "gad," which has been much used. Under the mass of copper, and in almost all the works lately opened, there are heaps of coals and ashes, showing that fire had much to do with their operations.

With these apparently inadequate means they have cut away a very tough, compact rock, that almost defies the skill of modern miners, and the strength of powder, for many miles in a continuous line, and in many places in two, three and four adjacent lines.

The great antiquity of these works is unequivocally proven by the size of timber now standing in the trenches. There must have been one generation of trees before the present since the mines were abandoned. How long they were wrought can only be conjectured by the slowness with which they must have advanced in such great excavations, with the use of such rude instruments.

This is a mystifying reference. Who wrote this passage? It is clearly someone who has firsthand knowledge of the Minesota Mine's location and Samuel Knapp's discovery. It seems that it is not Knapp himself, as he is referred to in the third person.

Based on the detailed information presented above, it seems the most likely candidate is (once again) William Hickok.

James Thatcher Hodge

Another observer of importance in the 1840s was the geologist James Thatcher Hodge. Hodge had a long involvement with the mines of the Copper Country, beginning with building a copper smelter on the Gratiot River in 1846 (Cooper 1901:23, 1903:463). At the time of the Minesota Mine revelations, Hodge was assistant editor for mining and metallurgy at the *American Railroad Journal* (ARJ). In addition to railroads, the journal also covered "Steam Navigation, Commerce, Mining, Manufactures." In the September 1, 1849, issue of the ARJ, he related details of his trip to the Minesota Mine and described the extant landscape and the curious ancient diggings.

This mine presents some of the most interesting features of all the extraordinary mines hitherto opened on the shores of Lake Superior. It is situated two miles south of the Ontonagon River, 20 miles above its mouth—in a straight line only about 12 miles. The trap range crosses the river in this vicinity; and its hills are as high and precipitous as on Keewena Point. As there, however, they afford on their more gentle slopes, south of the principal range of hills, tracts of great fertility, on which thrives a beautiful growth of sugar maple, birch and oak and other hard wood, intermixed with magnificent white pines. The mine now worked is on the southern slope of the northern of three parallel ridges, which here constitute the trap range. Here, though nearly up to the summit of this ridge, the surface inclines gently, so that the land around the mine is cultivated and occupied by the homes of the miners. So unbroken is the surface in general, the vein would probably have remained long undiscovered, but for some curious ancient diggings, which even in the thick woods could not fail to arrest the attention of one passing over them. They consist of lines of pits, now partially filled with rubbish and overgrown with trees, which pursue a course parallel with that of the ridge itself—North of East and South of West. The course is at right angles to that of the veins on Keewena Point. The walls of the pits are ledges of amygdaloidal trap, evidently the walls of a vein. They incline in towards the axis of the ridge, the dip being about 55° North. They may be traced with other indications of the vein for more than two miles in length; the actual extent of the workings themselves may not exceed half a mile. Their depth down to the rubbish does not exceed ten feet; how much deeper this may reach I cannot tell. In one instance I observed an arch of rock left standing, under which the unknown ancients had prosecuted their work along the vein. In the rubbish are found stone-hammers or picks, one end shaped like a wedge, the other rounded. A groove passes around them, as if for a withe to bind them to handles. They are of various sizes; some so small a boy could pound with them; others would seem to have required two men to use them. They are generally made of the hardest kind of greenstone trap, such as is found directly about the mines. The quantity of them is so great, that I was informed by Mr. Knapp, the agent of the company, many cartloads of them might be collected around the pits. Most of them are broken on the edge, as though used and then discarded. Two copper gads or wedges have been found, but excepting these, no other tools than these stone picks. These were each about 2 ½ inches long; one was a spike with a square head, the other a socket chisel. They are now with other relics in the possession of Wm. Hickok, Esq., 239 Water St., New York, who transacts the business of the company in this city. The use of iron seems to have been unknown to these miners. How their rude instruments could have been used to any effect in breaking down a rock about as hard as themselves—so hard that two good miners will be several hours in putting in a hole two feet deep with the best steel drills and mallets, and so tied together frequently with masses of copper that the drill will not penetrate, or if it penetrates, the powder will not break the rock—how such tools could have made way through it is more than we can comprehend. Possibly by the aid of fire the rock may have been rendered more brittle and easier to be removed; but so slow and tedious must have been their progress the copper was necessarily of more value to them than gold is to us.

From the extent of their operations it seems that these ancient men must have had excellent success at this locality; and much credit is due them for their skill in discovering so rich a vein so far back in the woods—if indeed the region was then as wild and untrodden by human feet as it now is.

In one of their pits, twelve feet below the surface, they encountered a mass of copper too large for them to remove. From under the rubbish of centuries it has again been brought to light, and with it the evidences of their unsuccessful toil. The ancient miners had succeeded in taking the mass out of the solid vein stone in which it originally lay, and in raising it about a foot upon two skids or timbers, one of oak and one of birch. These were found in tolerably good preservation, the wood, of which I have a piece before me, being partially converted into lignite. Some ashes were found under the gravel near the mass of copper. This mass they had worked over with their stone hammers, till they had removed every particle of veinstone, that adhered to it, and filled its ragged interstices. This appears to have been done for the purpose of lightening it as much as possible. But it was still heavy enough to defy all their exertions, and having now been taken out and cut up, is found to weigh 6 tons, 5 cwt. and 47 lbs. The bottom of this mass was about 25 feet below the surface; almost directly under it was the rich stamp work of the vein; over it had accumulated about 12 feet of gravel nearly filling the pit—and on this gravel were standing large hemlock trees, one of which, considerably smaller than some of the others, presented 397 rings of annular growth; and others also evidently still older lay rotting on the surface of the gravel. It was thought by men of good judgment in such matters, that the vegetable growth indicated at least 500 years since the pits were filled—and for the period of the ancient mining we must add to this the time taken for the gravel to slowly collect by natural causes to the depth of more than 12 feet; for we can hardly suppose these long lines of pits were filled by any other agency. The above data were given me by Mr. Hickok, who was upon the spot at the time the mass was found, and who can be relied upon as one careful in forming and expressing an opinion. (1849b:544–545)

Hodge's description largely supports the Knapp document and adds interesting details concerning the orientation of veins. This article is almost certainly the basis for a presentation Hodge made at a meeting of the American Association for the Advancement of Science held at Cambridge, Massachusetts, in August 1849 (Hodge 1850b). Note that William Hickok again puts in an appearance. (For someone whose office was in New York City, he certainly appears to have spent a lot of time in Ontonagon County and also seems to have assumed the role of curator of relics.)

This is one of the few contemporary descriptions of the Minesota Mine discoveries that are attributable to a knowledgeable reporter who was also a trained geologist. James Hodge's role in reporting the first discoveries of ancient mining has been largely overlooked until now (see the bibliography), probably because most of them were published in the *American Railroad Journal*, which is not generally viewed as a primary archaeological reference. His descriptions of roaming in an Arcadian landscape that was succumbing to inevitable development are duplicated by no other writer. Hodge continued his geological studies in the Upper Peninsula (among other places) until his death in the sinking of the steamer *R. G. Coburn* on October 20, 1871 (Wilson and Fiske 1898:224).

Later Reports

Regardless of who the author of the Knapp article was, the information in the original article and subsequent reprintings quickly became common knowledge. Two years later, the report (or possibly an interview with Knapp) formed the basis for a portion of Foster and Whitney's discussion of ancient copper mining in the seminal document, *Report on the Geology and Topography of a Portion of the Lake Superior Land District in the State of Michigan: Part I. Copper Lands* (1850:159–160). Based on their own field observations, they also offered some preliminary observations on the broad geographic occurrence.

From the northeast quarter of section 31, township 51, range 37, to section 5, township 49, range 40, a distance of nearly thirty miles, there is almost a continuous line of ancient

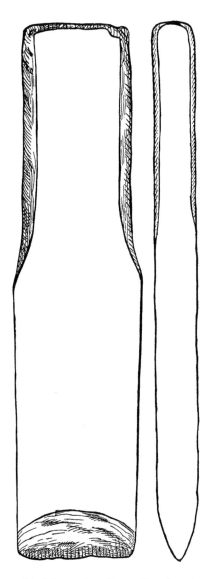

Figure 4.1. This copper chisel (two views), was found at the Minesota Mine.
Dimensions are given as 7¼ inches in length, 1¾ inches wide, and ⅝-inch thick. In
the text, Whittlesey describes this tool: "Towards the upper end the corners are
taken off, apparently for the purpose of being held in one hand, while it was struck
by a mallet with the other... None of the tools show signs of having been ground
to an edge on stone, but are beaten down roughly by hammers." (Reprinted from
Whittlesey 1863:19–20.)

Chapter 4

pits along the middle range of trap, though they are not exclusively confined to it.

Upon Keweenaw Point, they have been found extending from Eagle river eastward to range 28, a distance of twelve miles, along the base of the trap range. (1850:161)

But Isle Royale could not be ignored. A few years later Henry R. Schoolcraft reported the following discovery. Unfortunately, at this remove in time, the location or name of the mine or its operators cannot be identified.

In the beginning of 1855, discoveries were made on Isle Royal, Lake Superior, by persons mining for copper there, which denoted that antique labors of the same kind had been performed at the same place. A series of ancient pits were opened, on the line of the copper veins, to the depth of four or five feet. In these excavations, now filled with accumulations of soil, pieces of flattened copper were found, together with stone hammers, with the marks of hard usage. These old excavations in the trap rock seem to have been made by burning wood in contact with the rock, and then breaking it up with stone hammers. A large quantity of charred wood, coal, and ashes, is invariably found in these pits. A piece of oak wood, in the bottom of one of them, was, with a portion of the bank, in a good state of preservation. One end shows the marks of the instrument by which it was cut as plainly as if it had just been done. It is the most perfect specimen of the kind yet seen. The stick is about five inches in diameter, and seems to have been cut standing, by a right-handed person, with an instrument similar to an axe, having a bit at least two-and-a-half inches broad. The first blow penetrated, in the usual slanting direction, about three-fourths of an inch, cutting the bark smoothly, and leaving at its termination the mark of a sharp-edged tool. The antiquity of these excavations does not appear to be great—not probably anterior to the first arrival of the French in this lake (1855:111).

The Minesota Mine discoveries made public by Knapp started a landslide of reports. It seems that once observers knew what to look for, they started recognizing ancient diggings all over Copper Country. The following notice was initially published in the *New-York Tribune* and later published in the *Detroit Free Press*. It was also later published in toto (without crediting the original sources) by Ephraim Squier (1850:184–185) and in part by Sir John Lubbock, who claimed that it came from "Prof. W. W. Mather in a letter to Mr. Squier" (Lubbock 1875:256), the original source having long been forgotten.

A writer in the *New York Tribune* writing on American Antiquities says:

At the Copper Falls and Eagle River, as at the Vulcan and Minesota Mines, the ancient shafts are frequently discovered. Prof. W. W. MATHER, the eminent Geologist, in a private letter, referring to the two mines first named, says: 'On a hill south of the Copper Falls Mine is an excavation several feet in depth and several rods in length, extending along the course of the river. Fragments of rock, &c. thrown out of the excavation are piled up along its sides, the whole covered with soil, and overgrown with bushes and trees. On removing the accumulations from the excavation, stone axes of large size, made of green-stone, and shaped to receive withe-handles are found. Some large round green-stone masses, that had apparently been used for sledges, were also found. They had round holes bored in them, to the depth of several inches, which seemed to have been

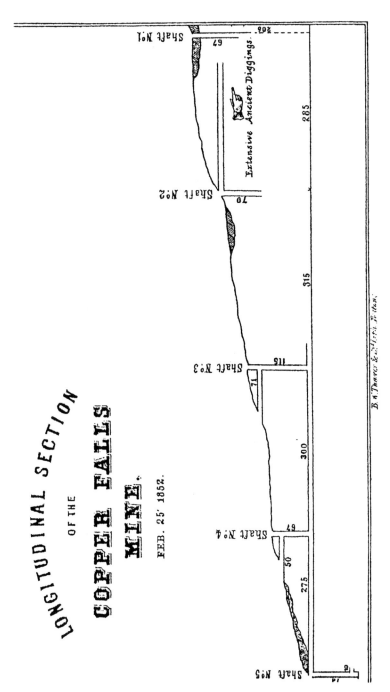

Figure 4.2. In an 1852 report to the directors of the Copper Falls Mining Company, Samuel Hill included this cross-section of the Copper Falls Mine in Keweenaw County. On this drawing, Hill recorded the presence of "Extensive Ancient Diggings." In the text, he also mentioned the "deep ancient pits and excavations" that marked the presence of rich copper veins. (Reprinted from Copper Falls Mining Company 1852:16–19.)

designed for wooden plugs, to which withe-handles might be attached, so that several men could swing them with sufficient force to break the rock and the projecting masses of Copper. Some of them were broken, and some of the projecting ends of rock exhibited marks of having been battered in the manner suggested. (Bagg and Harmon 1848b:2)

5 What Discoverers Saw in the Nineteenth Century

The high antiquity of this mining is inferred from these facts: That the trenches and pits were filled even with the surrounding surface, so that their existence was not suspected until many years after the region had been thrown open to active exploration; that upon the piles of rubbish were found growing trees which differed in no degree, as to size and character, from those in the adjacent forest; and that the nature of the materials with which the pits were filled, such as a fine-washed clay enveloping half-decayed leaves, and the bones of such quadrupeds as the bear, deer, and caribou, indicated the slow accumulation of years, rather than a deposit resulting from a torrent of water.

John W. Foster
1874

It is impossible to avoid a feeling of wonder that where these ancient diggings abound, there should always be found masses of copper. How is it they never, like us, were compelled to *trench* or dig to discover the veins of copper? Could it be that these works, which are found on this location [the Nebraska Mine] in abundance, were the result of the labor of those who inhabited this country *three or four centuries ago*? And might it not be that then but little or no soil was on the veins?

Edward Bradford
1855 (In Tenney and Leeds 1855b:189)

Chapter 5

The importance of Samuel Knapp's discovery at the Minesota Mine should never be discounted, but there were others who made what we see today as significant reports of "ancient diggings." One of the most important of these was Samuel W. Hill. In October 1847, Hill, in his role as a U.S. deputy surveyor, was working for Charles P. Jackson, who was at that time a U.S. geologist. Hill produced a map titled *Underground Works of the North West Mines*, which showed sectional views of three copper veins and a plan view of the North West Mine community in Keweenaw County. Clearly shown in the plan view are five spots designated as "Ind Works." The map would eventually be published in 1849 as an illustration in Jackson's controversial geological report. The lack of emphasis on the significance of the "Ind Works" would appear to indicate that such features were commonly known to surveyors, or at least to Sam Hill. The true significance of "Indian Works" as places to look for readily available copper would become clear with the Knapp discovery.

The evidence of prehistoric mining efforts was not always obvious. As the Canadian polymath Daniel Wilson noted:

> Here, in the neighborhood of the Minnesota Mine, are traces of the ancient mining operations, consisting of extensive trenches, which prove that the works must have been carried on for a long period and by considerable numbers. These excavations are partially filled up, and so overgrown during the long interval between their first excavation and their observation by recent explorers, that they would scarcely attract the attention of a traveler unprepared to find such evidences of former industry and art. (1856:227)

At least one early observer thought seriously about the steps that may have been necessary to successfully carry out bedrock copper mining using only rocks to free the copper.

> It is not to be presumed that these ancient people were unacquainted with the advantages of the division of labor. There were undoubtedly miners, bailers of water, and men whose part it was to manufacture tools and implements out of the pieces of rough native copper produced by the miners. Others were engaged in procuring and transporting food and other necessities of life, and still others were employed in collecting and transporting from the shores of the lake the rounded, water-worn boulders of diorite and porphyry, which were used by the miners as hammers and sledges. (Houghton 1876a:80, 1876b:168, 1879:142–143)

A common feature drawing the attention of the nineteenth-century miner was the presence of unnatural pits and trenches. After that came igneous cobble hammerstones weighing generally between 10 and 36 pounds. There can be little doubt that if there were pits or trenches visible, there were hammerstones present. Some of these stones were grooved for attachment to some kind of handle. In one case a cedar root was still twisted in the groove (Whittlesey 1863:10).

There is some limited information that suggests that the hammerstone question was complicated. Ephraim G. Squier cited a letter from geologist W. W. Mather:

Figure 5.1. William Henry Holmes in a prehistoric copper mining pit he excavated in 1892 near the Minong Mine, Isle Royale. Note the hammerstones visible at his feet and in the fill of the pit. (BAE 2273, Photo Lot 14, National Anthropological Archives, Smithsonian Institution.)

> On a hill, south of the Copper Falls Mine, is an excavation, several feet in depth and several rods in length, extending along the course of the river. Fragments of rock, etc., thrown out of the excavation, are piled up along its sides, the whole covered with soil, and overgrown with bushes and trees. On removing the accumulations from the excavation, stone axes of large size, made of green-stone, and shaped to receive withe handles, are found. Some large round green-stone masses, that had apparently been used for sledges, were also found. They had round holes bored in them to the depth of several inches, which seemed to have been designed for wooden plugs, to which withe handles might be attached, so that several men could swing them with sufficient force to break the rock and the projecting masses of copper. Some of them were broken, and some of the projecting ends or rock exhibited marks of having been battered in the manner here suggested. (Squier 1850:184–185)

J. H. Lathrop (1901a:6; 1901b:253–254) raised the possibility of specialized stone hammer "workshops." (See also Houghton 1876a:80–81, 1876b:168, 1879:143, and

Reed 1876:7.) Most of the stones, however, were unmodified, suggesting that they were used simply by grasping them with two hands and battering the trap rock enclosing the copper. However, Holmes opined that these would make "very ineffective tools" and that he believed it possible to successfully "withe-haft" hammerstones without grooving them (1901:694).

Explorers sometimes found complete hammerstones in the mines, but far more often they found incomplete specimens and fragments intermingled with the mining debris. Occasionally they reported broken specimens stacked in piles next to the pits. Hammerstones occurred in prodigious quantities, being measured, depending on the author, by the ton, by the cartload, or by the millions. (See Brine 1894:44; Clarke 1975a:8; Fox 1911:86–87; Hagar 1865:308–309; Holmes 1901:692–695; Whittlesey 1868:474; Wilson 1855:204, 1856:228; Winchell 1881a:604–605 and especially Martin 1999:96–107 for additional discussions of hammerstones.)

Collectors have continued to discover hammerstones, some apparently in undisturbed contexts, in relatively recent times (Straw 1962). Amateur mineral collectors have scoured the spoil piles at old mines for decades, looking for mineral specimens. If the pile is relatively shallow, it is possible that artifacts (most likely hammerstones) could be exposed, recognized, and taken by the collectors. Such discoveries and collection almost certainly would not be reported to state officials. Artifact collection of any kind on state lands is generally forbidden, but unfortunately this is next to impossible to enforce.

In addition to hammerstones, there were also mining tools made of copper. These include chisels, wedges, and gads (Anonymous 1890:180; Emerson 1862a; Gillman 1873:173; Houghton 1862; Whittlesey 1852a:28, 1863:11, 19, 21, 1868:475; Winchell 1881a:602). There were also rare examples of knives, "projectile points" (McKenzie 1864a; Whittlesey 1863:11, 1868:477, Plate 16, 2–2b), copper mauls or sledges (*Philadelphia Ledger* 1883; Whittlesey 1863:19), and small hammers (Cresson 1884; Leidy 1867). Conspicuous by their near-absence were probable or possible picks (Draper 1879b:47; Matson 1872:14; Whittlesey 1868:475), of which only a few were found. Perhaps the early miners quickly discovered that native copper, being relatively soft and pliable, did not make a very useful wedge or chisel.

Pits could be of surprising size. One pit at the Copper Falls Mine (20KE63), near Eagle Harbor on the Keweenaw Peninsula, was described as being 70 feet long and 37 feet deep (Kayner 1883b:330). Once the pits obtained such depths, gaining access must have become a major problem. Climbing the rock face was probably not always possible or desirable. Trunks of trees with the branches trimmed to leave rungs were reported from several sites (Anonymous 1850; West 1929:52; Whittlesey 1863:19).

Explorers often found significant quantities of charcoal in the pits. They presumed that the prehistoric miners had built wood fires against the mine walls and then doused the heated walls with water, causing the stone to crack and expose or free the embedded copper. However, others questioned the viability of this process, principally based on the logistical problem of having enough water readily available to do the dousing

Figure 5.2. Hammerstones from a prehistoric copper mining pit excavated by William Henry Holmes in 1892 near the Minong Mine, Isle Royale. Note the pick-axe in the foreground and the silver dollar (left center) for scale. (BAE 2294, Photo Lot 14, National Anthropological Archives, Smithsonian Institution.)

(Fox 1911:90–91; Martin 1999:93–96; Parker 1975:9, 11). The presence of charcoal in the mines is problematic. It seems ubiquitous, but could it simply be the byproduct of brushfires? Or did ancient miners burn off the ground cover, brush, and trees that were growing in the places they wanted to dig? Or perhaps the fires were simply to make the cold, damp pits a little warmer.

James Hodge reported the successful results of an early experiment carried out at the Minesota Mine.

> While I was recently at this mine an experiment was tried with fire upon one of the large masses of copper raised to the surface. The object was to remove from it the veinstone which adhered closely to and filled its interstices. After being surrounded by burning logs and wood for a day or two, cold water was thrown upon it, and the stone by this means was rendered much more fragile, so that the labor of many days was saved by the operation. Witnessing the group of men busy about this mass, one could not but be carried back to the time, when on this very spot another people were occupied with the same labor. No memorials remain of their existence but the vestiges of their work upon the surface. (Hodge 1849c)

Sometimes the actual process of mining could be observed frozen in time.

> A few unfinished jobs have been found in these ancient pits, which throw some light upon the manner in which the work was carried on. In two instances there were projecting masses somewhat resembling urns, or inverted short-necked bottles, and completely smoothed by hammering, especially at the thinner portion or neck. It appears that the ancient miners first removed the rock from around the veins of copper. This was done by building fires upon or about it, and when heated, crumbling it by throwing on water. By means of stone mauls the fragments were broken up and removed. When the vein was sufficiently exposed on all sides, a point was selected where the copper was thinner or narrower than the average of the vein. Here they commenced cutting off a mass, and by patient and long-continued hammering severed the two portions of the vein. (Hagar 1865: 311)

The document known as the Knapp report (see Chapter 4) described a mass of detached copper resting upon a cribwork of black oak logs 6 to 8 inches in diameter at the Minesota Mine (20ON1), in the Upper Peninsula near present-day Rockland, Michigan. The copper mass must have been raised several feet, presumably through the use of levers (Baldwin 1872:43–45; Foster and Whitney 1850:159; McCracken 1876a:513; Whittlesey 1852a: 27, 1863:19).

At the Waterbury Mine (20KE73), just west of present-day Delaware on Michigan's Keweenaw Peninsula, men reported finding blocks of stone weighing two to three tons that in prehistoric times were removed from the working face. These stones also must have required levers to move them (Allen 1885:395; Baldwin 1872:45; Whittlesey 1863:7). A 5,720-pound mass discovered in the Minong Mine on Isle Royale was reported to have poles underneath it, presumably for moving it (Winchell 1881a:602, 1881b:184).

Other reports indicate that a range of facilities were developed within the pits and crevices, such as a gutter or trough made of cedar bark to carry off water from the mine (Baldwin 1872:46; Whittlesey 1863:7), drains cut into bedrock to carry off rain or meltwater (Gillman 1873:173; Lane 1898:16), a scaffolding of white-birch poles that had been used as a landing to assist in removing the excavated rock (Lawton 1881:81; Wood 1907:288), and the placement of large granite boulders (300 to 400 pounds) to hold up the hanging ground (Davis 1875:370; Lane 1898:16). (Hanging ground is a term for rock or rock slabs in the "ceiling" above a tunnel floor that are not artificially supported by posts or beams. The presence of hanging ground was possibly the most dangerous aspect of mining for the ancient miners.)

Some scholars have argued that the miners didn't use water, as it was too costly in terms of labor. This is a question that is unlikely to be resolved at this late date, given that all known mining sites discussed in this study have been altered beyond recognition, if not obliterated. Clearly, it is next to impossible to know the structural, geological, and mineralogical characteristics of each mine. We do know, based on some descriptions made by mine operators, that the natural presence of excess water could be a deterrent to mining. For instance, at the Central Mine,

...a shaft was commenced to be sunk in the vein seventy feet to the north of the mass of copper found in the ancient excavation, when the rock was penetrated to the depth of six feet, water was found too quick to carry forward the work of sinking, until a shallow adit gallery should be made to connect with the shaft and drain off the surface water. (Tenney and Leeds 1855d:172)

In 1854, William H. Stevens, at that time agent for the Agate Harbor Mining Company, reported:

...There are numerous ancient pits supposed to have been made by some ancient miners. At several points we attempted to clear them of the rubbish and see what the appearance of the vein is in depth; several of them were sunk upon to the depth from ten to twenty feet. At one point we sunk thirty-two feet and came to an ancient level, where a man can pass back and forth over twenty feet between the walls of the vein. All of this ancient work has been done in a good and well-defined vein—one of sufficient width to work without breaking either wall at the points cleared out. It varies from three to five feet in width and contains native copper. We have traced the vein by these pits over one-fourth of a mile in a continuous line, where they are plainly marked; the indentations (at some points) before they were cleared out, were from four to ten feet deep.

At the above mentioned depths we came to a large pool of water that fills the old mine up to a certain level and were compelled to suspend working until we get up machinery to hoist or drive a level to drain the mine. (1854:1)

It appears that at an unknown number of mines, getting rid of water was a more pressing problem than obtaining it.

Removal of mining detritus was also a necessity. Presumably this required some equipment other than bare hands. On both Isle Royale and the Keweenaw Peninsula there were reports of wooden bowl-shaped utensils (Gillman 1873:173–174; Hagar 1865:310; Whittlesey 1863:20; Winchell 1881a:602, 1881b:184). At least one of these had "splintery pieces of rock and gravel embedded in its rim" (Foster and Whitney 1850:161). At several mines excavators found cedar wood objects shaped much like canoe paddles, but showing extensive wear, suggesting they were used as shovels (Allen 1885:395–396; *Detroit Free Press* 1893; Fox 1911:91; Hagar 1865:309–310; West 1929:52; Whittlesey 1852a:28, 1863:8, 10; Winchell 1881a:603, 1881b:184). Birch bark may also have been used to make bowls or boxes to remove soil excavated to reach buried lodes or boulders (*Detroit Free Press* 1893).

A unique specimen came from the Hilton Mine (20ON426) near Ontonagon: a leather bag, 11 inches long and 7 inches wide, was found with a mass of native copper deep in a prehistoric mine. Many believed this to be a carrying bag for mined pieces of copper (*Ashland Press* 1873; Hagar 1865:310). Given that a cubic foot of copper weighs about 559 pounds, the removal of small, easily portable pieces in a "tote bag" of this size makes eminent sense.

Another organic artifact of unknown purpose, preserved deep in a prehistoric mine, was "a piece of string, about a foot long, made of some raw-hide, supposed to be of the caribou, tied in the middle by 'a square knot and a half-hitch' " (Winchell 1881a:603–604, 1881b:184).

Prehistoric copper mining must have been an arduous, physically stressful undertaking, made even more unpleasant by the presence of black flies, midges, gnats, mosquitos, and other pests. James T. Hodge was highly conversant with these denizens.

…it is the insects…that most interfere with agricultural operations, and discourage men from undertaking to clear and cultivate the lands. These are the various kinds of flies common in the spring and early part of the summer all along our northern wild lands. The first that come are the common black flies—half as large as a house fly. They seem to rise out with the first flowers from under the snow, and in still weather, or at all times under the shelter of the woods, surround their poor victims like swarms of bees, attacking incessantly the hands and face, biting particularly the wrists under cover of the wristband, and the ears, and behind them under the edge of the hair, and round the neck, till the parts become severely sore and swollen, sometimes so much so that the cavities of the eye are covered over by the swollen flesh. Soon their forces are joined by mosquitos, which keep up their attacks all night long as well as all day, and in certain favorite spots far exceed in number the dense swarms of the black fly. The mosquitos have not been long at work, before one is surprised occasionally by a new sort of bite, a sharp prickling sensation, burning like fire, but without any visible agent to produce it. A repetition of the pain, however, makes the eye more watchful, and the two almost invisible wings of a very minute fly are discovered standing on end, as the head of the little rascal is turned down into the pores of the skin. These torments are called midges, or by the Indians in Maine, very appropriately *"no-see-ums."* They work, too, night and day, and encounter no serious obstacle in the meshes of a mosquito net. Next appears a fly similar to the common horse fly, which, as horses are not very plenty, seems driven to attack the human race—below generally treated by them with more respect—their bite is sharp, but not so often repeated as of the others. With them comes a yellow buzzing fly, which I have mistaken for the stinging "yellow jacket," whose appearance is later in the season. This fly hangs buzzing in the air in one spot directly before the face, and if frightened suddenly takes another position before it lights to take its meal. The bite of this fly is like that of a leech, but as the attack is not made by battalions it can be generally guarded against. The last of these pests is the great clumsy "moose fly," as he is called in Maine, much larger than the "deer fly," which I have not seen here. From his size he seldom gets a chance to bring his apparatus to bear upon the human person, and the pump never performs more than the first stroke before the whole machinery is knocked into utter destruction. But in connection with the two preceding species this fly is very severe in its attacks upon cattle and horses. (Hodge 1850c:434)

Given this description of the various insect plagues (which were probably no less severe in prehistoric times), it may be reasonable to propose that the mining season was scheduled so as to avoid the insect plagues.

Mining must have been a dangerous activity, especially in the deeper pits. However, historic excavators recovered remarkably few human remains from the pits. J. Venen Brown reported that Charles Whittlesey showed him "a piece of human skull and other bones, which have lately been found in the Ancient Indian Diggings on the Ontonagon River" (1850b, 1850c). James T. Hodge reported, from an unnamed site (or sites), "The only human bones yet found in these pits are those of the arm, which were colored green by the action of copper. It is hoped that these may prove a forerunner of more important relics" (1849b:545). Generally speaking, if there had been a catastrophic wall collapse in a deep pit, it would have been nearly impossible for the miners of that time to recover a seriously injured individual or a body. Historic-period miners faced the same dangers (see *Detroit Free Press* 1855b, 1855c).

The actual nature of some pits has been subject to alternative interpretations. For example, the Calumet Ancient Pit (20HO4) at the Hulbert Mine (later the Calumet & Hecla) has most commonly (and recently) been interpreted not as a copper mining pit, but rather a place where many native copper and organic artifacts (deerskin, birch bark baskets, and more) were cached (Griffin 1966a; Hulbert 1893, 1899; Lathrop 1901a:6–7, 1901b:255–258; Charles Moore 1915:449–453; Wood 1907:291–292). This was one of the most famous of the early "ancient diggings" discoveries.

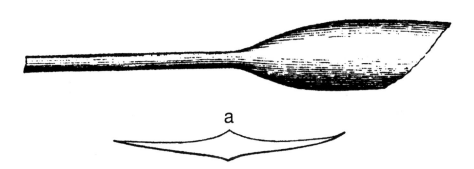

Figure 5.3. Newton H. Winchell's illustration and transverse section of a wooden paddle found near prehistoric copper mining pits on Isle Royale. The original caption reads: "At McCargoe's Cove, Captain William Jacka discovered a wooden shovel or paddle, which showed by its worn and battered side that it had been used in moving dirt. The blade was four and three quarters inches wide, and about twelve inches long. The handle had been broken, but still showed the length of about a foot. It was all perfectly wrought and smooth, and very true in form. A rounded ridge on the upper and lower sides of the blade extended along its middle, tapering off along the same sides of the shaft or handle upward. It was wet and swollen when found, but, on drying, it shrank to a width of three fourths of an inch, and curled out of shape...a represents the upper side of the blade, and the ridge, evidently designed to strengthen the instrument, extends to within an inch or two of the end, and gradually and smoothly sinks to the level of the surface." (Reprinted from Winchell 1881a:Figs. 2 and 3.)

Chapter 5

However, an earlier account (*Detroit Free Press* 1898; McKenzie 1865b) gives a very different impression.

In company with nearly all the mine agents of the district we visited this remarkable exploration on Tuesday last. The chief feature is the fact that the existence of copper in the conglomerate was well known to the ancient miners, who were never known to work in poor places, but the richer the deposit the more extensive their workings. In the belt of conglomerate crossing the Calumet property these old miners had made an immense excavation, and from appearance they were engaged on it every summer for at least six years! This is shown by the layers of leaves found in the mound of attle or waste rock thrown out of the pit. The pit has been cleared out to a depth of eighteen feet, and the bottom has not yet been reached, though it is probable it is not much deeper. For sixteen feet the pit is filled with decomposed vein matter, surface soil, boulders, etc, the whole mass most thoroughly impregnated with green carbonate. The next two feet of debris is entirely composed of finely broken conglomerate, fresh and clean, and rich in native [copper] and the carbonates and oxides of copper. Several large lumps of copper have been found here, and in clearing out the pit over two barrels of nuggets have been obtained. The attle heap forms a pear-shaped mound, about thirty by thirty feet, and six feet in height. Over this there is simply a layer of roots and decayed vegetable matter not more than four inches thick, and by digging through the attle the yellow drift sand is reached, and the layers of leaves passed through, showing that the rocks were not only covered with soil, but that there was as heavy growth of timber as at the present day. And yet there are growing on top of the mound a pine sixteen and a birch fourteen feet in diameter, and, still more, an immense birch root was found in place, but no trace of the tree, and a stump of pine almost entirely rotted away. These partially show the antiquity of the workings, though no exact period can be fixed. The rock of the waste heap is very much decomposed, and green as epidote from discoloration by carbonate of copper. It is quite evident the ancients mined here for coarse copper, as all the stamp and small nugget copper has been cast aside and is now found partially and entirely decomposed. It has not generally been supposed that heavy copper existed in the conglomerate, nor are we yet prepared to say it does, as we are inclined to the opinion that the heavy pieces were and are to be found at the junction of the conglomerate with a belt of amygdaloid associated with and underlying it. In a pit made several hundred feet north this feature is well known and some very rich rock taken out. It is seldom that the ancients have dug deep except for coarse, heavy copper, and when this pit is entirely cleaned out we think this will be found to be the fact there.

In reopening this old working several curious relics have been found: a birch bark basket, similar to those made by the Indians of the present day, a wooden shovel of hickory or ironwood and pieces of beaver, Indian tanned, with the fur as good as if just taken. Strange enough, however, not a single stone hammer has been found.

It is clear from this description that the pit *was* the result of mining for copper, especially with the inclusion of the detailed description of the large spoil pile adjacent to the pit, a critical feature ignored by Hulbert and Griffin.

Mine by mine synopses are provided in Appendix 3. There is extensive duplication of information between references for some of the better-known sites, a fact ruefully noted by John W. Foster in his *Pre-Historic Races of the United States of America.*

What Discoverers Saw in the Nineteenth Century

In the foregoing account of ancient mining on Lake Superior, I have availed myself, with the occasional omission of a sentence or the substitution of a word, of a chapter prepared by Mr. Whitney and myself, and contained in our "Report on the Geology of the Lake Superior Region," Part I, published by the authority of Congress, 1850. This was the first connected account of these ancient explorations. The facts embodied in that chapter have since been appropriated by subsequent writers, and often, I regret to say, without any acknowledgment of the source from which they were derived. (1874:269)

How Much Copper Was Obtained?

What was the yield from all this rock-bashing? There were imaginative and curious observers as mines were being created in the nineteenth century. One of them was experienced miner Alvinus Wood, who reported the following experiment at the Quincy Mine (20HO128) in 1855.

At the place where this shaft was sunk, barrel-copper (i.e., copper too coarse to go to the mill, yet not in pieces large enough to be shipped separately) was abundant. This class of copper was most valuable to the old miners; since large masses were too large for them to handle, and stamp-copper was too small to serve their uses. It is evident that they did not understand the art of smelting.

This excavation represents the removal of 41 tons of rock, and the ground at that place, judging from what could be seen in the walls of the shaft, carried fully 12 per cent. of coarse copper. This would represent 5 tons of copper actually extracted. From the amount of ancient work showing further south on the lode, it would not be too liberal to estimate that four times as much more copper, or say, 25 tons [90 cubic feet] in all, was obtained from this locality. The amount may have been much greater, for the ground under the workings was afterwards proved to be very rich in barrel-copper.

I still have in my possession a handful of fragments of copper, torn from the barrel-copper by blows from the stone hammers, each fragment showing the torn face caused by the blow. These were found in the bottom of the shaft. (1907:288–289)

The next year Mr. Wood conducted the following experiment at another location.

At the Huron mine [20HO275], next south on the Isle Royale lode, the first four shafts were started in 1854, in ground not suited to the ancient miners. In 1856 I took charge of the property, as agent, and pursued work in the first shafts. In 1857 I discovered signs of ancient workings to the south; but in this work, instead of digging pits, the lode had been stripped from north to south. The large stones had been piled up, and the dirt thrown back as the work progressed.

I caused the lode to be cleaned off, beginning at the north end of the old work, for a length of 30 ft. and a width of 16 ft. To the south there were abundant signs of ancient work for an additional distance of 80 ft. About 3 ft. in depth of the lode had been removed where it was uncovered. This surface had had the benefit of disintegration since the Ice Age.

To get an idea of the amount of copper which the ancients obtained, I had the uncovered surface—30 by 16 ft.—worked over with pick and bar, no powder being used. Ten tons of barrel-copper were taken up; the largest piece weighing 1,500 lb. The copper obtained was weighed and shipped to the copper smelting-works at Detroit. I should estimate that the ancients got twice as much as we did, or 20 tons, to say nothing of what they took from the additional ground worked by them to the south. (1907:290–291)

Wood also reported on the activities of other mine developers.

At about the same time, Mr. Charles H. Palmer, Agent, and Mr. Wm. B. Frue, his foreman, opened an ancient pit on the Pewabic mine location [20HO130], where the Pewabic main shaft was afterward sunk. The lode at that place was very rich in barrel-copper for a width of 20 ft. or more. The pit was somewhat irregular, about 6 ft. deep and 15 ft. wide, the length being 20 ft. across the lode, and representing the removal of 138 tons of rock, which carried about 10 per cent. of barrel copper, or 13.5 tons [48 cubic feet] of metallic copper.

The ancient work extended to the south more than 100 ft., but was not so deep. The ground under these pits proved to be very rich in barrel-copper, as is shown by the workings of the Pewabic Co. (This ground now belongs to the Quincy Co.) I think it is safe to estimate the copper taken from this south working by the ancients to be twice that taken from this shaft, or a total of 40.5 tons [145 cubic feet]—probably a great deal more. In the bottom of this shaft were found the remains of a large fire, the unconsumed brands of which had not been moved, indicating a sudden stoppage of work. One brand, a part of a young black-oak tree, 6 in. in diameter, showed, by the form of the cutting, that it had been cut at the stump while standing. Having been continuously under water, it was well preserved and of full form. It had evidently been hewed with a thin-edged tool, about 2 in. wide. A thin chip, which had been cut from the face left by a previous chip, was broken off at the base, about 1/8 in. thick; but the cutting tool had gone down 0.25 in. below the upper end of the remaining stub, leaving a gash into which I inserted my knife-blade. This proved that the cutting had been done by a thin, sharp blade, probably of copper, and not by a stone axe. (1907:289–290)

How many seasons of toil were expended by the ancient miners? This question intrigued some observers in the 1870s. Michigan newspaper publisher and printer S. B. McCracken offered the following statement.

Though it is probable that not one-tenth of these ancient excavations have so far been revealed, some idea of their extent may be arrived at, from the statement of a gentleman familiar with the mines, who has calculated that, at one point alone on three sections of land toward the north side of Isle Royale, the amount of labor performed by those ancient men, far exceeds that of one of our oldest mines on the south shore of lake Superior, a mine which has now been constantly worked with a large force for over twenty years. Or, stated in another form, that it would have required a force of one hundred thousand men fifty years (with their means of working) to do an equivalent amount of work. (1876b:58–59)

Number of Prehistoric Copper Mines

Another thing we do not know is how many mining pits there actually were. Drier and Du Temple (1961:15) estimated that there were "over 5000." George Quimby (1960b:52) said "thousands." Roy Drier (1961:7) stated: "On the surface of the Minong rock dumps a hundred or more hammers were visible. The 1870 estimate of

Figure 5.4. In 1862, at the Mesnard Mine in Keweenaw County, Michigan, a group of men uncovered this mass of native copper weighing 18 tons. Marks left by prehistoric miners showed that they had hammered off sections of it. (Courtesy of the Burton Historical Collection, Detroit Public Library.)

three thousand pits and one thousand tons of hammers does not seem to be out of line considering that hammers have been removed from the area for some eighty years." How these numbers were arrived at, we do not know. It is clear that, historically, no one, at any time, was keeping an accurate count of the number of pits and hammers. Certainly the sample size presented in this study is small, but it appears to be all we will ever have, given that most ancient pits were destroyed in the latter half of the 1800s and the first quarter of the 1900s. However, there is the possibility that some pits on the Keweenaw were less obvious to nineteenth-century explorers and remain undiscovered today. Even in ancient pits that were discovered and exploited, there are possibly sediments left in corners here and there that could yield datable charcoal and samples of hammer fragments.

When Did the Ancient Mining Begin?

Attempts at determining the age of the pits began early. Charles Whittlesey, perhaps the most important of the early observers, estimated the pits to be 1,200 to 1,700 years old. Most of these age-determination efforts in the mid-1800s were dependent on counting the growth rings of trees growing in the ancient pits. Some of those espousing tree-ring counts to establish a minimum age were Samuel O. Knapp, mine agent and first discoverer of the "ancient diggings" at the Minesota Mine (20ON1) (Foster and Whitney 1850:160), pioneering Lake Superior area missionary John Pitezel (1859:425–426), and mine executive William Hickok. All of these men understood that the actual "mining ages" of the pits were older (perhaps much older) than the trees growing in them. This passage by John W. Foster (1874) shows that the writer clearly recognized the great age of the ancient diggings.

> The high antiquity of this mining is inferred from these facts: That the trenches and pits were filled even with the surrounding surface, so that their existence was not suspected until many years after the region had been thrown open to active exploration; that upon the piles of rubbish were found growing trees which differed in no degree, as to size and character, from those in the adjacent forest; and that the nature of the materials with which the pits were filled, such as a fine-washed clay enveloping half-decayed leaves, and the bones of such quadrupeds as the bear, deer, and caribou, indicated the slow accumulation of years, rather than a deposit resulting from a torrent of water.

The relatively simple question of age is difficult to answer. To begin with, the first usage of native copper may have resulted not from bedrock mining but from the discovery and manipulation of pieces of drift or float copper (Halsey 2010; Salisbury 1885).

The prehistoric copper pits indicate a spectrum of abandonment strategies, ranging from backfilling the trench with attle (broken mine rock containing little or no metal) to simply walking away and allowing the excavated pit or trench to fill naturally with slumped attle, branches, twigs, leaves, silt, and even an occasional clumsy animal.

To date a pit, archaeologists need organic material from the actual period of use. This is because copper artifacts cannot be assessed by radiocarbon dating. Unfortunately, miners of the nineteenth century were highly efficient in destroying or dispersing the material that could have provided multiple radiocarbon dates: charcoal.

Archaeologist and author Susan Martin reports that the earliest known dates documenting copper usage are in the range of 7000 years ago (Martin 1995:122). These dates are based on radiocarbon dating of organic materials (one wooden spear haft and one charcoal sample) associated with copper artifacts. For more on the dates of these ancient mines, see also Beukens et al. 1992; Halsey 1966; Hill 2011, as well as the Afterword and Conclusion of this volume.

6 Who Were the Prehistoric Miners?

It is an error to suppose that any more civilized or superior race of people did this work, for the tools betray their true Chippeway origin, and are such as all Northern Indians made use of prior to the coming of Europeans.

I am perfectly convinced that most of the native copper veins, now opened and wrought by European and American miners on Lake Superior, were known and worked superficially by the red men, hundreds if not thousands of years before America was discovered by Columbus.

Charles T. Jackson
1850

That these ancient miners belonged to the present race of North American Indians is not at all probable. The mines seem from the unfinished jobs left to have been suddenly abandoned, as if caused by the dispersion of the people engaged in working them. The Indian race now inhabiting North America ... would have continued to use and perpetuate the knowledge derived from them, but when the Jesuit missionaries came among them in the sixteenth century they found no tradition even pointing to the existence of these mines, while copper ornaments and utensils were unknown.

Ashland Press
1873

Chapter 6

Charles T. Jackson was right. Time has shown that he was correct in every one of his statements above, even to his rough estimate of how long ago the mining took place. But he was in the minority. Few other people in the mid-nineteenth century were thinking about the age of the prehistoric mines in terms of thousands of years.

Jackson, a geologist, wrote at length about prehistoric mining and copper trading as he imagined them.

> Long anterior to the settlement of this country by white men the children of the forest were familiar with the use of native metals ... we find proofs in the ancient working of on Lake Superior, as well as in the accounts recorded by the ancient French Jesuits, who were the first Europeans that visited the lake, that the Indians built fires on and around the masses of native copper which were too large to be removed, and after softening the metal, cut off portions with their hatchets. They understood how to fashion the malleable native metals into all the weapons, ornaments, or tools employed by them, and manifested considerable ingenuity and skill in this handicraft, but no proofs have ever been discovered that they ever made any castings of metals fusible at a high temperature ... Native copper was an invaluable metal to the Indian tribes in North America. It must have given to the nations possessing it great power over their neighbors, and by exchanging of the metals for other articles the Indians must have carried on a most extensive trade. It seems to be proved by the fact that tools of native copper derived from Lake Superior are found in many of the western States, both in the soil and in ancient mounds...
>
> It has been suggested that a more civilized people than those who are supposed to have built mounds in Ohio, Kentucky, and some of the other western states, might have been the original copper miners of Lake Superior. It is certain that they had pieces of the metal in their possession, for it is found buried in the mounds; but it is more probable that they obtained it by trading with the Indian tribes living on the borders of the great lake ... That the mining was performed by the Indians having about the same degree of civilization as that of other northern tribes, cannot be doubted by those who see the rude tools dug up in the ancient excavations on Keweenaw point. The stone hammers made of oval polished shore pebbles, grooved in the centre for a with which served for a handle, are exactly such as the Indians of Maine, New Hampshire, Massachusetts, and Rhode Island employed anterior to the settlement of this country by Europeans. The Indians here used such hammers to break up the stones of which they made their arrow-heads, and on Lake Superior the native tribes employed them for similar purposes, and for breaking out pieces of native copper from decayed veins. The Indian miner assisted the operation by kindling a fire on the rocks he wished to break, and hence the origin of charcoal and charred firebrands which have been found in the ancient mines on the Ontonagon river.
>
> None of the ancient excavations thus far discovered extend to more than six feet in depth, and are such as Indians would be most likely to have made. Furthermore, we found half-finished scalping knives and spearheads in the soil near the Eagle river copper mine; and those instruments bear ample evidence of their Indian origin.
>
> We may therefore consider this question as settled, that the ancient Chippewas wrought many of the copper veins, as well as loose masses of native copper which they found on the shore of the lake and in the beds of the streams. (1849: 373–375)

However, there were others who were not convinced that the question was settled. Some decided the answer of identity was simply unknowable (*Chicago Daily Tribune* 1856; *Commercial Bulletin* 1849:1; Egleston 1879:282; Forster 1879:142; Foster and

Who Were the Prehistoric Miners?

Whitney 1850:162). One of these was Jackson's one-time employee and ultimately his replacement, Josiah F. Whitney. Whitney took the view that the prehistoric miners couldn't possibly have been the ancestors of the Native Americans of the nineteenth century. Below, his description of the Minesota Mine discovery.

> During the winter of 1847-'48 it was discovered that mining operations had been carried on at this location many hundred years previously; by what people or nation, it seems quite impossible to decide, ... Large quantities of stone hammers, or boulders of an ovoidal shape, with a groove cut around them near the middle, probably for the purpose of attaching a handle with a withe, were found buried in the rubbish which filled these ancient mines. In the principal excavation a large mass of copper, about ten feet in length by 3½ feet in breadth, and averaging twenty inches in thickness, estimated to weigh seven tons, was met with, as represented in the sketch, (fig. 4). This mass of copper had evidently been loosened from the rock in which it occurred, probably by the action of fire and sudden cooling, and with the aid of copper wedges and the stone hammers; and had been raised several feet above its native bed on blocks of wood, which were found still remaining under it quite decayed, yet preserving their original form. The whole surface of the mass of copper was hammered smooth by repeated blows ... The whole aspect of the excavations argues in those who made them a higher degree of perseverance and intelligence than is exhibited by the Indians of that region at the present day. Who were the real workers of these ancient mines is a question for the antiquary to study. (Whitney 1849:745–746)

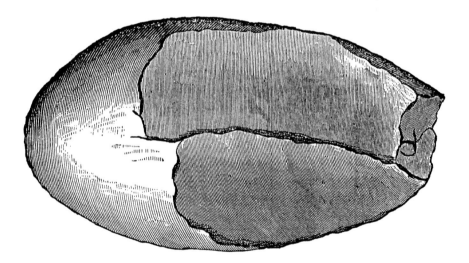

Fig. 6.1. This broken stone maul was found at the Central Mine on the Keweenaw Peninsula. (Reprinted from Whittlesey 1863:Fig. 7.)

Chapter 6

At the time, Jackson and Whitney had a very public disagreement, and Jackson wound up substantially the loser in public opinion. His assessment of Michigan's prehistoric copper mines has stood the test of time, but in his lifetime he was discredited.

An interesting aside: 136 years after Jackson's death, Ramon Martin and Sukumar Desai, Harvard Medical School specialists in anesthesia and pain management, studied Jackson's life and problems. Of his mining/surveying problems, they reported:

> When one of his former surveying assistants, Whitney, applied for membership to the American Academy of Arts and Sciences, Jackson opposed the application because Whitney had filed a complaint against Jackson's handling of the Michigan surveying expedition. When Jackson could not substantiate his claims against Whitney, the academy briefly considered revoking Jackson's membership. It ultimately did not but only out of respect for his prior scientific accomplishments. It was this negativity/hostility and perceived lying that made the public distrustful of Jackson's claims…Charles Jackson had a formative role in William Morton's demonstration of ether as an anesthetic. He lectured Morton about the use of ether and provided him with a bag of the gas. But, as at other points in his life, Jackson's ideas were taken and developed by others to fruition. Although he was a successful chemist, surveyor, and geologist, symptoms of ADHD [Attention Deficit Hyperactivity Disorder] limited his ability to follow through and be more productive. The symptoms of ODD [Oppositional Defiant Disorder] led him to be less than truthful about his follow-up actions after stating an idea and to relentlessly denounce those who did further develop "his" ideas. (2015)

More information on Charles T. Jackson and his legal troubles can be found in "Full Exposure of the Conduct of Dr. Charles T. Jackson, Leading to His Discharge from the Government Service, and Justice to Messrs. Foster and Whitney, U. S. Geologists," 1849, unknown publisher.

Jackson wasn't the only nineteenth-century writer to support the idea that the prehistoric miners were ancestors of Native Americans. Many prominent authors thought the same.

Charles Whittlesey was a no-nonsense man intimately familiar with the beliefs and practices of Native peoples. Below are his conclusions concerning ancient mining on the shores of Lake Superior. (These were written in 1856 but not published until 1863.)

> An ancient people extracted copper from the veins of Lake Superior of whom history gives no account. They did it in a rude way, by means of fire and the use of copper wedges or gads, and by stone mauls. They had only the simplest mechanical contrivances, and consequently penetrated the earth but a short distance.
>
> They do not appear to have acquired any skill in the art of metallurgy or of cutting masses of copper. For cutting tools they had chisels, and probably adzes or axes of copper. These tools are of pure copper, and hardened only by condensation or beating when cold.
>
> They sought chiefly for small masses and lumps, and not for large masses.
>
> No sepulchral mounds, defences, domicils, roads or canals are known to have been made by them. No evidences have been discovered of the cultivation of the soil.
>
> They had weapons of defence or of the chase, such as darts, spears, and daggers of copper. They must have been numerous, industrious, and persevering, and have occupied the country a long time. (1863:29)

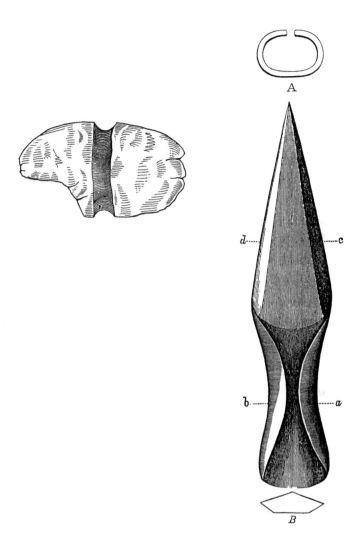

Fig. 6.2. These two prehistoric artifacts were found at the Copper Falls Mine on the Keweenaw Peninsula. On the left, a broken stone hammer or maul with a central groove for a withe, 7 inches long. On the right, a copper spear head (B=section of blade at c-d; A=section of blade at a-b). About the spearhead, Whittlesey wrote: "It was evidently formed by beating the metal while cold, probably between stones, having a rough and not a polished exterior...A piece of decayed wood was found in the socket...being apparently the remnant of the shaft, by which it was hurled. As the edges of the "shank" or socket are not soldered together, but only bent around the shaft, it was probably wound with some ligament to give it strength." (Reprinted from Whittlesey 1863:Figs. 4 and 5.)

73

Chapter 6

Henry R. Schoolcraft, who probably had more practical experience with Native Americans of the Lake Superior region than any other European individual, offered some trenchant observations on the probable practicalities of ancient copper mining.

It is apparent, that the ancient Red miners of Lake Superior supplied the demand [for copper], in its fullest extent. They probably received in exchange for it, the zea maize of the rich valleys of the Scioto and other parts of the West; the dried venison and jerked buffalo meat of the prairie tribes; and sea-shells of the open coasts of the Atlantic and Gulf. It is not improbable, indeed, when we examine the rocky character of much of the Lake Superior region, and the limited area of its alluvions and uplands, which appear ever to have been in cultivation, that parties of various tribes performed extensive journeys to this upper region, in the summer season, when relieved from their hunts, to dig copper, that it was a neutral territory; and having supplied their villages, in the manner the Iowa and Minnesota Indians still do, in relation to the red Pipe-stone quarries of the Coteau des Prairies, returned with their trophies of mining.

No tribes, indeed, whose history we know or can guess, possessed civilized arts to sustain themselves in this latitude during the winter solstice. The shores of the lake yield neither wild rice, nor Indian corn. They did not anciently cultivate the potato. They depended upon game and fish, and it is only necessary to have passed a single winter in the lake latitudes, to determine that a large body of miners could not have been kept together a long time for such a purpose, without a stock of provisions. On the contrary, as the theatre of summer mining, in a neutral country, or by self-dependent bands, hundreds of years may have passed in this desultory species of mining. (1851:99–100)

From the very beginning of the discovery of the ancient mines, however, there was a range of opinions as to who had done the prehistoric mining on the Keweenaw Peninsula and Isle Royale.

Some credited a mysterious lost race of Mound Builders; Aztecs or Toltecs; "a much older race (than) North American Indians" (Christian 1932:34); "some long forgotten race" (Greeley 1847); or "the unknown, mysterious ancient race of miners" (Greusel 1903:7).

The term "mound builders" was used as early as the mid-1840s by the intrepid mound explorers Ephraim G. Squier and Edwin H. Davis in their landmark publication, *Ancient Monuments of the Mississippi Valley*. Though the book was published in 1848, the manuscript had actually been accepted for publication in June 1847—before the Keweenaw/Isle Royale "ancient diggings" were widely publicized.

Systematic archaeological surveys—particularly on Isle Royale—have not turned up a shred of evidence that anyone but the prehistoric Native American inhabitants were involved in the exploitation of the native copper deposits (see Clark 1988, 1990, 1991; Martin 1988; and Martin, Martin, and Gregory 1994). Susan Martin (1995) presented a comprehensive and spirited response to claims that Europeans had contact with copper-mining Native Americans prior to the arrival of Columbus in the New World.

Historic Reports of Native American Comments on the Identity of Prehistoric Miners

In the first centuries of European contact, Native Americans often disavowed any knowledge of the copper mines and the identity of the prehistoric miners. This led some Europeans to argue that the miners could not possibly have been the tribes' ancestors. For how could such knowledge be totally lost?

Later, some scholars pointed out that Native Americans might have been trying to protect their resources by pretending not to know about copper sources or copper mining. Clearly they valued copper and believed it had spiritual significance, and more than one source indicates that there was a strict taboo against revealing to white men information about this precious metal.

Nineteenth-century author Sir Daniel Wilson interviewed Chippewa chiefs and wrote down their stories and comments concerning the ancient mines.

> Of Loonsfoot, an old Chippewa chief of Lake Superior, the improbable statement is made that he could trace back his ancestry by name, as hereditary chiefs of his tribe, for upwards of four hundred years. At the request of Mr. Whittlesey he was questioned by an educated half-breed, a nephew of his own, relative to the ancient copper mines, and his answer was in substance as follows:—"A long time ago the Indians were much better off than they are now. They had copper axes, arrowheads, and spears, and also stone axes. Until the French came here, and blasted the rocks with powder, we have no traditions of the copper mines being worked. Our forefathers used to build big canoes and cross the lake over to Isle Royale, where they found more copper than anywhere else. The stone hammers that are now found in the old diggings we know nothing about. The Indians were formerly much more numerous and happier. They had no such wars and troubles as they have now." (1876:219)

At La Pointe on Lake Superior, in what is now Wisconsin, Wilson met with Beshekee, or Buffalo, another elderly chief of the Chippewa. He reported Beshekee's comments (translated by the chief's grandson):

> "The white man thinks he is the superior of the Indian, but it is not so. The Red Indian was made by the Great Spirit, who made the forests and the game, and he needs no lessons from the white man how to live. If the same Great Spirit made the white man, he has made him of a different nature. Let him act according to his nature; it is the best for him; but for us it is no good. We had the red-iron before white men brought the black-iron amongst us; but if ever such works as you describe were carried on along these Lake shores before white men came here, the Great Spirit must once before have made men with a different nature from his red children, such as you white men have. As for us, we live as our forefathers have always done." (1876:220)

The principal entity connected with native copper in the beliefs of the Ojibwa people was the Underwater Manito. According to Christopher Vecsey:

The Underwater Manito possessed great and dangerous powers. It could cause rapids and stormy waters; it often sank canoes and drowned Indians, especially children. The Ojibwas associated it with the sudden squall waters of the Great Lakes which prevented fishing, even picturing it in the shape of Lake Superior itself (Kellogg 1917:104–105). Some Ojibwas thought of the Underwater Manito as a thoroughly malicious creature.

It was not totally evil, however. In some traditions it fed and sheltered those who fell through the winter ice. It offered medicinal powers to those who accepted it as a guardian. (Radin c. 1926:n.p.) It gave copper to the Indians, who cut the metal from the being's horns as it raised them above the surface of the water (Kellogg 1917:105). Those who attempted to take the copper without offering proper payment met severe punishment from the Underwater Manito. (1983:74–75)

Charles E. Brown, a pioneer in museum practices and co-founder of the Wisconsin Archaeological Society, published a brief summary of Native American myths and legends relating to copper. He depended primarily on accounts derived from the reports of Jesuit missionaries. In these he found stories about the sacred significance of copper, most of which appears to have been pieces of drift copper recovered from rivers and the shores of Lake Superior, but none relating to copper mining (Brown 1939).

From 1860, we have this detailed description from J. G. Kohl, peripatetic German traveler, historian, and geographer, regarding the significance to native people of copper (in this case drift copper) and the personal consequences of breaking taboos. (This account dates to 1855.)

Among the dead stuffs of nature, the dwellers on Lake Superior seem to feel the most superstitious reverence for copper, which is so often found on the surface-soil in a remarkable state of purity. They frequently carry small pieces of copper ore about with them in their medicine-bags; they are carefully wrapped up in paper, handed down from father to son, and wonderful power is ascribed to them.

Large masses of metallic copper are found at times in their forests. They lie like erratic blocks among the other rocks...

One of my acquaintances here, an ex-Indian fur-trader, a man of considerable intelligence and great experience among the savages, told me the following characteristic story of one of those lumps of copper:

"In the year 1827," he said to me, "I was trading at the mouth of the Ontonagon river, when the pleasant little town now existing was not thought of. The old Ojibbeway tribe, now known as the Ontonagon Band, lived there almost entirely independent, and Keatanang was the name of their chief.

"Keatanang, from whom I purchased many skins, and paid him fair prices, was well disposed towards me. He was often wont to say to me, 'I wish I could do thee some good. I would gladly give thee one of my daughters.' Once, when he spoke so kindly to me, and renewed his offer of his daughter, I said to him, 'Keatanang, thou knowest I cannot marry thy daughter, as I have a wife already, and the law forbids us Christians marrying several wives. But listen! Thou hast often told me of another treasure which thou possessest in thy family, a great lump of metal, which lies in thy forests. If thou really wishest me so well as thou sayest, and wouldst do me good, show me this lump of copper, and let me take it to my house. I will carry it to my countrymen, and if they find it good they will surely seek for other pieces of ore in thy country, and thou wilt soon have many lumps instead of one. If thou wilt show it to me I will pay thee any price thou mayst ask for it."

"Keatanang was silent for a long time after hearing my proposition. At length he began: 'Thou askest much from me, far more than if thou hadst demanded one of my daughters. The lump of copper in the forest is a great treasure for me. It was so to my father and grandfather. It is our hope and our protection. Through it I have caught many beavers, killed many bears. Through its magic assistance I have been victorious in all my battles, and with it I have killed our foes. Through it, too, I have always remained healthy, and reached that great age in which thou now findest me. But I love thee, and wish to prove my love. I cannot give thee a greater proof of my friendship than by showing thee the path to that treasure, and allowing thee to carry it away.'

" 'What dost thou ask for it, Keatanang?'

"After long bargaining, we agreed that I was to give him two yards of scarlet cloth, four yards of blue cloth, two yards of every colour in silk ribbons, thirty pairs of silver earrings, two new white blankets, and ten pounds of tobacco; and that when I had all this in readiness, he would show me the next night the road to the copper, and allow me to carry it off in my canoe. Still he made it a condition that this must be done very secretly, and neither any of his people nor of mine should hear a word of it. He proposed to come to me at midnight, and I would be ready for him.

"The next night, exactly at the appointed time, while lying in my tent, I heard a man creeping up gently through the grass, and felt his fingers touch my head. It was Keatanang.

" 'Art thou awake,' he said, 'and hast thou the goods ready?" I gave him all the articles one after the other, fastened the bundle up with the silk ribbons, placed half the tobacco on the top, and the rest in his belt. He took the packet under his arm, and off we started.

"We crossed a little meadow on the river bank, and reached a rock, behind which Keatanang's canoe lay in readiness.

"I offered to help him in paddling, but he would not allow it. He ordered me to sit with my back against the bow of the boat, and paddled along so noiselessly, never once lifting the paddle from the water, that we glided along the bank almost, I may say, like Manitous.

"In two hours we reached the spot we call the High Bluffs. From this point our path trended landwards. Keatanang took up the bundle, and when we had climbed the bluffs he turned quite silent, raised his eyes to the starry heavens, and prayed to the Great Spirit.

" 'Thou hast ever been kind to me,' he then said, in so loud a voice that I could plainly hear him. 'Thou hast given me a great present, which I ever valued highly, which has brought me much good fortune during my life, and which I still reverence. Be not wroth that I now surrender it to my friend, who desires it. I bring thee a great sacrifice for it!'

"Here he seized the heavy bale of goods with both hands, and hurled it into the river, where it soon sank.

" 'Now come,' he then said to me, 'my mind is at rest.' We walked to a tree which stood on a projecting space of the slope. 'Stay,' Keatanang said, 'here it is. Look down, thou art standing on its head.'

"We both commenced clearing away from between the roots the rotten leaves and earth, and the fresh herbs and flowers that had just sprung up, for we were then in spring. At length we came to several large pieces of botch bark. These, too, were removed, and I discovered under them a handsome lump of pure copper, about the size and shape of a hat. I tried to lift it, and it weighed a little more than half a hundred. I carried it out into the moonlight, and saw that the copper was streaked with a thick vein of silver.

"While I was examining the copper, Keatanang, who was evidently excited, and was trembling and quivering, laid the other five pounds of tobacco he had thrust in his belt as a conciliatory sacrifice in the place of the copper, and then covered it again with bark, leaves, and roots. I wrapped my lump in a blanket, and dragged it down to the canoe. We paddled down the river as noiselessly though more rapidly than we went up. Keatanang did not say a word, and, as we found everybody asleep in the encampment, we stepped into our tents again as unnoticed as we left them.

"The next day I loaded my treasure on my canoe, and set out. My specimen was ultimately sent to the authorities of the United States, and was one of the first objects to draw public attention to the metallic treasures of this remarkable district.

"Old Keatanang bitterly repented afterwards the deal he had with me, and ascribed many pieces of misfortune that fell on him to it. Still I always remained on a friendly footing with him, and gave him my support whenever I had an opportunity. Afterwards he became a Christian, and found peace." (1860:60–64)

Certainly Keatanang was not alone in his concerns about revealing the location of sacred lumps of copper. Bernard Peters details the tribulations of Wa-bish-kee-pe-nas (White Bird), a member of the Ontonagon Band of Ojibwa, "… whose behavior in relation to the Ontonagon Copper Boulder did not follow his culture's accepted norm" (1989:47).

Wa-bish-kee-pe-nas defied a sacred taboo by attempting to show the copper boulder to a white man and his life was forever changed as a result. His band banished him, and he attributed his subsequent bad luck to his failure to respect Gitchi Manitou and the god's sacred substance, copper. This shiny metal, torn from the rocks of Isle Royal and the Keeweenaw Peninsula by the Pleistocene ice sheet and dropped along the south shore of Lake Superior, was highly honored by the Native peoples of the Lakes. It was a religious symbol which affected their culture as much as Christian religious symbols affected Anglo-European culture. The European/American visitors to Lake Superior who observed the Native American worship of copper called it "superstition," and, in the ensuing years, devoted their energies (in some cases, their lives) to eradicating such "savage" beliefs. (1989:60)

7 The Fate of the Prehistoric Artifacts

...at the North West Companies' mines [20KE34], near Eagle Harbor, there was a depression of six feet, in the metalliferous lode, where the Indians had mined out the sheets of copper. Barrels full of hammers, much worn, were thrown out of this old excavation; and I have preserved a number of them for the Government collection.

Charles T. Jackson
1850

Obviously, there was a considerable amount of material culture present in Michigan's prehistoric copper mines. So what happened to it?

From the very beginning, the nineteenth-century explorers and miners appear to view the artifacts as interesting "curiosities" and not much more. The Horace Greeley reference cited above suggests that hammerstones were kept for a time in onsite mine offices, along with rock and ore samples. Daniel Wilson reported, "At the Cliff Mine (20KE46) some specimens of the ancient copper tools of the native metallurgists are preserved..." (1856:227).

Nevertheless, some people had artifacts from the pits and seemingly traveled with them and readily showed them to friends and newspapermen (e.g., Bagg and Harmon 1848; Ingersoll 1846, 1848).

Stone Tools and Organic Materials

At a very early date (but year unstated), geologist James Hodge donated a stone hammer from the Minesota Mine to the Cooper Union in New York City (Hodge 1873:55).

Joseph Henry of the Smithsonian Institution reported what might have been an early effort at onsite interpretation.

> The survey of the mineral land in the vicinity of Lake Superior, has disclosed the site of an ancient copper mine, whence, in all probability, the copper of the metal ornaments, instruments, &c., found in the mounds was derived. The remains of the implements and of the ore, as left by the ancient miners, are exhibited in place, and afford an interesting illustration of the state of arts among the mound builders. The geological surveyors have promised to make accurate measurements, and drawings of everything of interest connected with these works, and to present them, with suitable descriptions, to the Institution, for publication. (1851:22–23)

The timing of this discovery suggests that the surveyors were Foster and Whitney (1850), although what site is being discussed is unknown. The thought that one of these pits might be preserved in place was, in retrospect, highly naïve. It is extremely unlikely that the copper and the artifacts exhibited in place stayed that way very long.

In at least one case, hammerstones were put to a latter-day practical use. At the Minesota Mine location (20ON1), "The amount of ancient hammers found in this vicinity exceeded ten cart-loads, and Mr. K., [Samuel Knapp], with little reverence for the past, employed a portion of them in walling up a spring" (Foster and Whitney 1850:160). (It is unknown if the walled-up spring still survives.)

Charles T. Jackson, in his role as U. S. geologist, stated:

> ...at the North West Companies' mines [20KE34], near Eagle Harbor, there was a depression of six feet, in the metalliferous lode, where the Indians had mined out the sheets of copper. Barrels full of hammers, much worn, were thrown out of this old excavation; and I have preserved a number of them for the Government collection. (1850:284)

The Fate of the Prehistoric Artifacts

To Jackson's credit, when his collection of rocks, minerals, and ores were cataloged at the Smithsonian Institution in 1854, there was a total of six "Indian stone-hammers" from two different locations along with the hundreds of mineralogical specimens (Jackson 1855:342, 346).

Alexander Winchell, professor of geology, zoology, and botany and director of the Museum of Anthropology at the University of Michigan, led an expedition to the "mining region of Lake Superior" in 1867. Along with numerous geological specimens, Winchell reported the following additions to the Department of Ethnology and Relics.

Lake Superior Expedition—1. Sundry stone mauls, weighing from 5 to 15 pounds, collected in and about the ancient mines at Rockland, Ontonagon County. These, like all others heretofore found, are of diorite rock. 2. An elongated stone implement of hornblende rock. Bark of a pine tree which grew on the attle from an ancient mine, the stump of which contained 322 rings of annual growth, showing that the mine had been abandoned more than 322 years since—or before the year 1545. (1867:11–12)

Another who saw to permanent preservation of at least a few objects was Alexander's brother, Newton H. Winchell, Minnesota state geologist. In the *Ninth Annual Report*, he reported the presence in the "General Museum" of "Stone hammers," "Charred wood," and "Pounded copper flakes," all "from the ancient mines of Isle Royale" (Winchell 1881c:162). In the *Tenth Annual Report*, he reported his own donation of a "Small stone hammer, withed. From Michigan" (1882a). He also elaborated on earlier information, reporting that the "stone hammers" and "battered copper" from the Minong Mine on Isle Royale and that "wood" (presumably charred) came from "the ancient dumping heaps" (1882b:54). Winchell also reported that "Ship-loads of these stones [hammerstones] are transported from the north shore of Lake Superior for paving streets in Chicago and other cities" (1881a:605, 1881b:184).

Peripatetic mound surveyor T. H. Lewis included an apparent copper "chisel" from "an ancient mining pit" on the property of the Knowlton Mine (20ON270) in his collection deposited in the Macalester College Museum of History and Archaeology (Lewis 1890:180).

Figure 7.1. Newton H. Winchell sketched these Isle Royale hammerstones, which were found at the Minong Mine. He notes that on Isle Royale, the ancient miners used "beachwrought hammers"—that is, stones they found on the beach. "Their battered and even fractured extremities are the only sign of the agency of man in giving them shape." (Reprinted from Winchell 1881a:Fig. 6.)

The State Historical Society of Wisconsin aggressively collected native copper artifacts and on at least one occasion took in hammerstones from the Aztec (20ON271) and Mass mines (20ON274/275), donated by the aforementioned Daniel Beaser (Draper 1882a:19).

Perhaps the most important private collector was Charles Whittlesey. In his early writings, he specifically uses "I had" and "I have" in reference to pieces of timber from the Minesota (20ON1) and Waterbury (20KE73) mines (1852a:27–28), as well as a "shovel." He gave an unknown number of these artifacts to the Western Reserve Historical Society (Whittlesey 1871:40). Two of them, a hammerstone and a wooden "shovel," were described and illustrated by MacLean (1901:207–210).

Whittlesey mentions a "spear-head" in the possession of S. W. Hill, as well as a grooved hammerstone (1852a:28–29; 1868:477).

One of the most puzzling statements was made by Jacob Houghton.

> A copper chisel, stone hammers with bands hollowed out for the reception of handles of twisted twigs, and other relics of the miners of an extinct race, have been found in these workings and most of them carefully preserved. (1862)

Unfortunately, Houghton gives no clue as to where, how, and by whom these relics were "carefully preserved."

Occasionally an artifact was considered to be so rare that it might have some financial value. Albert Hagar reported one of these instances.

> At the Hilton Mine [20ON426] in the Ontonagon district, in October, 1863, as the men were removing the vegetable mould that had accumulated in one of the old pits, they found at the depth of about nine feet a leather bag, which was eleven inches long and seven inches wide. It was lying upon a mass of native copper which the ancient miners had unsuccessfully attempted to remove from its parent vein. The bag was in a remarkable state of preservation, the leather being quite pliable and as tough as sheepskin. It was made up with the hair inside, was sewed across the bottom and up one side with a leather string, and near the top holes were cut and a leather string inserted to close the mouth by drawing it together. The bag was empty, but from its appearance I judged that it had been used for transporting copper or other mineral,—the leather in places showing marks of much service, and the hair being almost entirely worn off. I was unable to determine what kind of skin it was, but inclined to the belief that it was from the walrus, as the short, stubby hairs more closely resembled those of that animal than of any other with which I am acquainted. At the time I saw the bag,—the day after it was discovered,—it was in the possession of C. M. Anderson, Esq., the agent of the Knowlton Mine; but I hear it has since been taken to Boston and sold. (1865:310)

There were some primitive attempts to preserve some of the organic materials and speculations about how they had survived for so long. From the very earliest days of discovery we have this example from the Minesota Mine (20ON1).

> We learn from Capt. J. W. Hunter, who bro't down last week a piece of this stick about four feet in length and to whom we are indebted for a small piece of the same, that when

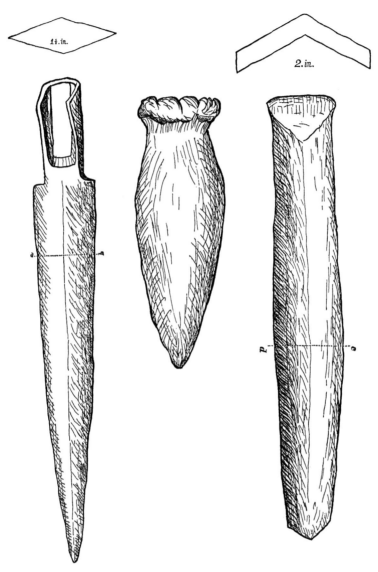

Figure 7.2. Sketches of (left to right) a copper spearhead from Ontonagon (14 inches long); a copper gad from the Minesota Mine; and a copper chisel from Ontonagon (13 inches long). Of the chisel, Whittlesey notes: "It is bent up longitudinally from near each end in the manner shown by the cross section in c d. The object in giving it this form must have been to stiffen it and thus save metal. This contrivance speaks well for the ingenuity of the maker." (Reprinted from Whittlesey 1863:Figs. 12, 13, and 14.)

taken out of the mine it could easily be pulled in pieces by the hands, but by carefully drying it in the sun it became strong and hard and very much 'season cracked' by the exposure to the air and sun.

How long would oak timber probably remain in this state of preservation under such circumstances? and would the presence of copper around it have any effect on the preservation of it? These are questions to which we should be glad to have more satisfactory answers than we are able to give them. (Brown 1850b:2)

Most organic materials taken from the pits survived only because they had become permanently waterlogged in the undrained pits. The existence or availability of appropriate conservation materials and techniques on the copper-mining frontier seems highly unlikely. Once removed, the terminal fates of wooden and bark artifacts, in particular, were sealed.

These [wooden shovels] and wooden bowls for bailing water from the pits had been preserved from decay by lying covered with water and fell to pieces as soon as they became dry from exposure to the air. (*Ashland Press* 1873)

George A. West, harking back to a "visit to the mining region on the South Shore in 1875" reported that he

…saw a ladder, two paddles and fragments of a birch bark receptacle, taken from one of the ancient pits. These were all in a good state of preservation, excepting that the birch bark when dried was very brittle. This ladder, like many others that were found in that district, was merely a trunk of small tree with limbs cut away, leaving stumps of about six inches in length. This is the well-known form of Indian ladder. The one seen was cut off by hacking, with some instrument, clear around the portion to be cut and not cutting from two sides, as a white man does. The instrument used was reasonably sharp and might have been a copper or stone axe. (1929:52)

This is a remarkable observation, considering that West was only 16 years old when this "visit" took place. He also stated that during this "visit" he "conducted investigations along this Copper belt" (1929:43). It is unfortunate that he did not identify what mine or mines contained these materials.

Large Masses of Copper

Some of the monster masses of copper found in the pits, particularly on Isle Royale in the 1870s, were given brief reprieves before they were melted down.

About 1872, Mr. A. C. Davis worked, on the island, a property called the Minong, and brought to Detroit a very interesting piece of ancient work—a mass of copper of 4 or 5 tons, taken from the bottom of an ancient trench along a vein. This mass had been deprived of all portions of copper that could be detached by pounding with stone hammers, and the whole surface of the mass had been pounded into deep pits and ridges in attempts to

The Fate of the Prehistoric Artifacts

detach other pieces. It showed an immense amount of labor done. Mr. John R. Grout, of the Detroit copper-smelting works, shipped the mass to the Michigan University at Ann Arbor, and offered to sell it to the University at half its cash value, but the Regents did not make the purchase. Perhaps they lacked money; perhaps they lacked appreciation. The mass was shipped back to the smelting-works and went into the furnace. Thus a thing of very great interest was lost. Had the University taken it, it would be the most interesting in all its collections, and of special interest to Michigan. (Wood 1907:293)

A Canadian, Arthur Harvey, also commented on a large mass of copper that had a shot at immortality.

One pit was found on Isle Royale in 1870, filled up with the accumulations of leaf mould for centuries on which pine trees 2 feet in diameter were growing. It was 100 feet across, nearly circular, 20 feet deep. At the bottom was a mass of copper, raised upon skids, and weighing over 16 tons. Lying in the hole were handspikes, 7 or 8 feet long. Both skids and bars were thoroughly impregnated with copper solutions. The two skids were 10 or 12 feet long, 8 or 9 inches thick; the marks of knife or hatchet were visible on them, and on the handspikes which are now in the Detroit Museum. The copper mass showed signs of having been hammered all over with stone hammers, of which dozens were lying around. This mass was raised, sent by steamer to Detroit, and offered to the Council to be put up as a monumental base in front of City Hall, but the proposition was not entertained and it was sent to Wyandotte to be smelted. Mr. Shortiss of this city saw it on Isle Royale, and saw it with the stamp mark of its weight at Wyandotte. (Harvey 1890:224)

Yet another mass from Isle Royale had a depressingly familiar story.

The most remarkable specimen of copper was exhibited by the Minong Mining Company, of Isle Royale. It weighs three tons, and was discovered in the summer of 1875 by A. C. Davis, of Detroit, on the lands of that company lying directly upon one of the so-called copper belts of the Island, but perfectly detached from the parent rock. Underneath it lay a handspike [a heavy bar used as a lever], 6 ½ feet in length, made of white cedar, and in a remarkable state of preservation. Some 16 ½ feet of soil filled with particles of charcoal, rolls of birch bark and stone hammer lay above it. Its surface is covered with indentations and depressions caused, undoubtedly, by the efforts of the pre-historic people who labored there to separate from the large mass such projecting pieces as might be made useful to them.* Such traces of the labor of this people in search of copper are found in direct connection for about three miles, while in other portions of the island other evidences of their labors have been found.
　　*This specimen is at present in charge of the Detroit Scientific Association. Its value, about $1,100, is more than its owners care to donate, and its final disposition is uncertain. It ought to be saved from the furnace. (McCracken 1876a:513)

The *Detroit Free Press* (1893) offered this follow-up on the famous Minong Mine mass (with some rather startling differences and omissions):

The locality which shows the greatest evidence of these early miners is the property of the Minong & Cove Mining Company, at McCargo's Cove, on Isle Royal. All along the outcrop of the belt for at least two miles the soil has been turned over in the eager search

Figure 7.3. In 1874, historic miners found this hammered chunk of copper at a depth of 16½ feet in the Minong Mine on Isle Royale, Michigan. Prehistoric miners used hammerstones to remove small sections of copper—it was this work that created the overlapping marks on top. Shown here on a dock in Detroit. (Courtesy of the Burton Historical Collection, Detroit Public Library.)

Figure 7.4. Another view of the copper chunk on a Detroit dock. James B. Griffin included this photograph in Lake Superior Copper and the Indians: Miscellaneous Studies of Great Lakes Prehistory (1961d: Plates V and VI), noting that the chunk did not survive long in this form. He notes: "After being exhibited in Detroit it was shown at the Philadelphia Centennial Exposition in 1876. It was then smelted." (Courtesy of the Burton Historical Collection, Detroit Public Library.)

for copper. There was found in 1874 a celebrated mass of copper which was exhibited in the Centennial [the 1876 Philadelphia Centennial International Exhibition]. When found this mass was covered with seven or eight feet of soil. It was marked all over by the hammers of the early workers who had removed every projecting point of metal. Near it was found a wooden paddle or shovel which was used to throw the dirt aside; also, a great deal of charcoal and fragments of birch bark which had once been baskets for handling the earth. Many tons of hammers were found in that locality. After the Centennial this mass of copper was melted. This is generally considered a great misfortune, as it should have been preserved by the state as valuable relic, but it was offered to the state at the mere value of the copper, and this offer was not accepted. The mining company to whom it belonged was not particularly thriving and did not care to go to the expense of preserving a relic of bygone days in mining, and accordingly melted the mass. This mass would have been a valuable addition to the exhibition at the world's fair, showing the manner in which the earliest workers in copper strove to penetrate the earth in their search for metal.

More details may be found in Michigan State Board of Centennial Managers (1876:57). In fact, 14 of the 23 mines that "contributed" copper specimens for display at the Philadelphia Centennial International Exhibition had originally been discovered by the ancient miners. The "1874 mass" was also reported by Gillman 1876:330–331 and Winchell 1881a:602.

Lack of foresight and imagination—and stinginess at university, municipal, and state levels—cost the people of Michigan multiple opportunities to purchase geological specimens whose size would have put the more famous, but much smaller (3,708 pounds or 1.85 tons) "Ontonagon Boulder" in the shade. However, the historical significance of the Ontonagon Boulder in drawing attention to the copper resources of the Keweenaw Peninsula will never be surpassed.

Paradoxically, some of the enormous copper masses discovered in the historic mining era (though not necessarily in the prehistoric pits) survived meltdown only because they were sent to exhibitions, both in the United States and abroad, as "ambassadors" of the fantastic mineral wealth to be found in the wilds of Michigan.

According to Ralph Williams, perhaps the first of the native mass copper exporters was John "Old Blind" Hays, discoverer of the copper deposits that would become the Cliff Mine.

As the Cliff mine was running a large amount of copper in masses from one ton up to eighty-one tons, it became necessary to erect smelting works, so as to put it in marketable condition, that is to cast it into ingots weighing ten pounds, more or less. For this purpose Hays went to England to examine English furnaces, carrying with him samples of copper, one piece weighing 3,852 pounds, and others from one up to ten pounds. The large piece was sold to King's College on the Strand and the smaller specimens were given to the British Museum. There were no furnaces in England, however, for smelting mass copper. The English obtained their copper from ore combined with sulphur known as sulphate of copper, which had to be crushed and then washed, yielding but five per cent of copper. The specimens that Hays left in England created great excitement among scientific men, especially geologists, and did much to enlist the interest of capitalists in the wonderful mineral region. (Williams 1907:17)

The Fate of the Prehistoric Artifacts

This version omits the role apparently played by one Alonzo W. Brockway in facilitating the removal and transport of the copper mass from the Cliff Mine to New York (Speer 1888:327–328).

Entrepreneur Henry Pinkham wrote:

> The first great impulse given to the mining interests in Ontonagon County was through the discovery of the Minesota Mine, in 1847, by Mr. S. O. Knapp. He was led to the discovery by examining the indentations, which were plainly discernible along the surface outcrop of the vein, and which proved to be ancient mining excavations, in some of which were found enormous masses of nearly pure copper, one weighing upwards of six tons. Accounts of these monster masses were everywhere published, and the reality almost staggered belief. Masses of copper weighing thousands of pounds were purchased and taken to Europe to be exhibited as curiosities. (1888:29–30)

According to James Ritchie (writing in 1858), overseas belief in the purity of the great masses of Lake Superior copper was a hard sell.

> Several years ago, a block of copper from Lake Superior was sent to London as a specimen; the geologists there could not be convinced, at first, but that it was a Yankee trick—they had never heard of copper being found in such a pure state, and supposed the block had been cast for the purpose of exhibition. (1858:193)

Among the early museological activities were those of Pierre Louis Antoine Cordier, prominent French geologist and founder of the French Geological Survey.

> A specimen exhibited to the academy [French Academy of Sciences] by M. Cordier weighed upwards of 112 lbs., and contained very little gangue. This specimen, as well as another more than ten times its weight, formed part of a cargo of several tons which recently arrived at Havre; it is intended to be placed in the Museum of Natural History. (1849:679)

Supplementary information in the article reveals that this copper came from the Copper Falls Mine (20KE63) in Keweenaw County (1849:680).

From the Rockland Mine (20ON427) "...one mass, weighing 6,075 pounds, taken from an ancient pit and bearing the marks of the rude mining tools of a former period and an unknown race, was sold and sent to England as an object of great curiosity" (Pinkham 1888:36).

Apparently, the source of Pinkham's knowledge of the Rockland Mine discovery was a notice credited to "an intelligent correspondent of the *Pontiac Jacksonian*" that appeared in *The Mining Magazine*.

> Here was the largest mass [of copper] found that has ever been discovered so near the surface. It was found on opening one of the ancient pits, and within ten feet of the surface. At the head of the mass the ancient had endeavored to get off a piece of copper, and the end was rounded with a ring, or neck piece, formed by the hammering, but which was not separated from the mass. This measures enough for 100 tons. Already some

20 tons have been taken from it, and the mass gives more thickness and breadth as it is exposed. The mass cut from the top, containing the neck, was sold upon the ground, to go to Liverpool for exhibition for 22 cents per pound—weight 6,075 lbs. It was on the wagon to go to the landing. (Tenney 1855:253–254)

The New York Exhibition of the Industry of All Nations (perhaps better known as the New York Crystal Palace) provided the venue for displays of minerals and crystals from around the world (Russell 2008). Michigan proudly displayed native copper.

The State of Michigan exhibited a mass of native copper, cut from the lode of one [of] the North American Mining Company's (No. 210, Class I.), weighing 6,300 pounds. This mass was cut into a rectangular form. Portions of the epidotic gangue or veinstone were adhered to the upper surface, but the sides were clean-cut surfaces of pure copper, upon one of which was engraved the locality and weight. Many other very large masses of copper were exhibited by different mines in the Lake Superior region, of which that from the Minnesota Mining Company weighed over 5,000 pounds. (Silliman 1854:597)

This specimen (or perhaps a different one) was so impressive that one of the British commissioners for the London Crystal Palace Exhibition offered to buy it for that exhibition (Wilson 1853). Further confusing the issue is a brief notice originally published in the *New York Journal of Commerce* and picked up by the *Detroit Free Press* (1855a), announcing that large masses of native copper from the Minesota and Rockland mines were lying in front of the Merchants' Exchange and that they were "on their way to England as a sample of the mass copper from Lake Superior." The Rockland specimen was said to show signs of working with stone hammers. These were quite likely the same specimens described earlier. Frustratingly, the writer did not give the ultimate resting place(s) of these specimens, both in the 200- to 300-ton range.

In the end, it seems that the excavation methods employed by historic miners resulted in many (or most) artifacts being roughly evicted from their resting places and tossed onto a spoil pile, as suggested by Edwin Hulbert's lament cited at the beginning of this paper. The quantities of mining artifacts may be only guessed at, but Theodore Lewis gives us an idea of the density of artifacts with observations he made in 1888 at the remains of a "clean out" at the Caledonia Mine (20ON255) in Ontonagon County.

…where the debris had been dumped in a heap, I noticed a few hammers protruding, and in less than an hour, by using a sharp stick, I exposed 132 grooved hammers, ranging in size from a hen's egg to twelve or fifteen pounds in weight, only a few of which were unbroken, and also a broken stone ax and several large hand hammers. (1889:295)

Absence of Images

As we have seen, many of the artifacts associated with the prehistoric mines were lost through one means or another, often before their likeness was recorded. Foster

Drawn by J. C. Tidball U. S. A.

Figure 7.5. Henry Rowe Schoolcraft commissioned a soldier, John C. Tidball, to create this engraving of a prehistoric copper miner at work. Schoolcraft published the image (entitled "Ancient Mining on Lake Superior") in Part V of his series on tribes of the United States. (Schoolcraft 1855:Plate 16, Fig. 1. Image courtesy of University of Michigan Special Collections Library.)

and Whitney offer rather primitive representations of an unexcavated pit and two hammerstones (1850:Figures 30 and 31). The importance of Charles Whittlesey's 1863 monograph and the images of artifacts and mine sections contained in it cannot be overemphasized. It could never be improved upon because, as noted previously, the "ancient diggings" had been more or less systematically destroyed. The few who cared were unable to preserve the fragile organics.

Slightly later, Henry R. Schoolcraft published an engraving by John C. Tidball, then a young career soldier. The image was titled "Ancient Mining on Lake Superior" (Schoolcraft 1855:Plate 16, Fig. 1). It depicts an adult male climbing out of a mining pit using a tree-trunk ladder and carrying a lump of copper in his right arm. On the floor of the pit a fire is burning against the pit wall. Also on the pit floor lay a hammerstone attached to a handle, an unhafted hammerstone, a chisel, and an axe, the latter two both apparently made of copper. Virtually all the significant artifacts and mining techniques known at that time are present. It is unclear if Tidball ever saw any actual copper-mining artifacts or if he was acting on Schoolcraft's verbal descriptions.

> For a higher grade of drawing I received thirty dollars per plate. These were what Mr. Schoolcraft called historical subjects. The mechanical execution of these came very easy, but they required considerable draft upon my imagination to make them historically correct. Another, and the highest order, were intended to illustrate characteristic traits and customs. Schoolcraft was very particular about these. He had in his mind an ideal difficult for me to grasp, but by sketching in as nearly as possible what I supposed was his idea and getting him to remark upon it I could work into it what he imagined it should be. (Tidball 2002:150)

It would appear that "Ancient Mining on Lake Superior" was of "the higher grade of drawing."

Nothing approaching the scope of information provided in this image would be created until the Milwaukee Public Museum's early twentieth-century diorama of copper mining, visible yet today on collector postcards. Indeed, Tidball's engraving was still the image of choice 155 years later, when it was chosen for inclusion in Howard and Lucille Sloane's *A Pictorial History of American Mining* (1970:1).

Conclusion

That discovery of Mr. Knapp at the Minnesota location, and the success to which it led, infused new energy into that and other enterprises of the kind wherever other 'ancient pits' could be found.

La Fayette Mining Company
1863

Conclusion

Thousands of years ago, people mined the bedrock copper of Michigan's Upper Peninsula. Today we continue the process of trying to understand the details of how they mined, worked, transported, and used copper. What can we take away from this summary of Euro-American observations of the ancient diggings of the Keweenaw Peninsula and Isle Royale?

Discovery and Destruction of the Prehistoric Bedrock Copper Mines

The information base for ancient diggings is extremely small. Had not Charles Whittlesey and Samuel W. Hill been on the scene, for example, there would have been virtually no primary illustrations of artifacts or mine cross-sections. By the early twentieth century, there were hardly any new reports of ancient diggings on the Keweenaw Peninsula (although discussion continued regarding Isle Royale). Comments were limited to reminiscent pieces such as "A Mysterious Race" (Greusel 1903); also, George Brinton Phillips (1923, 1925a, 1925b, 1926) made some useful contributions with presentations at home and abroad.

The early notices of prehistoric mining had a major economic impact that essentially ensured the early destruction of the known ancient pits as sources of archaeological information. Because of the irregular distribution of native copper deposits and the vast distances from the sources of supplies, mines active in the late 1840s were on the verge of being abandoned. According to Cyrus Mendenhall, a well-known early mining engineer:

> The fact is, had it not been for the discovery of the 'Ancient Mining Pits'—made early in 1848 by Mr. Knapp of the Minnesota Mine—*that, and all other* mining in the Ontonagon district, would, under this general discouragement, want of success, and distrust of value, added to the great expense and difficulty of prosecuting such work at that date, have been abandoned before the close of that year.
>
> That discovery of Mr. Knapp at the Minnesota location, and the success to which it led, infused new energy into that and other enterprises of the kind wherever other 'ancient pits' could be found. (La Fayette Mining Company 1863:16–17) [Emphasis in the original.]

Twelve years later, F. G. White of the Osceola Consolidated Mining Company echoed Mendenhall on the significance of ancient diggings as a guide to economic well-being.

> The ancient mining upon this belt was also very extensive, which, taken with the fact long ago developed in Ontonagon County, that in most cases this mining was done in the vicinity, and usually in contact with copper in quantity, gives of itself a warrant of promise to investors in the enterprise. (Minong Mining Company 1875:14)

Taken all together, these nineteenth-century accounts present a reasonably detailed picture of what tools, facilities, and mining techniques made a fully operational prehistoric copper mine. Of course, now we understand that what the observers of that time might have seen actually represented the terminal phases of prehistoric activity in that mine. Archaeologists continue to analyze prehistoric objects and manufacturing techniques around the country (e.g. Ehrhardt 2005; LaRonge 2001; Leader 1985, 1988; Peterson 2003a, 2003b; 2004).

The prehistoric mining of copper on the Keweenaw Peninsula and Isle Royale undoubtedly spanned more than a thousand years, but had ceased before European contact. Given the time span and geographic extent of these ancient mines, they arguably represent one of the most extensive mineral extractive activities ever undertaken by any prehistoric people in North America (exclusive of flint quarries). The terminal fates of the ancient diggings as sources of archaeological information were sealed from the moment of their discovery (Brown 1850a). And as was the case with the discovery and fate of the Hohokam canals of Phoenix (Bostwick 2008:95ff), they are bitter testimony to the price that some classes of archaeological sites paid for being signposts to economic advantage.

Discovery and Significance of Prehistoric Copper Caches

While it is unlikely that we will learn much more about prehistoric mining at the locales of the mine sites themselves, we are learning more all the time about what happened to the copper after it was mined. Finished copper artifacts were stored in caches, often associated with partially processed copper nuggets. Unfortunately, very few of these caches have had radiocarbon-datable material (e. g., wood or leather) in them.

Long-time Wisconsin archaeologist Robert Ritzenthaler was perhaps the first to address the question of what to call a copper nugget that has undergone a certain level of processing yet is not really identifiable as a formal artifact. Referring to a lump of copper 2½ x 1¾ x ⅜ inches found in Dodge County, Wisconsin, he used the term "ingot," citing Webster's dictionary definition of ingot as "a mass of metal cast into some convenient shape for transportation or storage." Ritzenthaler went on to say:

> This piece was not cast, but the possibility of its having been beaten into a more compact shape and eliminating any sharp edges for convenient transportation seems worthy of consideration. The paucity of worked implements near the great quarries such as Isle Royale suggests that the Indians mined the native copper and took it to their respective villages to work it.
>
> A weakness of the ingot theory is that if this were a consistent practice, one would expect to find more examples of it. This is the only example that the writer has encountered. Obviously most of them would end up as artifacts and thus disappear from the record, but even so, it would be suspected that other examples would have been forth coming by now as having been lost in transit or in a workman's stockpile, if this theory is to be given credence.

Conclusion

These pieces should not be confused with float copper, which occasionally shows evidence of battering, but we would be interested in hearing from any collector who might have a specimen similar to this which could lend further support to the ingot theory. (1963:215–216)

Ritzenthaler's prescience in the second paragraph as to what should be found would be borne out in later years.

Ingots, or blanks, as they are also known, have sometimes been a misunderstood product. However, metallurgist William Vernon clearly recognizes their actual importance:

The blanks used in this study seem to represent an important intermediate stage in the production of artifacts, a level somewhere between mining the raw material and producing a finished artifact. Their nondescript appearance has not invited critical analysis on any stylistic level in the past, yet the blank is functionally important, if not critical in the fabrication of many artifacts. It is a starting point for fashioning any of a number of different artifacts from prepared material. (1985:161; 1990:509)

I believe there is a high probability that many caches may have been "high-graded" when they were found in modern times. That is, artifacts such as mini-ingots, if they were included in the cache, were discarded as being of little or no interest or value. Caching activities extended well into historical times. Examples of finished caches (not to be confused with burial/grave goods) are described in Brown (1905a:207, 212; 1905b:90, 92, 93; 1906:296, 299, 361, 362, 369, 388, 391, 413; 1907:66–67).

What follows is a summary, in order of discovery, of what I believe to be travel caches found from the late nineteenth to the twenty-first century. This listing most certainly should not be considered exhaustive of cache discoveries. There are many more caches that I believe are not travel caches, but assemblages of finished artifacts (not necessarily all made of copper) that were for unknown reasons never retrieved by the owner.

I will start with a cache of copper artifacts discovered near Sault Ste. Marie, Michigan, and first reported (apparently) in 1883.

It contained twenty-three pieces consisting of six awls, the largest one about six inches in length, the smallest, three inches, five knives of various sizes, and thirteen pieces composed of axes, hammers and chisels. They were found lying piled together, straight and close, encircled by a little pile of stones, and are supposed to be the treasure of some ancient miner, as some of them show marks of very considerable age. (Anonymous 1883b)

Now, more than 130 years later, we do not know who found the cache or what became of it. More than likely it was quickly dispersed. This may have been what I have called a travel cache, but in the absence of information about whether there were partly worked nuggets and whether any of the tools were unfinished or damaged by use, it is impossible to be certain.

In 1946, Gerald Haltiner discovered the following in a sand blow-out in the city of Alpena, Michigan: a 4-pound unworked nugget of copper; 31 partially worked pieces of copper scrap totaling 1¼ pounds; 2 socketed piercing implements; and a rectangular cross-sectioned awl with tapering ends (Binford 1961). Apparently there was no newspaper or other account at the time the discovery was made. Today the Haltiner Archaeology collection is curated by the Besser Museum for Northeast Michigan in Alpena.

While disking farm land near the bank of the Sandy River in Aitkin County in central Minnesota, Ed Borg unearthed:

> ...an assortment of nine pieces of tempered copper implements consisting of what appears to be an axe, two tomahawks, two instruments which were apparently used as needles, being about 4 inches in length, squared and tapered to a point from about ⅛ inch diameter. One piece has the general form of an arrowhead, but unfinished. Three other articles of hardened copper which are 3½ to 4 inches in length, have not been identified. (Anonymous 1954)

While the terms used here to describe the various artifacts lack the precision and commonality of descriptions we see today in the discussions of copper caches, it appears rather obvious that the Sandy River discovery was a cache, probably consisting of three adzes or celts of various sizes, two rectangular-section awls, one projectile preform, and several worked mini-ingots: in sum, a travel cache as defined in this work. The fate of the Sandy River cache is unknown.

An extremely important account of the discovery of "an ancient copper transportation and manufacturing site in northern Wisconsin" (Kent 1991) is perhaps the earliest report of such a site. The author (but not the discoverer of this site) did not, in this article, reveal the location of the finds, apparently at the behest of the actual finder. Nevertheless, the location is described as being

> ...on a high neck of land, ranging from one-quarter to one-half mile in width, which extends in a roughly north-south direction between two small lakes. A short, narrow creek cuts through the neck of land in an east-west direction, connecting the two lakes. The copper site, located near the midpoint of the width of the neck of land on a high ridge over the south bank of the creek, is on rolling forest land, with sandy-gravel soil. It has never been disturbed by plowing, but was logged over at one time and is now covered with second growth mixed forest. (1991:124)

It appears that some excavation was done in this area, as 148 tiny worked fragments of native copper were found, ranging from a fleck to one-half inch in diameter. The fragments were concentrated in 11 areas, each roughly the size of the floor of a dwelling or a work area. The 11 areas, interpreted as work areas or domestic dwellings, were scattered along the ridge over an area roughly 25 to 30 yards wide by about a quarter of a mile long.

South of this area began another area of copper finds about 150 yards wide in six vaguely concentrated areas, parallel to the direction of the creek and roughly a quarter of a mile in length. In it were found 34 completed copper implements, 1 unfinished conical point, 14 scattered worked copper pieces, and pieces of bone, pottery sherds, and flint-knapping chips. These artifacts were found 4 to 8 inches below the present surface. There also were worked pieces of copper scattered within the area of the finished artifacts, including 2 irregular nuggets (approximately 2 inches by 3 inches), 5 smaller pieces (diameters ranging between ¾ and 1½ inches), and 7 small irregular pieces (diameter between ¼ and ½ inch).

And there was more:

> Directly to the north, across the creek from the above copper site, was discovered a companion copper site: it covers an area on the crown of the ridge above the creek bank and parallel to the direction of the creek about twenty-five yards wide by eighty yards long.
>   Here were found nineteen completed copper implements of the same assemblage of artifact types as found south of the creek. This site also yielded an apparent supply of copper nuggets, mined and transported from the source regions farther north for the manufacture of implements and ornaments. Two large, irregular masses of copper, each roughly six inches in diameter, were found apparently cached together fourteen inches below the present surface. In addition, nine irregularly shaped pieces ranging from two to 4 ½ inches in diameter were found, within the four to eight inch depth. The completed implements and nuggets were found in scattered locations over the site, in seven vague areas of minor concentration. No unfinished implements or small copper-working debris pieces were found on the site. (1991:125)

A copper repair band on a ceramic vessel, theoretically the theme of the Kent article, is insignificant compared to the rest of the data conveyed. This article sets the stage for beginning to understand the transport and usage of both mined and drift copper. (And it almost certainly took place in Vilas County, Wisconsin, the scene of many other copper transport discoveries described below.) What I find of particular interest here is the cavalier attitude of the ancients expressed in the leaving of finished artifacts, as if there was no concern about someone else coming along and collecting the finished implements. Was copper really that readily available?

Susan R. Martin, an archaeologist who for decades excavated in Michigan and studied prehistoric copper, reported on a copper workshop and other finds—notably a cache of artifacts—made by an avocational archaeologist with a metal detector at a site in Keweenaw County, Michigan (1993, 1994). According to Martin:

> The cache consisted of a carefully curated bundle or deposit of copper tools, partially worked nuggets and raw copper concretions in association with a woven textile and pieces of what appear to be leather. According to the collector's recollection, the cache was found near the ground surface in a heavily wooded area a number of meters above the current elevation above Lake Superior. The cache may have been contained in a leather bag or wrapper. Near the bottom of the cache were 30 or so pounds of raw and partially worked copper nuggets, above which lay a piece of aboriginal textile or woven material...The materials appeared to be carefully arranged with the leather [missing word?] Moreover, the tiny nuggets of copper, concreted together as wads of oxidation-encrusted material, were clearly collected and stored for some intended use. In addition to this material,

the cache included 43 awls, 1 crescent knife, 2 triangular points, 1 hammered nugget of unknown intended function, and more than 300 copper beads, some strung on original cord. (1993:154)

In Martin's 1994 paper, she broached the possibility that the cache had belonged to a woman. This suggestion apparently was based largely on the presence of beads and bead raw material.

Many avocational archaeologists have been pivotal in bringing to light more recently discovered caches. E. J. Neiburger, a dentist from Waukegan, Illinois, is one of these. It is unclear what Neiburger's role was in the actual discovery of caches in northern Wisconsin. Nevertheless, apparently he had access to the recovered artifacts and used photographs of the caches to illustrate articles in the Central States Archaeological Journal without divulging the specific locations of the discoveries or the names of the collectors (Neiburger 2008; Neiburger and Livernash 2009a, 2009b; Neiburger and Spohn 2007). In 2013 Neiburger published (2013a, 2013b) the first "comprehensive" illustrations of Vilas County caches. The contents of caches became commonly used illustrations after that.

Two collectors from Vilas County, Wisconsin, have used metal detectors to make significant discoveries of four caches of native copper artifacts (knives, awls, conical projectile points) and lightly processed copper nuggets or mini-ingots (Neiburger 2016a). The nuggets all appear to have been condensed from their original size by hammering. Many are in the 1 x 1 inch range. Neiburger believes that many of them already are or could be easily made into small knives (2016b). The presence of what we have defined as mini-ingots, and perhaps of partly finished tools, suggests that these are travel caches, as defined above.

Professional archaeologists might ask, "How many of these caches have been found? How representative are the Vilas County discoveries?" In a 2014 article, Neiburger offered the following tantalizing information:

Figure 3a. Frame of Old Copper Culture artifacts found at one camp site by metal detectors. Note the numerous flattened mini-ingots, the fish hook, awls, harpoon tip, knife, conical points including matching mandrel (for manufacture). This collection, one of over 100, was found in Vilas County, Wisconsin. (2014:131)

Later he states:

I propose that the copper trade, from the Great Lakes region to points south, was very active and consisted of many individuals transporting and trading relatively small amounts of copper (a kilo or so). A traveler could easily transport such amounts and hide them while camping for the night. The immense number of copper caches (100+) found in Vilas County, Wisconsin in the last few years are an example of the considerable volume yet to be identified. There may be hundreds of thousands of such caches. (2014:132)

Conclusion

An accidental discovery that bears a strong resemblance to the Vilas County discoveries took place in an unnamed lake in the vicinity of Marquette, Michigan (Pepin 2001). Jayme Petersen was using a screened catch basket and a submersible metal detector to find old fishing lures when she heard "a sharp electronic beep." Scooping with her basket, she found something in about 18 inches of mud and 3 feet of water: a copper spear point with a notched base. Additional scooping brought up 40 more pieces of copper, including a number of copper pins and about 30 fragments of copper, which were like the mini-ingots described earlier in this paper.

The discoveries to date seem to indicate that there was a regular copper trade across the western Upper Peninsula of Michigan and the northern counties of Wisconsin, and it is clear that the region holds many copper caches yet to be discovered.

Neiburger (2013b:78) also raises the issue of theft of copper nuggets, comparing it to modern criminal systems extorting payments or the presence of roaming gangs. He concludes that the great number of known and probable lost/abandoned caches is due to this problem:

> Trading valuable copper appears to have been a relatively lawless and hazard prone occupation. Considering the dangers encountered in long distance trade, the profits must have been enormous…even for a prehistoric native.

In the same article, he goes on to discuss lost caches.

> Wisconsin and Michigan are covered with caches of copper. These deposits amount to more than a pound (1/2 kg) or less of copper. The majority of this copper is in the form of ingots, though finished, preformed and partially worked tools are represented. There is seldom any pottery, stone points or other implements found. The lack of basket or clay materials imply that the copper was kept in a skin or woven pouch which degenerated over the years. Most caches are found buried in soil less than a foot deep. Like the Roman coin hoards being found in Europe today, these caches give all the indications that they were once secretly but temporarily buried and eventually abandoned. There is rarely any evidence of human/animal burials, buildings or trash heaps; the signs of long term habitation. The interesting thing about these copper caches is that there are so many of them and they all have the same signature: small amounts of copper (mostly ingots), no other materials like bone, wood or stone, no surviving containers and a very shallow burial.
>
> What does this mean? I propose a theory: These caches were one night camps made by small-time traders who briefly hid their "goods" in shallow pits and were unable to retrieve them. It is possible that they were surprised by bandits and ran away, but more likely killed/captured, since such a valuable hoard would later be retrieved when bandits left the area. This explains the lack of pottery, stone, etc. for they were one night camps made by traveling traders who traveled light and produced little trash. The caches were in shallow pits because there was no need to securely hide them, just to keep the copper out of sight for a night. The valuable copper was in small quantities because of the need to travel quickly (smuggling), and the reality that a small-time trader usually did not have the wealth to accumulate and transport larger loads. We find the caches only because the owner could not retrieve them. (2013b:78–79)

Needless to say, likening the ethnology of the Lake Superior region thousands of years ago to contemporary norms raises many questions.

The late Canadian artifact collector, Oliver Anttila, made significant discoveries in northwestern Ontario (to the northwest of Isle Royale), both on the mainland and on offshore islands (Anttila 2009). The topography of this region has been significantly disturbed by gravel extraction, timber harvesting, and the alteration of the original levels of many interior Ontario lakes. One particularly interesting find—interpreted as a possible canoe spill—was made up of 4 knives, 2 awls, 1 flaking baton head, and many copper preforms and lightly worked nuggets.

The collecting activities of Anttila, a long-time resident of the Thunder Bay area, have added to our understanding of Old Copper activities, particularly around Dog Lake. According to Anttila, most of his metal-detecting activities took place in the water and on water-washed shorelines. In an illustrated article published in *Indian Artifact Magazine*, he reported the discovery of three caches of copper nuggets and implements: Cache 1 consisted of 34 hammered nuggets; Cache 2 had 10 conical points, 2 tanged points, 2 partially completed tanged points, 3 copper bars, and 18 nuggets, 10 of which were hammered flat; and Cache 3 contained 2 conical points, 49 copper nuggets, and 2 taconite chert scrapers (Anttila 2000).

In a later article (2006), he described the discovery of another cache on the shore of Dog Lake. The site, which he discovered on an eroding shoreline (the lake had been dammed in 1905, raising the lake level about 10 feet), yielded 194 pieces, consisting of 11 formal artifacts and 183 mini-ingots.

In another 2006 article, Anttila provided an illustration (Figure 5) of a cache of 52 raw and hammered preforms and a grouping of 4 "surgical tools" (Figure 7) on the lakeside shoreline of Surgeon Bay (Anttila 2006).

One of Anttila's artifact-hunting partners, known in publications only as "Terry," apparently discovered a cache of 27 gaff hooks, 11 copper ingots, and possibly other tools or weapon heads (Anttila 2002:Photo 8, 45). Even from this small and nearly always accidental quantity of sampling, it is apparent that a relatively broad selection of goods was being transported.

Finding copper caches that could be dated has been difficult. However, researchers have had some success dating sites that appear to be copper-tool workshops. The Martin-Bird site (DbJm-5), located southwest of Thunder Bay, 50 kilometers inland, is one such site. K. C. A. Dawson reports:

> In Sq. 9 a copper cache occurred at 5" (7 cm) extending down 14" (35 cm) as a small irregular pit roughly 14" (35 cm) across.
> Marked by black ash and boulders it contained a hammerstone, a copper awl and 35 copper fragments. Black ash from this feature yielded a radiocarbon date of A.D. 805 ± 55 (S-891). The pit may have supplied heat during the annealing process and together with the hammerstones and worn abrading stones clearly indicates the manufacture of copper tools on the site. The only other record of this process in the immediate region comes from the north lakes area of Wisconsin directly across Lake Superior (Salzer

Conclusion

1974:48). There it was dated on a comparative base to the earlier Initial Woodland period. (Dawson 1987:49)

This could be the remains of a tool-manufacturing site. The tools and condition of the site are reminiscent of Hunzicker's description of experimental studies that aimed to replicate prehistoric metalworking (2002:58–59).

Copper tool production is not a difficult process but it requires significant amounts of both time and fuel. Copper can be worked hot or cold, but due to the difficulty of manipulating the metal when it is hot, it is likely that all but the very early stages of shaping a copper tool utilized cold forging. When cold forging, copper must be annealed [heated then cooled] frequently to relieve stresses that build up in the metal (LaRonge 2001). Personal experiments indicate that it takes 5-10 cycles of forging and annealing to produce a simple awl from an irregular copper nugget, and as many as 20-30 cycles to complete a larger, more complex form such as a socketed projectile point. LaRonge (2001) was able to forge a toggle head harpoon point in ten annealing cycles and Crucefix (2001) reported 30-67 cycles and an average of 33 hours to produce a copper axe. Given that sustained temperatures of at least 200-225 C, (Bastian 1961; LaRonge 2001), and preferably closer to 500-600 C, (Ormandy 1968, Schroeder and Ruhl 1968, Timothy Lloyd, personal communication 2000, Crucefix 2001), are necessary to satisfactorily anneal copper, considerable effort would be required to maintain a fire in addition to the actual work of manufacturing copper tools (Bastian 1961, Clark and Purdy 1982, Vernon 1990, Crucefix 2001). Experimental studies by Crucefix (2001) found that approximately 50 kg of wood would be consumed in the production of a single copper axe. The simultaneous annealing of many copper pieces would increase the efficiency of the process, as well as the potential for loss of individual pieces in a fire. By this explanation, the two concentrations of partially worked copper pieces at the Little Rice Lake site may indicate the locations of fires used in the copper tool manufacturing process.

There are a few fundamental techniques to cold forging copper that become clear through experimentation with the material. One is that heavy hammering is superior to soft hammering, especially in the early stages of shaping a copper tool. While soft hammering is useful for minor surface modifications, heavy hammering is evidenced at the Little Rice Lake site by deep hammer marks on some phase I and II copper pieces and the large concoidal fracture on the anvil stone.

Also of primary importance in the manufacture of copper tools is the maintenance of thickness. The importance of maintaining thickness is that once a piece of copper becomes too thin to be of use, the only way to thicken it is by folding, which inherently weakens any finished tool made from the piece because layers of copper cannot be fused through the process of cold forging. Rectangularity was used in personal experiments, as well as at the Little Rice Lake site, to achieve major changes in the shape of a copper piece while minimizing the reduction of thickness. The process of lengthening a piece, for example, involves hammering on one surface, rotating ninety degrees, hammering, rotating again, and repeating the process until the desired length is attained. The prehistoric use of rectangularity during copper tool manufacture is demonstrated at the Little Rice Lake site by the numerous elongate and triangular blank forms as well as some of the unfinished tools.

The copper pieces defined as blanks deserve additional comment. Artifacts similar to elongate blanks have been reported from other sites including the Haltiner Copper Cache described by Binford (1961), and pieces resembling triangular blanks have been found at the Reigh site where they were classed as wedges of chisels (Ritzenthaler et al. 1957), and at site 20KE20,(Martin et al. 1993). Forged copper pieces resembling both elongate and triangular blanks were recovered from the Chautauqua Grounds site (Pleger 1992). Blank forms were also reported by Moffat and Speth (1999) from the Rainbow Dam sites in Oneida County. These artifacts have undergone preliminary stages of manufacture and may have served as trade items or simply as a convenient form of storage of raw material reserved for later use. Copper blanks may in fact represent an important part of the Old Copper complex artifact assemblage that has been largely overlooked in the past.

The Robert Salzer discussion cited by Dawson reads as follows:

One of the most distinctive features of the Nokomis phase [Early and Middle Woodland] is the impressive extent to which copper tools were made and used. Cairn-type hearths almost certainly supplied heat during the annealing process and anvil-hammerstone combinations indicate copper toolmaking at some Nokomis phase sites. Well-worn abrading stones are also present and presumably functioned in the toolmaking process. The tools include simple awls, what may be elements of compound fishhooks, conical awls or harpoons, flat-stemmed projectile points with beveled edges, rolled conical projectile points, small chisels, punches, rolled beads, and what are probably ingots or blanks. Copper wastage is usually found throughout Nokomis phase deposits and, with the abundant discarded tools and features associated with copper processing, collectively reflects the significance of this industry in the local technology. (1974:48–49)

Given the description of "abundant discarded tools and features associated with copper processing" found close to bedrock copper sources, it appears that by Woodland times, at least in Wisconsin, copper and the artifacts made from it had few or no sacred properties.

It also seems clear that no matter what you call them—blanks, wedges, or mini-ingots—they were probably the principal sources of workable copper in distant places.

Researchers have published a significant study of ancient copper use in western arctic and subarctic Canada (Franklin et al. 1981), but the study appears to have no connection with developments in the Great Lakes area.

In summary, here is what I think we know about these caches and their contents.

1) With the exception of the Haltiner cache, they occur within roughly 100 miles of the bedrock copper sources in Minnesota, Wisconsin, and Ontario.

2) They do not seem to have occurred very far to the south in Wisconsin. Several caches of copper implements were found further south than Vilas County in Wisconsin, but they appear to be principally deposits of large, heavy, well-made spear points, pikes, axes, harpoons, and crescents (Brown 1907:66–67).

3) The caches are typically buried about 1 foot under the surface.

4) The cache bundles are light enough to be the burden of one person.

5) They have been found in areas that are today largely nonagricultural, i.e., less susceptible to destruction from agricultural activities. How

often plowing redistributed mini-ingots—particularly in Wisconsin, the final resting place of tons of drift copper fragments, among which the contents of many plowed-out caches could easily be "lost" (Lawson 1906:128–129)—is impossible to say.

6) Virtually all have been found through the use of metal detectors.

7) They contain a limited mixture of finished weapons and tools, unfinished items, and lightly worked copper nodules or mini-ingots.

8) The persistent presence of mini-ingots sets these caches apart from those caches that consisted of finished and generally higher-end weapons and tools (Titus 1914; Vansteen 2007).

9) In my opinion, the caches possibly began life at small copper-processing localities such as the Duck Lake site in Ontonagon County (Hill 2006).

10) Time for study may have already run out. As of the summer of 2016, at least some of the Vilas County artifact caches were being broken up and the specimens sold individually.

These caches may help us understand the prehistoric transportation of copper across land, a process long absent from the archaeological record in the Upper Midwest and northern shore of Lake Superior. We do not know how many (if any) caches were transported by canoe.

Charles E. Brown, the indefatigable director of the State Historical Society of Wisconsin's museum for 36 years, kept a close watch on the discovery of copper caches across Wisconsin. If one reads Brown carefully, it appears that none of the reported "Caches of Metal Implements" (1907:66–67) contained mini-ingots.

In a later report on copper implements in northern Wisconsin, Brown reported 163 copper implements found in 14 counties (including Vilas), but none were found in caches (Brown 1923b). The difference between 1923 and today is astounding: many more copper artifacts have been found in recent times. Perhaps more than anything else, the difference is due to the intensive use and significance of metal detectors.

It is possible, however, that the mini-ingots were seen but not recognized. For example, in Sheboygan County, Wisconsin, Alphonse Gerend reported:

We are apt to better realize the appropriateness of the application of the word "work-shops" to certain portions of this area, when we reflect upon the fact that the greater number of the objects found here were also manufactured here. Thus flakes and cast off particles of copper occur in great quantity, while larger partly finished objects are also quite common. Among the latter are flat oblong strips, bars, and blades of various shapes plainly exhibiting what appear to be the marks of the stone hammer and cutting implements. Large lumps of unworked native copper, although not discovered in recent years, have also been found. (1902:17)

See also Robert Hruska's account (1967). In another case, an interesting but not uncommon discovery of a cache was made at Portage Entry, Keweenaw County, Michigan. According to artifact collectors Walter C. and Edward F. Wyman:

Knives and spears found at Portage ship canal lighthouse by the light keeper while digging a stump on the lighthouse grounds, when at a depth of fourteen inches. They were all found together during the summer of 1889 and obtained from William McGue, the lighthouse keeper. (Quimby and Spaulding 1957:190)

This discovery consisted of 2 spear points, 3 knives, a fragmentary awl, and "one small fragment of worked copper." I think it is likely that this was a traveling cache of the type discussed above, and that there may have been other lightly worked copper fragments, so common in those caches, that were discarded as unworthy of being saved. Why one was saved is unknown.

Native Copper Use in North America

It appears obvious that the significance of native copper and its use by native peoples changed greatly over time. The summaries presented below show that vividly. The main focus in this segment is on the Old Copper period, because it is during that time that we can be reasonably certain that many native copper artifacts were being manufactured with mined rather than drift copper. However, as far as I know, no one has yet developed a technological system for differentiating a finished mined copper artifact from a drift copper artifact.

As the reader progresses through time and distance, the number of references relating to discoveries in Minnesota, Wisconsin, Illinois, Indiana, and Ohio, overwhelms the number of references to Michigan discoveries. What happened? The answer is simple. Long before the ancient miners had discovered and exploited bedrock exposures of native copper and other valuable materials such as chert, glaciers expanding and moving south scraped up untold tons of copper and deposited these materials (pieces ranging in size from pebbles to tons) south of the western Lake Superior basin, predominantly in Wisconsin. The Lower Peninsula of Michigan lay east of the main distribution of bedrock occurrences of copper. (See Hill 2009:61–76 for a more detailed treatment.)

It appears now that there were no firm cultural boundaries between the north and the southeast when it came to the use of copper. Research done by Sharon Goad has shown that some Great Lakes copper was used in the manufacture of artifacts discovered at Middle Woodland period sites in Alabama, Georgia, and Tennessee. There appeared to be no correlation between artifact type or style and ore source (Goad 1979). That is, we do not know how early the comingling of northern and southern copper sources began. It is clear that native copper was just one of the many materials sought after by the participants in the Middle Woodland Hopewell Interaction Sphere (Spence 1983).

Unfortunately, there have been too few extensive, thoughtful summaries of native copper's role in North American prehistory, although I acknowledge the efforts of Thomas (1898:109–113); Martin, Quimby, and Collier (1947:40–46) and, in the more restricted frame of Mississippian societies, Goodman (1984). For those who want to understand the basics of native copper, there is no better place to start than Frank Hamilton Cushing's pioneering study "Primitive Copper Working: An Experimental Study" (1894a, b).

Since some time periods experienced much greater usage of copper than others, I have created an appendix of chronological bibliographic references (Appendix 5), in which sources are arranged alphabetically within time periods.

When did the first use of Great Lakes native copper take place? Copper may have been used as early as the Paleoindian/Early Archaic period (ca. 8000 BC). Copper projectile points with the distinctive Plano and perhaps even stemmed forms are known (Mayer-Oakes 1951:315, Fig. 101, 30; Pettipas 1983:66–73, 1996:55, 59–61, Fig. 37, Fig. 48; Steinbring 1968, 1970:55–60, 1974:65, 68, and 1991; Woolworth and Woolworth 1963: Plate 7). It should be noted that Mayer-Oakes was writing at a time when "Steubenville Stemmed" points were believed to be Early Archaic. In fact, in 1958, Don Dragoo demonstrated that the Steubenville site (which Mayer-Oakes excavated) was Late Archaic. East of the Appalachians, its Fox Creek/Selby Bay lookalike dates to the Middle Woodland, as demonstrated by Henry Wright in 1973.

David Hunzicker has suggested that, "The Little Rice Lake site may belong to this possible transitional period of early copper working toward the end of the Late Paleo-Indian stage" (2002:58). It may or may not be significant that finished copper artifacts, blanks, and worked pieces excavated at the Little Rice Lake site are similar to what was found in the caches of Vilas County discussed above. Alas, very few copper artifacts have been found with datable wood shafts, handles, or the like.

From the Middle Archaic well into the Late Archaic period (ca. 5000–1000 BC; Pleger and Stoltman 2009:707–712), coppersmiths of the Old Copper Culture (OCC) created some copper tool and weapon types that had no obvious local antecedents in stone. These then disappeared, never to be seen in succeeding cultural periods. At first copper was used extensively (if not solely) in the manufacture of tools for domestic use and weapons for the hunt. The earliest workers in copper produced at least sixteen varieties of projectile points, five varieties of straight-bladed knives, nine varieties of crescent knives or ulus, four varieties of socketed axes or adzes, awls, punches, eyed-needles, pikes, drills, celts, chisels, wedges, gouges, fishhooks, gorges, harpoons, gaffs, spatulas, bracelets, and beads. In areas where copper was less common—e.g. southernmost Illinois, Iowa, and Tennessee—mostly small items have been found, such as awls and ornaments. The majority of copper artifacts found in eastern Wisconsin and Michigan's western Upper Peninsula may be reasonably attributed to the OCC.

In a small cornfield in southeastern Minnesota, property owners discovered a quantity of native copper artifacts. According to the abstract of the lengthy article:

> The Robert and Debra Neubauer site on Mission Creek in Pine County has yielded extensive materials including Late Paleo-Indian, Archaic and Woodland diagnostics. The majority of the diagnostics are from Middle and Late Archaic periods. One of the most significant activities appears to have been copper tool manufacturing. Experimental

archaeology has replicated a copper clad thought to have been used as a digging stick tip. In addition, different patterns of use-wear on hammerstones have been related to knapping, copper hammering and cutting, pecking and bone work. The site represents a copper tool manufacturing site probably associated with the "Old Copper" industry of the Middle and Late Archaic. (Romano and Mulholland 2000:120)

One wonders how many more "Neubauer-class" sites there are across the Upper Great Lakes.

Outside of a significant number of discoveries made in New York State (Beauchamp 1884, 1902), few OCC artifacts have been found in the Northeast and Middle Atlantic states (Abbott 1881:411–422). A few have been found far from the copper sources. For example, a copper spearhead was found in far northern Saskatchewan (Meyer 1979, 1983:157–158), an adze in northwest Missouri (Easterla 2007), spear points from North Dakota (Spiss 1968), a crescent-shaped knife in South Dakota (Gant 1963), a variety of artifacts in eastern Pennsylvania (Berlin 1885), and two spear points in Maryland (Curry 2002).

Of particular interest for this study are discoveries made in Canada in the form of grave lots with significant quantities of copper artifacts. These sites, which are probably of disparate ages, are the McCollum site, south of Lake Nipigon in western Ontario (Griffin and Quimby 1961; Wright 1972b:20 and facing page, 1995:287–288), and the Farquar Lake Cache, Harcourt Township, Haliburton County, southern Ontario (Popham and Emerson 1954:5–9; Wright 1995:287, 289).

Much further east, lying just east of Pembroke, Ontario, are the rapids of the Ottawa River. Avocational archaeologist Clyde Kennedy excavated the Morrison's Island site (Clermont and Chapdelaine 1998) and the Allumettes Island site (Clermont, Chapdelaine, and Cinq-Mars 2003), both lying just barely in Quebec, in the 1960s. Both sites contained numerous burials with both stone and copper artifacts. At Morrison's Island, there were stemmed copper projectile points, celts, socketed adzes, bipointed awls, fishhooks, harpoons, gorges, needles, awls, bracelets, punches, knives, rings, discs, and "50 pièces difformes dont une pourrait être de la matière première brute peu travaillée" (Clermont and Chapdelaine 1998:120, Planche 79), undoubtedly from the illustration of one specimen, a mini-ingot. Radiocarbon dates obtained from three burials were 4620 ± 40 BP, 4630 ± 40 BP, and 4860 ± 50 BP (Clermont and Chapdelaine 1998:25, Tableau 3).

The Allumettes Island site burials yielded conical projectile points, stemmed projectile points, rat-tailed points, socketed adzes, gorges, fishhooks, harpoons, awls, punches, needles, crescents, beads, pendants, and knives. In regard to recovered copper not made into artifacts, Chapdelaine reported, "Nous avons distingué trois types de déchets : les fragments difformes (N=129), des fragments de feuilles (N=4290 et des fragments de tiges (N=343)," (Clermont, Chapdelaine, and Cinq-Mars 2003:246). There are no illustrations of these waste pieces, but I assume that the term "difformes" here has the same meaning as it did in the Morrison's Island report; thus there were 129 fragments that might be classified as mini-ingots. Virtually no discussion is given

Conclusion

to the age of the site, but in Figure 1.7 there is a map of excavation units and burials. Four burials supplied dates: 4680 ± 40 BP, 5240 ± 80 BP, 5270 ± 40 BP, and 5440 ± 40 BP (Clermont, Chapdelaine, and Cinq-Mars 2003:39), making it about 600 years earlier than the Morrison's Island site. These sites are not to be confused with numerous ancient and historic human burial locations in eastern Canada that contain copper tools, weapons, and beads, but never mini-ingots (Jury 1965, 1973).

There was also significant "leakage" of OCC artifacts north into the Canadian Prairie Provinces, although there is little in the way of substantial sites there (Buchner 1979:35–36, 57, 86–93, 1980:13; Forbis 1970:15, 43; Meyer 1979, 1983; Pettipas 1983, 1996; Steinbring 1966, 1967, 1968, 1970a, 1970b, 1971b, 1974, 1975, 1980:64–66, 178–243, 1991) or later (Brink 1988; Dyck 1983:96). The elaborate burial of a child near International Falls, Minnesota, accompanied by a pair of diagnostic Old Copper projectile points attached to dart shafts, is of particular interest (Steinbring 1971a).

We must pause here and consider the likelihood that copper almost certainly held different meanings for members of societies far removed from the bedrock sources. An example of this was a discovery made in the course of excavations at the Carrier Mills Archaeological Project in the Saline Valley of southern Illinois (Early to Late Woodland). In Burial 190B (an adult male), excavators discovered a copper wedge placed where the missing skull should have been (Jefferies and Butler 1982:Vol. 1, 632, Plate 37, Vol. 2, Table 410, Fig. 318, Plate 132). Since the rest of the skeleton is largely intact, it would seem that this individual (and the wedge) must have occupied unique statuses in that society.

Several mortuary complexes that flourished near the end of the OCC and the Late Archaic period and into the Early Woodland present interesting conundrums in regard to copper. The Red Ocher Culture had, as one of its nuclear traits, worked copper beads and occasional copper tools. The Red Ocher people apparently had nothing to do with raw, mined copper or mini-ingots, but drift copper would have been abundant for sites and finds throughout Wisconsin, eastern Iowa, Illinois, Michigan, and Ontario.

A seldom-cited description of a copper discovery presented in an article by pioneering nineteenth-century avocational archaeologist John F. Snyder gives a breathtaking look at the amount and variety of copper artifacts discovered in a Red Ocher grave beneath the later Hopewellian Hemphill Mound in Brown County, Illinois.

> With only one of the entombed bodies had been interred worldly possessions of a kind that survived the lapse of ages. We are at liberty to imagine that this one was a distinguished personage, and the other seven, his wives or slaves, slain at his death to attend him in the other world. Let that be as it may; if in his day the finances were based upon a single copper standard, he was reasonably well fixed. Near his head was a nodular nugget of pure, native copper—unwrought raw material—weighing 24 pounds; and along his sides were ranged ten copper axes. Around his neck were three necklaces; one of oblong, large beads, made from the columella of marine shells, perforated longitudinally; another of over 200 incisor teeth of the squirrel bored at the root, shown, with one of the beads in Fig. 9, and the third was composed of 283 globular, copper beads, solid, and smooth as if moulded and then polished. The largest ones, in the middle of the necklace, are half

an inch in diameter, and they gradually decrease in size at the ends to the quarter of an inch. The cord that suspended them, a two-strand, twisted twine, apparently of hemp, was still in place, but crumbles at the touch. Across his breast, and following each other an inch apart, were five plates cut out of fluor spar, each six inches in length, two and a half inches wide, square-cornered, and the fourth of an inch in thickness, as smooth as glass, and in the sunlight as resplendent as burnished silver. Each was perforated with two holes, one two inches from either end, for attachment to the dress. The copper axes are of three types, three of them of the thin, hammer-marked sort, Fig. 18, three inches wide and seven, nine and ten inches long respectively. Three are of the celt shape, Fig. 25, compact, very smooth and sharp-edged; and three, four and four and a quarter inches long. The other four are flaring at the edge, Fig. 26, heavy, with even, well-finished surfaces, weighing from two to four pounds each, and are ornamented by cuts a line in depth and an inch to an inch and a half in length, on both side sides, at irregular intervals of half an inch or more, seemingly made with a cold chisel or other edged tool. (1898:22–23).

The name of the Glacial Kame Culture of southern Michigan, southwestern Ontario, northern Ohio, and Indiana was derived from the practice of burying the dead on the glacially deposited gravel hills or kames. Copper beads, shell gorgets, and bracelets are found in Glacial Kame burials, but are not common. Again, caches of finished artifacts with mini-ingots are completely unknown.

Further south, during the Early Woodland period, the use of copper changed and became more restricted in usage—for decoration as opposed to tools and weapons. For example, in the Adena culture of the Ohio River Valley, copper was used mainly for rectangular gorgets and plates, bracelets, finger rings, crescents, pins, celts, awls, boatstones, antler headdresses, bars, foil covering for certain varieties of artifacts (such as bear canines or effigy bear canines), and above all, beads.

In the northeastern United States and eastern Canada, in some groups, especially those related to or influenced by Adena, or the Meadowood Interaction Sphere (Brose 1971:12; Flanders 1986:78; Garland 1986:75–76; MacNeish 1952:48; Ritchie 1955; Ritchie and Dragoo 1960; Salkin 1986:110; Spence and Fox 1986:23, 26, 31–33; Taché 2011), copper appears, but it is almost exclusively in the form of beads and small celts. Analyses of northeastern copper artifacts (e.g. Lattanzi 2013; Levine 1996) suggest that, after the Late Archaic, the sources of copper in the northeastern Early Woodland appear to be almost entirely local. Therefore, I will draw the line here and not venture into northeast and east coast sites with copper.

There was a precipitous decline in copper use from the Archaic to the Early Woodland. In the Middle Woodland period, Hopewellian coppersmiths from Michigan to Alabama exceeded all previous works in complexity of artifacts and quality of workmanship. Flat artifacts were sometimes made of several sheets of copper artfully riveted together, hinting at the possibility that other techniques, such as cold welding, were employed (Gadus 1979). Earspools, breastplates, and headdresses with moveable parts were the most elaborate products of a repertoire that also produced ceremonial axes (one weighing more than 38 pounds), panpipe jackets, geometric cutouts, reel-shaped gorgets, repoussé plates of birds and fish, effigies of deer antlers, bighorn sheep horns, turtles, bats, poison mushrooms, and even prosthetic noses for corpses.

Many of these items were probably ritual equipment or adornments used by shamans or headmen. There can be little doubt, for example, that the elaborate headdress depicted by Charles C. Willoughby (1935:Fig. 3) was part of a larger costume representing Michabo, the Great Hare, an important Algonquian culture hero. The sophistication of construction and the restricted range of artifact types strongly suggest that craft specialists were responsible for their manufacture.

Societies contemporary with Hopewell in the northern Upper Great Lakes (Laurel and Northern Tier or Lake Forest Middle Woodland) made ready use of copper, but mostly for utilitarian objects (Brose 1970a:129–137; Janzen 1968:69; Ronald J. Mason 1966:68–69, 1967:320–323; Richner 1973:56–59; Webster 1973:104–105). Even at the furthest extent of moundbuilding in Manitoba, copper was to be found (Bryce 1887:5, 1890:345, 1904:21–22; Capes 1963:45, 52, Plate VI, 101–102).

During the Late Woodland, Fort Ancient, Upper Mississippian, and Oneota periods, copper usage almost ceased, with some or perhaps all of the raw material probably obtained from the glacial drift. At some sites, the only way we know copper was part of the cultural assemblage is the presence of copper stains on bone in burials (Oehler 1973:32; Wood and Brock 1984:18, 33, 41, 51, 66, 101, 112). Stylized bird and snake (feather?) effigy pendants, beads, awls, rings, and foil covering for wooden or bone earspools represent virtually the entire inventory. In habitation remains excavated at the Juntunen site (AD 800 to 1400) on the west end of Bois Blanc Island in the Straits of Mackinac was a concentration of copper products in the form of unworked lumps, thin sheets, beads, awls, and a "butter knife" (McPherron 1967:Plates XXXVIII and XXXIX). (See this volume's Afterword for more on Juntunen.) Alan McPherron, the excavator and author, posed the question:

> Why, then, were no large pieces of raw material encountered at the Juntunen site? Why were all fragments, both those derived from [the] working process and the essentially unaltered raw pieces, smaller than any artifact, and why did most artifacts look as if they had been made up of several pieces of copper? (1967:165–166)

One answer would be that they may have only been working with drift copper all along, possibly from sources at the northeastern end of Lake Superior, perhaps near Mamainse Harbour (Conway 1981:49) or Michipicoten Island. However, if the Juntunen craftspeople were working drift copper, one would expect some lumps and mini-ingots. They might have recycled tiny bits of copper scrap from earlier archaeological sites. There were several earlier sites nearby. Just east of Juntunen was Arrowhead Drive, a Lake Forest Middle Woodland site. On a higher beach ridge above and to the east of the Juntunen site, a corner notched point was found, similar to Feheely Corner notched points and to the point from Duck Lake. See Chaput 1969 for a brief history of this mysterious island.

Given the possibility of Appalachian sources and the many copper finds from the region, I will comment only minimally on copper use in southeastern North America (see Bishop and Canouts 1993:163–166; Goad 1978; Goad and Noakes 1978; and Hurst and Larson 1958 for more on this very complicated topic). Mississippian societies in the Deep South may have made use of northern copper sources (Goad 1978:210).

Craft specialists of the Middle Mississippian cultures (post–AD 1000) of southeastern North America and the central Mississippi Valley, where copper could have been obtained through trade or finding drift copper, produced stunning works of art in copper that reflected a complex, highly developed and war-soaked cosmology. (See this volume's Afterword for more on the copper workshop at Cahokia, extensively excavated recently by Kelly and Brown. Even though it is in a huge center with craft specialists, the debris is reminiscent of the small working area at the Great Lakes fishing villages on Juntunen, with many small flakes of copper being annealed together.) In Mississippian society, only those in the uppermost strata of society had access to the symbols of power made of copper (e. g. Peebles and Kus 1977:438–439). Thin plates or plaques of repoussé copper carry images of warriors or priests dressed in feathered costumes carrying maces and severed human heads, of dancing human figures, stylized raptorial birds, and geometric figures. Along with bilobed arrows, effigy feathers, and small, perforated, arrowhead-shaped pieces of copper, the smaller of these plaques were probably components of elaborate headdresses (Black 1967; James A. Brown and Rogers 1989; James A. Brown et al. 1996; King 2007a:111, 2007b:255–256; Pustmueller 1950). Larger specimens, based on their relationships to the skeletons discovered in graves, apparently functioned as ceremonial breastplates (e.g. Jones 1982; Kelly and Neitzel 1961:52–53; Thomas 1894:161). Large axe blades, long-nosed "god" maskettes, maces, circular breastplates, pendants, beads, long awls or needles, effigy tortoise shells, copper-foil-covered wood effigies, and stone, wood, and clay earspools and beads make up most of the rest of the inventory.

Very late prehistoric and early historic-period Indian use of copper was minor and largely restricted to clothing decoration or accents such as eyes on animal skin medicine bags (e.g. Groce 1980). Nevertheless, it is clear that in some areas, particularly the southeastern United States, native copper and increasingly imported European copper were important materials for ritual artifacts, both new and heirloom, even up through the 1860s.

The uses of copper and the control of its availability were critical features of earliest Native American and English relations in Virginia (Hodges 1993:36–39; Lurie 1959:43–44; Mallios and Emmett 2004; Potter 1989; Rountree 1989; Rountree and Turner 2002; Trevelyan 1985). At the sources of native copper on the Keweenaw Peninsula and Isle Royale, at the time of first contact, local inhabitants believed large pieces of copper to have miraculous powers and regarded them as divinities (Charles E. Brown 1939; Kohl 1860:61–62; Thwaites 1959:265–267). At first contact there may have been social sanctions against those who showed outsiders the locations of copper and iron sources (Buell 1906:38). However, it is clear that through the 1820s there was progressive loss in the belief of spiritual power inherent in native copper (Kohl 1860:60–64; Williams 1888:142–143). By the 1840s, such beliefs had largely

Conclusion

lost their effectiveness (Whittlesey 1852c:333–335) and copper became just another useful metal, devoid of any spiritual connotations.

Recent Research on Native Copper and Artifacts

Establishing the date of Native American copper mining activities has proved difficult. The dating of copper-working features in campsites, discussed above, indicates some copper working in the Keweenaw area before 6000 BC, but most of the features date to between about 1800 and 1400 BC. We do not, however, know whether the copper worked at these sites was mined from bedrock sources or found on the surface of glacial deposits. An obvious approach would be to date charcoal or wooden artifacts from the mines, but modern copper mining has destroyed all prehistoric known mines on the Keweenaw Peninsula. Only on Isle Royale has direct dating of mines been successfully attempted. Carbon samples collected by Tyler Bastian in 1961 from prehistoric mine shafts and debris deposits indicate some Isle Royale mines were exploited between 3340 and 1450 BC (Crane and Griffin 1964:7, 1965:126–27).

Recent studies have provided provocative new information on the beginning (and cessation) of copper mining in the western Lake Superior region. A geochemical research team led by David Pompeani, of the University of Pittsburgh, took sediment cores from three lakes near copper mines in the Keweenaw Peninsula.

> [The team]…analyzed the concentration of lead, titanium, magnesium, iron and organic matter in sediment cores recovered from three lakes located near mine pits to investigate the timing, location, and magnitude of ancient copper mining pollution. Nearly simultaneous lead enrichments occurred at Lake Manganese and Copper Falls Lake ~8000 and 7000 years before present (yr BP), indicating that copper extraction occurred concurrently in at least two locations on the peninsula. The poor temporal coherence among the lead enrichments from ~6300 to 5000 yr BP at each lake suggests that the focus of copper mining and annealing shifted through time. In sediment younger than ~5000 yr BP, lead concentrations remain at back ground levels at all three lakes, excluding historic lead increases starting ~150 year BP. Our work demonstrates that lead emissions associated with both the historic and Old Copper Complex tradition are detectable and can be used to determine the temporal and geographic pattern of metal pollution. (Pompeani et al. 2013:5545)

Two years later, the same University of Pittsburgh group studied sediment cores from McCargoe Cove, a long, narrow inlet on the north shore of Isle Royale. They found that there were two periods of increased metal pollution.

> …concentrations of lead, copper, and potassium increase in the sediments after AD 1860 and between 6500 and 5400 years before AD 1950 (yr BP). Metal pollution increases at McCargoe Cove exceed natural (or background) levels and coincide with radiocarbon dates associated with copper artifacts and existing lead pollution reconstructions from lakes on the Keweenaw Peninsula…a coherent cessation of lead emissions at multiple

study sites after ~5400yr BP coincides with the onset of dry conditions found in regional paleoclimate proxy records. After ~5000 yr BP, lead concentrations on both Isle Royale and the Keweenaw Peninsula remain at background levels until the onset of modern lead pollution ~AD 1860. (Pompeani et al. 2015:253)

A rough calibration of the radiocarbon dates of the Pompeani team indicates there was an increase in lead pollution between 7200 and 5700 BC in the Keweenaw lakes and between 5600 and 4000 BC in similar Isle Royale deposits. These interesting results are puzzling not simply because the direct dating of increased copper exploitation generally yields later dates, but because there is very little lead in Great Lakes copper, and how it might be deposited into lake sediments in the absence of actual smelting is unclear. In fact, a quick survey of lead records in sediment sequences in the northern hemisphere showed contemporary Middle Holocene lead peaks (and thus prior to metal smelting anywhere) in other areas, so this may be a planet-wide phenomenon unrelated to mining. The researchers go on to cautiously suggest that the period of inferred copper exploitation could have been a consequence of prolonged dry conditions.

> …the final cessation of Pb [lead] emissions at multiple lakes coincides with the onset of sustained dry conditions detected at several sites around the Great Lakes (Kirby et al., 2002; Smith et al., 1997, 2002; Yu et al., 1997), raising the possibility that metalworking activity associated with hunter-gatherer economies in North America was susceptible to influences of climate change (Pompeani et al. 2015:261).

The arrival of long-term drought could have brought an attendant loss of the wildlife and vegetation on which the miners depended for their livelihood. It might make sense that they would abandon their mines (or use them much less) during such periods. Mining had perhaps always been secondary to collecting and working drift copper, which was available in endless sizes, shapes, and weights on the land and in the ponds, lakes, and rivers of the Midwest ever since the retreat of the last glaciers (Halsey 2010). There was probably a prolonged "shutdown" rather than a sudden termination of hard-rock mining.

Taking great liberties with the time spans offered by the two Pittsburgh studies (and calibrating them from the radiocarbon calendar to the historical calendar), we can propose that the prehistoric copper explorers had opened and worked mines over at least 3000 years (7200–4000 BC). During these millennia, the people of the Lake Superior region discovered and utilized the copper resources there. If indeed there was no copper mining during the next period—the more than 5000 years between 4000 BC and AD 1860 (during which copper artifacts were likely created from drift copper)—it wouldn't be surprising if the people eventually forgot about the long-ago endeavors of their ancestors.

Clearly, we still have very few studies providing reliable dates of early copper working and mining. In the future, we hope that researchers will find and date more prehistoric

Conclusion

mines that have survived nineteenth- and twentieth-century destruction and will undertake more studies of sediments in lakes, not only near the mines but in areas far from early copper sources.

Afterword: Recent Work on Copper-Processing Sites

As recounted in the preceding chapters of this book, pioneering researchers of the nineteenth century made extraordinary advances in understanding prehistoric copper miners. These advances were followed by many decades of limited research. There were several reasons for this. First, the copper boom in the Lake Superior region tapered off. The surface mines, which produced so much evidence of ancient miners, were replaced by deep mines, made possible by constant pumping, which plumbed depths the prehistoric miners never reached. Second, two World Wars and the Great Depression focused interest and resources on other endeavors. Third, when archaeology was federally funded during the New Deal under the Works Projects Administration (WPA), as a way of providing jobs to unemployed residents of small towns and rural areas, this archaeology was necessarily concentrated in the areas where there were more unemployed people. Few WPA funds went to the states bordering the Upper Great Lakes.

After World War II, there was a renaissance of research on the prehistoric copper miners. There were several reasons for this. First, new technologies were available that could answer questions that the earliest archaeological investigators asked, but could not answer. The most important of these were new methods of figuring out when archaeological and geological sites were occupied. The first constructive use of nuclear physics was not a new way of generating energy or a new medical treatment, but a new way of telling how old things were, based on the regular decay of unstable radioactive isotopes. Of greatest use to archaeologists was the decay of the isotope carbon-14. Along with large amounts of ordinary, stable carbon-12, all living organisms accumulate small amounts of carbon-14, which is created by the impact of cosmic rays on our planet's upper atmosphere. When the organism dies, the unstable carbon-14 decays into carbon-12 at a known rate. After about 5700 years, half of the carbon-14 in formerly living tissues (such as wood, seeds, bones, and shells) has decayed. The first crude radiocarbon dates, based on the ratio of carbon-14 to carbon-12, were measured at the University of Chicago in 1949, four years after the bombing with nuclear weapons of Hiroshima and Nagasaki. By 1960, after many adjustments and improvements, reliable dates could be estimated for most archaeological sites less than 40,000 years old. Another use of nuclear technology was the identification of tiny amounts of trace elements and isotopes in artifacts of pottery, stone, copper, and other materials, and the assignment of these to natural sources of these materials. It became possible to document vanished systems of procurement and exchange. In archaeology, the first application of such methods was during the 1960s, when University of Michigan researcher Adon Gordus used neuron activation analysis to identify obsidian from eastern North American archaeological sites as having been carried from far-away Yellowstone Park in Wyoming.

Afterword

Second, under the influence of social and ecological anthropology, archaeologists became interested not just in the age of different artifacts, but in how they were used and how the people who made and used them actually lived. These interests developed rapidly after 1950, as archaeologists borrowed from such fields as zoology, botany, ethnography, and human biology, respectively, to develop a host of new methods to infer ancient lifeways from animal bones, carbonized seeds and wood, houses, and human skeletons. Whole new disciplines, such as ethnobotany, ethnozoology, and archaeogeology, became more and more important.

Third, archaeologists realized that we were discovering complex systems that could not be represented by traditional narrative accounts or deterministic models. Archaeologists begin to think in terms of formal models that could represent many interconnected actors and activities controlling diverse variables. This "processual" approach is still developing, but even the earliest steps toward this more complex paradigm during the 1960s influenced the ways archaeologists worked in the Upper Great Lakes.

These new technical and theoretical developments in archaeology were first applied in the Upper Great Lakes during 1958 and 1959, in a project designed by James B. Griffin, then director of the University of Michigan's Museum of Anthropology. The project was to study prehistoric copper mines and associated settlements on Isle Royale, on the northwest side of Lake Superior. Tyler Bastian, who went on to become a long-serving state archaeologist of Maryland, directed the actual fieldwork. It was thought that nineteenth-century mining was not so intensive on Isle Royale as on the Keweenaw Peninsula, and that excavators could find and date prehistoric mines there. This proved to be the case. As noted in previous chapters, Bastian and his team of graduate students recovered a carbon sample from a prehistoric mine dating to 3360–2880 BC and later (Bastian 1962; Crane and Griffin 1965:126–128). While some of the Isle Royale mining pits date to Late Archaic times, the occupational sites tested by the team had only Middle and Late Woodland dates and ceramics, so little could be said about early settlements and copper processing activities associated with the Isle Royale mines.

Late twentieth-century researchers have contributed to our understanding of the Native American copper industries of the Upper Great Lakes by applying these new methods and perspectives to workshop and occupational sites near the Keweenaw copper sources or on the routes connecting copper sources in the northern forests to the larger population centers farther south. By considering several such projects from the Upper Great Lakes, we can begin to trace the development of copper working in ancient Native American communities.

The earliest native copper tools found so far date to 7000 BC, the Early Archaic Period, but the earliest sites with direct evidence of copper working excavated in the Upper Great Lakes date to the Late Archaic Period. We will first consider the sites of the Late Archaic Period, then a site of the Middle Woodland Period, and then sites of the Late Woodland and Mississippian periods.

Lac La Belle

One Late Archaic site is Lac La Belle, a site in Keweenaw County on a protected bay near the northwestern tip of the Keweenaw Peninsula. In the trinomial numbering system used by all of the United States, Lac La Belle is 20KE20. In 1987, this site came to the attention of Patrick and Susan Martin—at the time teachers at Michigan Technical University and the only professional archaeologists on Michigan's Upper Peninsula—because of reports of finds made with metal detectors, most outstandingly a small cache pit with many copper nodules, artifacts, and even traces of fiber items preserved by copper salts. Residential development threatened the locality, so the Martins, with a team of archaeology students as well as local volunteers, undertook excavation in 1988 with support of the university, the National Geographic Society, and the National Science Foundation. A useful comprehensive report of the site was published in *The Michigan Archaeologist* (Martin 1993).

Lac La Belle is actually a complex of sites scattered over about 8 ha of sandy beach ridges of earlier stages of Lake Superior, uplifted by postglacial rebound after the weight of the glacial ice was removed. During occupation, the site would have looked south over a protected bay. Occupational traces are common on both the higher beach ridges, now raised to about 195–198 m above sea level (thought to result from a phase of Upper Great Lakes development called the Nippising stage, dating between 3500 and 1560 BC), and the lower beach ridges, now raised to about 189–195 m above sea level (the lowest of which are sometimes called the Algoma stage, which dates to between 1490 and 1015 BC). Modern Lake Superior averages about 183 m above sea level. The occupants would have had access to the food resources of the pine, spruce, hemlock, and birch forests, such as deer, moose, squirrels, rabbits, and bears, and the wild rice, tubers, reeds, turtles, and fish of the lakes and wetlands. Material resources such as wood and bark from the forest and cobbles of basalt, quartz, and chert were close. However, it was the rich seams of copper from Mount Bohemia, less than a mile to the north, that most likely attracted prehistoric people to this area.

It is hard to know how to approach a site with such a low density of artifacts occurring in discrete clusters, especially given the removal of many copper items found by avocational archaeologists with metal detectors. Given the size of the site, the few trained workers, and the limited funds, the Martins decided on the use of five different procedures.

1. Before doing any fieldwork, they consulted with the avocational archaeologists who had worked on the site, and tried to place their finds on the site map.

2. Using a proton magnetometer, they searched the path of the planned road for magnetic anomalies that might mark hearths. This was not particularly useful.

3. With a more sensitive metal detector, they systematically passed over the sand ridges where amateur metal detector enthusiasts had reported copper, looking for small pieces of copper missed by previous searchers. This was useful in better defining the artifact clusters.

4. They made shovel tests in areas where finds had been made (to define the boundaries of clusters) and also where road construction was planned. This was useful in finding stone flakes and other artifacts not detectable with metal detectors, which helped to better define the clusters.

5. They excavated larger test units (each composed of 2 to 10 1-m squares) in levels. They sieved all sediment through 6-mm screen, and recorded all plans and sections. The hoped-for post molds and floors marking structures had not survived, but two clusters revealed probable hearths.

The fieldwork documented four areas that could be absolutely dated with the radiocarbon method; all four areas had been occupied at several different times. All dates were made at the Beta Analytic Laboratory in Coral Gables, Florida.

To the north, above the beach that is 195 m above sea level, in Test Unit 12 in Cluster B, charcoal from a deep burned red area with associated basalt flakes and a copper fragment was 95 percent certain to date between 7585 and 6070 BC. This is somewhat puzzling, because the Nippising beach is thought to have begun to form about 3500 BC. Though it is possible that old driftwood was swept up by rising lake waters, it seems more likely that the Nippising beach has incorporated remnants of an earlier beach ridge. Though an actual working area was not found, the date indicates Early Archaic occupation at Lac La Belle.

To the south, above the lower beach ridge, in Test Unit 10 in Cluster C, charcoal from a hearth area with associated chert flakes was 95 percent certain to date between 1760 and 1450 BC.

To the southwest, also above lower beach ridge, in Test Unit 50 in Cluster A, scattered charcoal was 95 percent certain to date between 1730 and 1410 BC. These dates are statistically indistinguishable and fall in the earlier part of the Late Archaic period. They indicate that during this period there was a large campsite extending at least 150 m along the old beach, perhaps with several contemporary small camps for several subgroups, or perhaps with several visits during the long time span of the earlier Late Archaic period.

To the southeast, also above the lower beach ridge, in Test Unit 4, close to Cluster D, the discoverer of the site found a pit containing a copper cache, for which a piece of a woven fabric or bag yielded a date 95 percent certain to fall between AD 300 and 655, which is during the later Middle Woodland Period. If Cluster D is also of this period, this too would have been a campsite extending about 500 feet along the beach. Obviously more dates would be useful, but archaeologists are limited by the material they find. Though we cannot detect subgroups, we can learn something from the structure of the cache and from the structuring of copper manufacturing, to which we now turn.

The pit in which many copper items were cached was only 25 cm in diameter and 30 cm deep. It had five distinctive layers. The deepest and earliest layer was composed of about 14 kg of copper nuggets and hammered pieces. Above this was a folded fabric, perhaps a bag, partially preserved by copper salts. Above this was an unusual crescent-shaped copper object, and above that was a layer with more than 300 copper beads, some still strung on cord preserved by the copper salts. Finally, there was a layer with 43 finished copper awls, square in section.

An overview of all the copper items found at the site indicates the following four steps in the processing of copper.

1. Residents collected nuggets. Individually measured nuggets range from .2 grams to 20.6 grams, with almost half less than 2.5 grams. More than 70 percent of these have been hammered, whether in the course of removal from bedrock copper seams or to test the hardness of the metal, we do not know. Some of these, however, were hammered until they were flattened rectangular and triangular pieces—apparently preforms to make beads, points, etc.

2. They separated nuggets and hammered pieces into size categories destined for different uses: Careful examination yielded no evidence that larger items were by made heating and hammering together small pieces, so each artifact had to be made from an appropriately sized nugget. Beads from controlled excavations range from .2 to 2.6 grams and would have been made with the smallest nuggets. Tiny pins range from .7 to 2.6 grams and would also have been made from these small nuggets. The two awls for which we have weights are 5.6 and 10.5 grams and the two points for which we have weights are 5.2 and 11.5 grams. These would have been made from intermediate-sized nuggets. Unfortunately, we do not have the weights of the majority of copper artifacts that are not from controlled samples, and thus we cannot precisely specify the types of nuggets used for each type of finished tool.

3. Some tools would not have been made directly from hammered nuggets, but from the intermediate stage of a blank or preform. Thus, beads are made from flat rectangular pieces rolled around a cord, and conical points are made from flat triangular preforms rolled into a cone.

4. They manufactured standardized forms of finished artifacts from nuggets or flattened preforms. The most common finished items in the Middle Woodland cache in Test Unit 4 were hundreds of beads. Next most common are awls, of which there were 43. There is only one crescent-shaped item—it was possibly a knife. Elsewhere on the site, however, such distinctive forms as socketed points, stemmed points, and tanged knives appear to have been made, though none were found in caches. A spatial separation in the making of different copper types implies that individual craftspeople had special skills.

The final class of artifacts considered here is the flaked stone assemblage, from which we can learn about the broader regional contacts of the Lac La Belle occupants. Avocational archaeologists in this area rarely collect such tools, so (in contrast to the copper) we can be sure that we have a sample representing the full range of stone artifacts. All of the stone tools seem to have been made from rounded cobbles with battered and weathered outer surfaces, such as are found locally on beaches and in stream beds. The overwhelming majority, about 79 percent, are small flakes of locally available quartz, none retouched or evidently utilized, which probably served as expedient cutting tools. Cherts, carried by the glaciers from bedrock sources in Canada, were used for 11 percent of the tools. In contrast to the quartz flakes, some cherts were used. Many are flakes discarded in the course of making bifacial tools. A number of these were made into small scrapers, most designed to be socketed into handles. There is one biface: a side-notched projectile point of a type not known in Midwest Archaic sites to the south. The rest of the assemblage is made up of quartzite, basalt, and diorite, one of which was a utilized flake and one which has the characteristic battering of a hammer stone. All these stone types can be found nearby in glacial deposits on the Keweenaw Peninsula. As we will see, this strongly local character is in contrast to the flaked stone from the one other excavated Late Archaic occupational site in the area. Unfortunately, the table with lithic occurrences in the different test units was left out of the publication, so we cannot assess differences in activities in different parts of the Late Archaic Clusters A and C or during the development from Early Archaic in Cluster B to Late Archaic and perhaps Middle Woodland in Cluster D.

Duck Lake

Another Late Archaic site where people worked copper is the Duck Lake Site (20ON20) in Ontonagon County, Michigan. This site is 22 km southeast, at the closest, from the main bedrock copper deposits and 42 km southeast of the present shore of Lake Superior, on a sandy outwash plain in the interior of the Upper Peninsula. The site is immediately south of the deeply entrenched East Branch of the Ontonagon River and immediately north of some small wetlands in low areas. The site came to the attention of Mark Hill when avocational archaeologists found cultural materials in the forest access road crossing the site. Under the direction of Hill, a team from the Ottawa National Forest, aided by volunteers, commenced a field study in 1994; work continued in 1996 and 1997. An exemplary comprehensive report was published in the *Midcontinental Journal of Archaeology* (Hill 2006).

The site is an irregular shape draped over about 7 ha of low hills of sandy gravel between the river valley and the wetlands. The forest today is composed of pine, sugar maple, birch, and aspen, with a ground cover of ferns and berries. The charcoal from prehistoric features includes spruce, pine, maple, and red oak. The lack of birch and aspen and the presence of oak charcoal could represent a deliberate selection of woods suitable for hotter, longer-lasting fires, rather than a difference in climate. Today, the Ottawa National Forest is home to deer, bear, beavers, and a variety of small mammals, birds, and fish. Burnt animal bones from cultural features were exclusively from mammals: a few are deer-sized and one is a smaller mammal, perhaps a marten, but precise identification has not been possible. While the area has the same woods for technical applications as Lac La Belle, it is poor in stone useful for making flaked tools. Only quartz and basalt are common, and the latter was rarely used. Surprisingly, most of the flaked stone from Duck Lake is exotic chert from the Mississippi Valley, as is further discussed below.

The excavation procedures were similar to those used by the Martins at Lac La Belle. The team began with a systematic program of shovel testing. They conducted controlled metal detector surveys to delimit denser concentrations of copper items within the larger localities defined by shovel testing. This was followed up with a program of block excavations. Excavators sieved sediment through 6-mm screen.

Researchers defined three larger concentrations, called localities A, B, and C, based on copper found in shovel tests and with metal detectors. Of these, only one—Locality B to the northwest—is presented in detail in the report. Locality B extended 90 m from northwest to southeast and averaged about 10 m wide. Within the cluster, there were two concentrations: a smaller one to the southeast and a larger one to the northwest, in which a large block excavation was placed. This block produced three closely spaced cultural features, also termed A, B, and C.

Feature A is a small oval concentration of red soil and charcoal with copper scrap and copper preforms. It is interpreted as a surface hearth. Beta Analytic reports a radiocarbon date made using the then-new accelerator method. It is 95 percent certain that this feature dates between 1880 and 1620 BC.

Feature B, only 2 m south of A, is a roughly rectangular concentration of charcoal, blackened sand, and fire-broken rocks in a shallow pit with an anvil, hammered copper, and scraps. Beta Analytic reports a conventional radiocarbon date 95 percent certain to date between 1970 and 1450 BC.

Feature C, between A and B, is an irregular concentration of burned bone, worked copper, and stone flakes, which was not dated.

The dates from Duck Lake are statistically indistinguishable and fall in the earlier part of the Late Archaic period, very close to the dates obtained for Clusters A and C at Lac La Belle. Unfortunately, the fact that the radiocarbon dates from Duck Lake are from very close together in the northeast extremity of Locality B means that we have no grounds to decide whether the locality was occupied by one large group, as may be the case for Clusters A and C at Lac La Belle, or rather was visited on separate occasions by different groups, during the long time span of the earlier Late Archaic period.

We can, however, learn something from the structuring of copper manufacturing at the site. Unfortunately, no caches were found at Duck Lake, but 87 copper items were scattered around Locality B. Of these, 21 are unworked small copper nuggets with a mean weight of 2.2 grams; 42 are hammered copper pieces with a mean weight of 5.1 grams; 2 are large shaped pieces with a mean weight of 118 grams; 8 are termed scrap, with a mean weight of .3 grams; and 5 are various unfinished preforms for stemmed points, conical points, and beads. Only 9 items were finished copper tools—very small beads (2) with a mean weight of .1 gram, small pins (5) with a mean weight of .7 gram, and a tanged knife weighing 9.4 grams. There were no awls or finished stemmed points. A hammered copper piece was found on an anvil stone next to Feature B. It is clear that unworked copper was brought to the Duck Lake Site and used to produce preforms. In brief, we have the same four steps in production—collection, sorting, preform manufacture, and final finishing—found at Lac La Belle. Unfortunately, since most of the work was in one cluster, we do not have a clear idea of whether the clusters in one locality at Duck Lake were contemporary, and whether there was any degree of specialization in production among the clusters.

The final class of artifacts considered from Duck Lake is the flaked stone assemblage. The overwhelming majority, about 80 percent, are artifacts and flakes of imported chert, most of it Prairie du Chien-Galena chert from the Mississippi Valley in northeastern Illinois and adjacent Iowa or in southern Minnesota, 320 km or more southwest of Duck Lake. Locally available quartz, none retouched or evidently utilized, which probably served as expedient cutting tools, constitutes only 9 percent of the flaked stone items. There is a small amount of locally available glacially transported chert from Canada. There is one chert point, a corner-notched projectile point similar to several Late Archaic types known to the southwest and southeast in the Upper Midwest (Justice 2001).

It must be noted that other Late Archaic sites lacking copper but with diverse flaked stone inventories have been excavated in the Ontonagon area. One of these, the Ottawa North site, is 55 km south-southwest of Duck Lake and has an assemblage almost exclusively of quartz flakes (Hill 1994:14–29). The other, the Alligator Eye site, is 72 km west-southwest of Duck Lake and was a bedrock quarry for quartz, but with a diverse assemblage including various types of scrapers and bifaces (Hill 1994:15–44). We should not, however, jump to the conclusion that Duck Lake was occupied by strangers en route back from a visit to the copper sources, though this is one possibility. Many years ago, the late Howard Winters demonstrated that there were cycles of copper movement during the Late Archaic (1968), and it is possible that Duck Lake was occupied during a time of intense long-range exchange, while Lac La Belle, Ottawa North, and Alligator Eye were occupied during a period of little exchange and very local use of copper.

These two well-excavated Late Archaic occupational sites have little evidence of organized copper working, the feature at Duck Lake covering a small area suitable for at most one worker.

At both sites, the manufacturing process involved the hammering of a suitable-sized nodule of copper into a preform and thence into a tool. There is no evidence of the building up of copper preforms or tools from many small pieces, which becomes common in later times.

So far, researchers have found no site of the Early Woodland Period (800–200 BC) with evidence of copper working. However, a number of sites of the Middle Period Woodland (200 BC–AD 650) have been excavated, of which the best excavated and best reported is summarized below.

Summer Island

Summer Island is the northernmost of a line of islands at the mouth of Green Bay on the northwest side of Lake Michigan. This line of islands connects Wisconsin's Door Peninsula on the south and Michigan's Garden Peninsula on the north. Today Summer Island is near the southern limit of the Canadian Forest biome. The island is small, only about 6 km^2, and has a protected bay with sandy beaches called Summer Harbor on the northeast side. On the beach is a small occupational site (20DE4) covering about .8 ha, with evidence of a major occupation of the Lake Forest Middle Woodland Culture—which concerns us here—followed by limited occupations in Late Woodland and Historic times. A number of sites with this Middle Woodland cultural pattern have been excavated in the Upper Great Lakes, but this is one of the few with direct evidence of Middle Woodland copper working, and it is certainly the best reported of these sites. The site was first reported in the early twentieth century.

In 1967, James Edward Fitting, then curator of Great Lakes Archaeology at the University of Michigan Museum of Anthropology, received a grant from the U.S. National Science Foundation to study prehistoric settlements. A team from the museum, led by David Brose and composed of five trained students, aided by up to 20 younger students from a program called Summer Science, excavated extensively on Summer Island in July and August 1967. Coring and test squares revealed well-preserved, richly organic midden deposits, so the team focused on the block excavation of complete midden areas. They mapped the site in English units. Excavation was by successive natural layers within 5-foot squares. Team members made plans of the top of each layer and section drawings of the walls of the square. They screened all midden deposits through a 3-mm mesh, and floated some in water before passing them through a very fine screen to concentrate and recover carbonized plant remains. This work was documented in an exemplary site report published a few years later (Brose 1970a).

The Summer Island report shows that between AD 150 and AD 290, a small community of several oval houses occupied a sandy ridge between the lake and the dense mixed hemlock, fir, beech, sugar maple, and birch forest of Summer Island. In the past there would have been more pine, beech, and maple, but repeated logging since the 1970s has removed many of these (Brose 1970a:10–12). The most common large mammal bone discarded by Summer Islanders was bear, but they also hunted Virginia deer and moose. There was some hunting or trapping of smaller animals, such as beaver, otter, muskrats, rabbits, turtle, and migratory waterfowl. The island sustains deer and rabbits today, but the bear would require a larger territory, and the moose, beaver, otter, muskrats, and turtle would prefer more humid bogs and streams, and were probably found on the Garden Peninsula rather than Summer Island. However, the bones of terrestrial fauna were far less common than those of fish. The most common among these were the remains of sturgeon, but there was some taking of walleye, bass, and other fish. Sturgeon and walleye spawn in the shallows off Summer Island in the spring (Brose 1970a:145–147). Excavators recovered remains of some

Afterword

wild berries and nuts. There was no evidence that residents collected wild rice and or raised domesticates such as maize or squash (Brose 1970a:147–148). The animal and plant remains indicate the site was occupied from spring into mid-autumn. During this warm-season occupation, people lived in oval houses varying from 26 to 49 m², each probably with a few related smaller families (Brose 1970a:41–45). Based on modern observations of the rotting of posts in the area, Brose suggests the site was occupied for only three or four seasons. However, these are not fence posts subjected to harsh freezing and thawing, but posts protected from moisture by the walls and roofs of the houses. The rebuilding of hearths and many alterations of internal racks and walls suggest a somewhat longer occupation, perhaps a decade or even two. In the absence of detailed house-by-house artifact inventories, it is also hard to say whether there was one occupation with about 35 people in four houses or two occupations with with each occupation about 17 people in two houses.

Residents pursued craft industries such as pottery making, wood and fiber working, hide working, flint working, and copper working. Brose presents a commendably comprehensive study of the copper items, with counts, measurements, and weights. Summer Island is 270 km southeast by trail and canoe traverses from the bedrock copper sources in the Keweenaw area, but by canoe only 50 km from major areas of float copper in Wisconsin and 90 km from those in Michigan (Rapp et al. 2000). However, the excavations yielded two fragments of copper adhering to Keweenaw basalt, indicating some exploitation of the distant bedrock sources. The fragments termed scrap include no unworked nuggets, but some partly hammered pieces similar to the earlier-mentioned micro-ingots. Most of this scrap is flattened pieces, which had been repeatedly folded, heated, and hammered to make flat preforms that could be cut into knives or rolled into beads. Copper workers made awls by heating, rolling, and hammering the sheets to make a small, relatively solid, quadrilateral bar. Alternatively, they could repeatedly fold, hammer, and anneal copper sheets, a method that differed from the simple heating and hammering of larger nuggets reported from earlier sites such as Lac La Belle and Duck Lake.

Brose counted, measured, and weighed the 70 copper items from the Summer Island excavations (Brose 1970a:130–134, Table 14). Of these, 41 are called copper scrap, which here includes hammered nuggets, flat pieces, and flat flakes. Together they have a total weight of 79.6 grams and a mean weight of 1.9 grams. Unfortunately we do not have separate counts and weights for different types of scrap. About 60 percent of the excavated items were finished, and together these have a total weight of 98.4 grams. These are as follows:

Beads (13): total weight of 10.4 grams and a mean weight of .81 grams; made by rolling a folded and hammered flattened piece around a pommel.

Hooks without barbs (2): total weight of 5.3 grams and a mean weight of 2.6 grams; perhaps fishhooks; both hammered from a single small nugget.

Flattened bipointed pieces (3): total weight of 27.6 grams and a mean weight of 9.23 grams; two are notched to fix a central cord; all are perhaps gorges designed to

124

catch fish by lodging in their throats; Brose does not suggest a method of manufacture for these.

Awls (4): total weight of 28.9 grams and a mean weight of 6.9 grams; all square in section and all made by hammering together and heating flat pieces, rather than hammered from a single large nugget as at the earlier sites; most of these have one end sharpened or scratched from use.

Knives (2): one is small, like the ubiquitous "butter knives," weighing 2.9 grams; one is large and curved, more like the crescents of other sites, and weighs 12.2 grams; both appear to have been from hammered from a single nugget; neither were sharpened nor had they any signs of use, so we have no confirmation of their proposed use.

Small chisels (2): total weight of 9.10 grams and a mean weight of 4.6 grams; made by folding and hammering, as were the awls, rather than hammered from a single nugget; given their weight and method of manufacture, they may well be broken, recycled awls, but the squared ends do show use scratches, confirming their use as chisels.

Finally, there are two unique pointed items: a punch, perhaps used in flint flaking, which weighs 1.4 grams, and a piece in the shape of a bird claw (perhaps a piece of ceremonial paraphernalia) weighing .6 grams. The latter was made from a single copper nugget.

The small amount of scrap and predominance of finished items by both weight and count in the excavated area of the site suggest that Summer Island copper was mostly used for local consumption.

An important discovery at the Summer Island site was a copper-working area outside one of the houses (Brose 1970a:162). This was a small area of packed clay covering only 2.8 m². Excavators found 14 pieces of copper, including one piece of unworked copper still adhering to its basalt matrix, eight pieces of scrap, three beads, a finished awl, and a finished bipointed gorge, all concentrated in an area of only 1 m². Also associated were an anvil, two hammer stones, and two abraders. There is also much flaked chert debris, including four tools, 14 bipolar cores and 370 grams of small chert flakes, which could have been struck with the hammer stones, perhaps on the anvil, suggesting that the feature was not solely a specialized copper-working area, but one that was also used for flaking chert. This evidence shows that this is a small, unspecialized working area with some finished items and many flakes or flattened copper pieces lost in the course of hammering. Such loss is most logically interpreted as an indication that work was episodic, so that misplaced or cached pieces could easily become hidden. While the small and unspecialized nature of this area is similar to the Late Archaic working area at Duck Lake, the working process, which involves the making of tools and preforms from both small single nodules and small copper flakes that have been hammered and annealed together, is intermediate between Late Archaic techniques and Late Woodland techniques.

Afterword

Juntunen

We are fortunate to have excavation evidence from the Juntunen Site, a Late Woodland Period site with much evidence of copper working. Juntunen (20MK1) is located on the west end of Bois Blanc Island in the Straits of Mackinac. It is the only site in the Upper Great Lakes with direct evidence of Late Prehistoric copper working, which makes it the latest known survival of the tradition of copper working that began millennia before, in the Archaic Period. It was found not as a result of academic curiosity about prehistory, a training program for students, or an effort by either local museums or government agencies to salvage evidence of the past being destroyed by modern development. Rather, it was part of a worldwide effort to solve a grand problem of cultural ecology—the long-term relation between changing social organization and changing environments—and funded by the newly founded U.S. National Science Foundation. Teams of specialists, not only in archaeology but in geology, zoology, botany, and human demography and health, were at work in the Great Lakes, the arid Southwest, the jungles of Yucatan, and the mountain foothills of the Middle East.

A team from the University of Michigan Museum of Anthropology, led by Alan McPherron, excavated extensively at Juntunen from 1961 to 1963. The large size and stratigraphic complexity of the site required a complicated judgmental sampling scheme. The team surveyed the site in English units of measure. The actual excavation was similar to that described above for the Summer Island site. McPherron summarized their findings in a classic site report published a few years later (McPherron 1967). They showed that, for six centuries, the site occupied a sandy ridge between Lake Huron and the dense, mixed, old-growth forest of Bois Blanc Island (very unlike the cut-over and oft-burned forests of today). The common prey of Native American hunters in the eastern forests to the south, the Virginia deer, does not thrive in such mature forests and was rarely hunted. There was some hunting of animals such as moose, beaver, and migratory waterfowl, but the primary food quest was the fishing of sturgeon and whitefish (Cleland 1966). The people collected some wild berries and nuts, and consumed maize, though it is unclear whether there were local gardens or the maize was obtained through trade with people farther south (Yarnell 1964). At least in the later phases, people lived much of the year in large longhouses, each probably with a number of related smaller families.

As on Summer Island, local industries included pottery making, wood and fiber working, hide working, flint working, and copper working. It was not the intent of the team to study a copper industry, but archaeology is a discovery discipline, and they discovered copper in quantity, even though Juntunen is far from sources of bedrock and float copper. The site is 400 km by trail and 450 km by canoe eastward from the bedrock copper sources in the Keweenaw area. It is 230 km from major areas of float copper in Wisconsin and 300 km from float copper in Michigan (Rapp et al. 2000).

Excavations, whether scientific or not, destroy context, and they are only useful if the contextual information is available in publications. McPherron summarized an unpublished study of the Juntunen copper by Barry Kent, who became the state archaeologist of Pennsylvania, and Ronald Vanderwal, who became a prominent archaeologist in the Caribbean, Australia, and New Guinea (Kent and Vanderwal 1962). Study of the carefully screened samples shows that centuries after a brief Middle Woodland occupation with few indications of copper usage, village occupation revived at Juntunen about AD 800. At first, there is little evidence of copper working: primarily a few discarded finished tools. About AD 1200, however, there was a three-fold increase in the number of discarded tools, but a ten-fold increase in unfinished copper debris. This suggests that in the thirteenth and fourteenth centuries AD, people at Juntunen were making tools that they sent to other sites. The illustrated fragments show mostly unworked irregular nuggets and only one partly hammered piece that is similar to a micro-ingot. Much of the debris is flattened pieces, which were not only cut into knives and points or rolled into beads and cones as before, but were repeatedly layered, heated, and hammered to make flat sheets. Small tanged knives could be cut directly from these sheets. Awls could be made by heating, rolling, and hammering the sheets to make a small, relatively solid, quadrilateral bar. Both the knives and awls were finished and sharpened by grinding. This method for making awls is quite different from the simple heating and hammering of larger nuggets reported from earlier sites such as Lac La Belle and Duck Lake. Copper workers made the Juntunen awls by hammering together rolled flattened pieces. They were seemingly less durable than awls made from nuggets in earlier times, as they tended to break at the junctures of sheets (McPherron 1967:164–175).

An important discovery at the Juntunen site was a copper-working area within a longhouse. The longhouse contrasted with the oval houses known from earlier times, being 8 m wide and at least 17 m long (McPherron 1967: 234–239, Fig. 36). A working area near the eastern end of the longhouse had 405 pieces of copper, concentrated in an area of about 15 m^2 (ibid., Fig. 24c), which was centered on a large hearth.

Conserved in the University of Michigan's Museum of Anthropological Archaeology (formerly the Museum of Anthropology) are 104 copper items from a screened portion of the hearth. They could be counted, weighed, and compared to similar items from the Late Archaic working areas at Duck Lake and from the Middle Woodland working area at Summer Island. Of these, 17 (or 16 percent) are unworked small copper nuggets and fragments with a mean weight of 1.5 grams; six are hammered copper nuggets with a mean weight of 9.5 grams; five are larger shaped flat pieces with a mean weight of 3.0 grams; 17 are flat, folded pieces in the process of being hammered into larger flat pieces, with a mean weight of 2.1 grams; 56 (or 54 percent) are small flat flakes with a mean weight of only .8 grams; and three are unfinished preforms for awls with a mean weight of 5.3 grams. None of the items are finished copper tools. This evidence points to a specialized workshop area with a few unused nuggets (most small), more flattened pieces set aside in the course of hammering and never finished, a majority of small flakes, and a few finished items, which were perhaps misplaced in the sandy working floor.

Afterword

A Copper Workshop at Cahokia

The large amount and proportion of scrap in comparison to finished tools around the Juntunen longhouse implies that the products of Juntunen copper workers were sent elsewhere. The most parsimonious inference is that these tools were exchanged with nearby villagers in the Great Lakes region. However, we have few statistics on copper from such sites, and no evidence of comparable workshops. It is interesting, however, to compare the Juntunen workshop with a recently discovered copper workshop at Cahokia. The largest center in eastern North America, Cahokia is on the American Bottoms of southwestern Illinois, close to the junctions of the Missouri and Ohio rivers with the Mississippi River. Cahokia is the largest early Mississippian center, with more than 120 mounds in an area of about 9 km^2. In general, finished copper items are found only in ritual deposits in large Mississippian centers. Though Cahokia was a large center in the thirteenth century, excavations have not revealed any such ritual deposits from that date. However, researchers have found at least one Mississippian copper workshop.

This workshop is under a small mound laconically called Mound 34 on the Ramey Plaza, east of the central pyramidal mound called Monk's Mound, which is the largest earthwork built by Native Americans north of Mexico. Excavations in 1950 by James B. Griffin and Albert C. Spaulding with students from the University of Michigan revealed evidence of the specialized engraving of marine shell imported from the Gulf Coast. In 1956, this led Gregory Perino, working for the Gilcrease Foundation (later the Gilcrease Museum) of Tulsa, Oklahoma, to expand the Mound 34 excavations, revealing more evidence of shell working and suggestions of copper working. This in turn led John Kelly of nearby Washington University and James Brown of Northwestern University to work further in the Mound 34 area. Under and beside the mound, they found evidence of a number of rectangular post-and-beam structures. From 2007 to 2010, they focused on Feature 82, a single building dating to about AD 1250. It was 5.7 m on its long axis, a few degrees west of north, by 4.4 m (about 25 m^2), with tiny flecks of copper tinting parts of its floor green. The excavation and recording methods used are among the most precise and complete yet developed in North American archaeology. The floor was shaved and brushed so that even the smallest copper flecks could be precisely measured in three dimensions. Analysis of these data is still underway, but we are fortunate to have a preliminary study by Emily Coco (2014) written under the direction of John Kelly.

The copper-working building has a small refuse-filled pit slightly south of its geometric center. East of this is a small hearth. To the south of the central pit are four smaller pits, which the excavators suggested might be emplacements for anvil stones. The copper remains themselves show little variation. While two hammered copper nuggets were found nearby, none were found on the building's floor or in features cut into the floor. Coco studied and measured 96 of the largest pieces—all those with a measurable length, width, and thickness—among the 408 copper pieces in the building whose location had been exactly measured. These were all annealed and hammered flat pieces with a mean thickness of .007 cm and a mean weight of .23 grams. Copper hammered to such thinness is too thin to make plates, but could be made into a foil meant to cover wooden artifacts. Some of these had been folded several times and further hammered. A few had a straight edge, perhaps from bending to a break point or perhaps from cutting by hammering on an anvil edge and grinding. A few possible basalt anvil fragments and sandstone grinder fragments were found. There were no finished tools such as awls or knives and no beads made by rolling thin sheets. If such items were made, this was done elsewhere. Coco also undertook an innovative study of the distribution of copper and other items on the floor of the building. She found three larger concentrations of tiny copper flakes: one just north of the hearth (associated with possible basalt anvil fragments), one in the central pit, and one to the west of the central pit, in an area without features. There is a scattering of smaller concentrations in the southwest quadrant of the building and several on the west edge of the westernmost possible anvil emplacement (also associated with possible basalt anvil fragments). It is notable that some food remains were found—mostly fishbone, but also some of birds, small mammals, and amphibians or reptiles. Many were stained green by copper oxide, suggesting a close association with the copper workers. We can suggest that during the lifespan of the copper-working building, the focus of work changed several times. However, further study of excavation records is needed to assess the sequence of copper-working activities in the building.

Though archaeologists excavated the near-contemporary Juntunen and Cahokia copper-working areas at different times with different methods, we can usefully compare the two facilities. In contrast to earlier copper-working sites, both were focused on the production of flattened copper plates or sheets by a process of repeated annealing and hammering. An anvil and an annealing hearth are associated with Late Archaic Duck Lake, so the basic techniques is not new, but a focus on making thin pieces that are preforms for finished items appears to be typical only of the last one or two thousand years of Native American copper working. At Juntunen, in a dedicated area inside a residential structure, these sheets were transformed into awls, beads, and other items. At Cahokia's Mound 34, the sheets do not seem to have been finished in the copper-working building. Presumably they were passed on to other specialists, who covered items of wood or other materials with the thin sheets, but they could have folded and hammered them into other artifacts as well. The most striking contrast is how little was misplaced at the Cahokia facility and how small the lost pieces were in comparison with those from Juntunen and earlier sites in the Upper Great Lakes region. At Juntunen, there are large flattened pieces, and the small, flat flakes had a mean weight of .8 grams; at the copper-working building at Cahokia there are no larger flat pieces and the small, flat flakes have a mean weight of .2 grams. Was this simply because copper at Cahokia, far from major sources, was more valuable and more carefully conserved? Or was it because Cahokia society supported a greater degree of craft specialization, in which productive processes were more carefully organized? These are questions for future researchers.

Table 1. Counts and Mean Weights (gm) of Selected Classes of Copper Artifacts.

Site		Unworked nuggets	Hammered nuggets	Flakes	Small beads	Small pins	Awls	Points	Knives
Cahokia workshop	count	—	—	96	—	—	—	—	—
	mean weight			0.2					
Juntunen hearth	count	17	6	56	—	—	3^a	—	—
	mean weight	1.5	9.5	0.8			5.3		
Summer Island work area	count	$—^b$	41^c		13	—	4	—	2
	mean weight		1.9		0.8		6.9		7.5
Duck Lake work area	count	21	42	8^d	2	5	—	2^a	1
	mean weight	2.2	5.1	0.3	0.1	0.7		8.9	9.4
Lac LaBelle	count	277	186	?	38^e	5	2	2	—
	mean weight	0.3	2.0		1.4	1.8	8.0	8.3	

a = unfinished preforms; b = unworked nuggets not reported; c = nuggets and flakes combined; d = termed "scrap"; e = beads are perhaps Middle Woodland.

Summary

Stepping back from the details of individual sites, we can see some broad trends through time on Table 1. Any such effort is tentative: not only do we have very few sites from any one period, but the remains at each site are not what was brought to the site, but rather what was brought and not taken away. It is only what was discarded, lost, or cached. Also, the sites were not excavated in exactly the same way, and the copper items were not reported in comparable ways.

1. The Late Archaic contexts studied at Lac La Belle and Duck Lake have, among the finished items, many nuggets, particularly the hammered nuggets, but they are notably small. There is a diversity of finished tools of which small pins, awls, and stemmed points are more common, but a variety of unique items were also found. Since these two sites had varying amounts of collecting using metal detectors, there may be a bias against larger items.

2. The Middle Woodland working area studied on Summer Island has relatively few unfinished items, but we cannot comment on the types of debris. Finished tools are represented by small beads, awls, and tanged knives. Stemmed points were found elsewhere. Only small pins appear to be absent.

3. The Late Woodland hearth in the working area inside the Juntunen longhouse had relatively few nuggets, but a larger number of small, flat flakes. Awls are represented by unfinished preforms. Tanged knives and beads are represented elsewhere on the Juntunen site. Small beads are rare on the site, but longer tubular beads are more common. Stemmed copper points were not found anywhere on the site. Small pins were absent.

There thus seems to have been no consistent change in the size of nuggets, though their frequency may decrease. There was, however, an increase in small flake debris in the work areas through time. Finished tools decrease in diversity on the sites, and working areas have fewer and fewer finished items.

The Mississippian copper-working building at Cahokia can be said to take these trends farther. There are few nuggets and many small, thin flakes. There are no finished items. There may have been secondary workshops concerned with the production of finished items. If so, Cahokia had a much more complex system of crafting than contemporary Late Woodland sites working within the ancient traditions of the Upper Great Lakes region.

Henry T. Wright
Curator of Near Eastern Archaeology
University of Michigan Museum of Anthropological Archaeology

131

Appendix 1: Reporters and Discoverers

Name	Occupation	Year	Mine	Reference
Bauerman, Hilary	Geologist	1865	Central	Bauerman 1866:457
	Geologist	1865	Flint Steel	Bauerman 1866:459
Bawden, Henry	Captain	1851	Bohemian	Bohemian Mining Co. 1851:14
Beaser, Daniel	?	1861	La Fayette	La Fayette Mining Company 1863:10-11
Blake, "Dr." [John H.?]	Agent	1852?	Waterbury	Whittlesey 1852a: 27-28; 1863:6-8
Bradford, Edward	Newspaperman	1855	Nebraska	Tenney and Leeds 1855b:189
Brown, J. V.	Newspaperman	1850	Minesota	Brown 1850
	Newspaperman	1850	Copper Falls	Brown 1851
	Newspaperman	1852	Phoenix	Brown 1852
	Newspaperman	1853	Portage	Brown 1853a
	Newspaperman	1853	Waterbury	Brown 1853b
Buzzo, Thomas W.	Clerk (National Mine)	1862	Great Western	Great Western Copper Mining Company 1863b:9-10
Byerly, A. H.	Superintendent	1855	Cliff	Pittsburgh & Boston Mining Company 1855:16-19
"C., E."	?	1849	Minesota	Lake Superior News and Mining Journal 1849
	?	1850	Minesota	Lake Superior News and Mining Journal 1850
Christian, Sarah Barr	Housewife	1875	Minong	Christian 1932:33-35
Chynoweth, John	Captain (National Mine)	1862	Great Western	Great Western Copper Mining Company 1863b:11
	Captain	1882	Aztec	Lawton 1883:170
Clarke, L. W.	?	1850	Aztec	Aztec Mining Company 1851:11
Coburn, Augustus	?	1850	Aztec	Aztec Mining Company 1851:13
Coulter, Joseph	Agent	1864	Penn	Coulter Mining Company 1864:7

Name	Occupation	Year	Mine	Reference
Curnow, Mathew	Captain	1854	Siskowit	Brown 1854
Davis, A. C.	Agent	1863	Hudson	Hudson Mining Company 1864:4
	Superintendent	1874	Minong	Davis 1875
	Superintendent	1874	Minong	Minong Mining Company 1875:15
	Superintendent	1874	Minong	Lane 1898:18
	Superintendent	1874	Minong	Wood 1907:292–293
	Superintendent	1876	Minong	Swineford 1876:52
Day, S. M.	Secretary	1854	North West	North West Mining Company 1854:3
Douglass, C. C.	?	1851	Isle Royale Quincy	Whittlesey 1863: 16–17
Duffield, George	Minister	1858	Quincy	Detroit Free Press 1858a, 1858b
Edwards, Richard	Captain (Albion Mine)	1850	Aztec	Aztec Mining Company 1851:12
Emerson, George D.	Newspaperman	1855	Mass	Mass Mining Company 1856:10–11
	"	1856	National	Tenney 1856b:309
	"	1857	Pewabic	Emerson 1857a
	"	1857	Ohio Trap Rock	Emerson 1857b
	"	1857	Minesota	Emerson 1857c
	"	1857	Copper Falls	Emerson 1857d
	"	1857	Superior	Emerson 1857e
	"	1857	Nebraska	Emerson 1857f
	"	1857	Evergreen Bluff	Emerson 1857g
	"	1857	Adventure	Emerson 1857h
	"	1857	Nebraska	Emerson 1857i
	"	1857	Rockland	Emerson 1857j
	"	1857	Adventure	Emerson 1857j
	"	1857	Nebraska	Emerson 1857k
	"	1857	Minesota	Emerson 1857k
	"	1857	Ogima	Emerson 1857l
	"	1857	Minesota	Emerson 1857m
	"	1858	Minesota	Emerson 1858c
	"	1862	Bohemian	Emerson 1862a
	"	1862	Mesnard	Emerson 1862b
	"	1863	Portage	Emerson and McKenzie 1863

Appendix 1

Name	Occupation	Year	Mine	Reference
Emerson, L. G.	Mining Engineer	1862	Great Western	Great Western Copper Mining Company 1863b:9-10
Forster, John H.	Agent	1860	Columbian	Columbian Mining Company 1862:8
	Explorer	1862	Mesnard	Forster 1891:21-22
	Explorer	1862	Mesnard	Packard 1893:182
	?	?	South Pewabic	Packard 1893a:183; 1893b:75
	?	?	Arcadian	Forster 1879:158
Frue, William B.	Foreman	ca. 1855	Pewabic	Wood 1907:289
	?	1862	Columbian	Columbian Mining Company 1862:11
	?	1862	Ogima	Ogima Mining Company 1862:8
Gailey, G. K.	?	1879	Minong	Winchell 1881:603
Gillman, Henry	?	1873	Triangle Island	Gillman 1874:387-389
Greeley, Horace	Director	1847	Northwest	Williams 1950:123
Grout, John R.	General Agent	1857	Eagle Harbor	Eagle Harbor & Waterbury Mining Companies 1858:6
Hancock, James	Captain	1849	Siskowit	Siskowit Mining Company 1850:18
Hanna, Joshua	President	1853	Ohio Trap Rock	Tenney 1853c:635
Hardie, George	?	1874	Minong	Minong Mining Company 1875:12-13
Harris, William	?	1863	Hilton	Hilton Mining Company 1863:12
Hart, ?	Miner	1870s	Minong?	Fox 1911:86
Hickok, William	"Owner"	1848	Minnesota	Ingersoll 1848; Trowbridge 1881:122
Hill, Samuel W.	Superintendent	1850	Copper Falls	Brown 1851

Name	Occupation	Year	Mine	Reference
	Superintendent	1851	Copper Falls	Hodge 1851a:338, 1851b:255; Kayner 1883b:330
	Superintendent	1852	Copper Falls	Copper Falls Mining Company 1852:16-18; Whittlesey 1863:10
	Superintendent	1853	Copper Falls	Tenney 1853b:172
	Superintendent	1854	Copper Falls	Copper Falls Mining Company 1854:10; Tenney and Partz 1854e:669
	?	1855	Central	Tenney and Leeds 1855a: 291-292
	Geologist	1855	Garden City	Garden City Mining Company 1856:6-7
	?	1857	Eagle Harbor	Eagle Harbor & Waterbury Mining Companies 1858:18
	?	1857	North Cliff	North Cliff Mining Company 1858:19-20
	?	1872	Minong	Foster 1874:268
	?	1874	Minong	Minong Mining Company 1875:13-14
	?	1874	Minong	Swineford 1876:52
	?	1874	Minong	Lane 1898:18
Hodge, James T.	Geologist	1849	Minesota	Hodge 1849a-c
	Geologist	1849	Adventure	Hodge 1849c:752
	Geologist	1849	Aztec	Hodge 1849c:752
	Geologist	1850	Minesota	Hodge 1850:307
	Geologist	1850	Adventure	Hodge 1850c:434
	Geologist	1850	Ridge	Hodge 1850c:434
	Geologist	1850	Aztec	Hodge 1850c:434
	Geologist	1851	Peninsula	Hodge 1851c:353
	Geologist	1856	Evergreen Bluff	Evergreen Bluff Mining Company 1857:8, 1858
Hogan, Thomas H.	?	1852	Forest	Forest Mining Company 1852:14, 20
Holmes, W. H.	Archaeologist	1892	Minong	Holmes 1901
Houghton, Jacob	Agent	1862	Pontiac	Houghton 1862
Hulbert, Edwin J.	?	1857	North Cliff	Lathrop 1901a:5, 1901b:250; 1961:121
	Explorer	1860-61	"Hulbert Tract" (Natick)	Thomson and Co. 1864:10

137

Appendix 1

Name	Occupation	Year	Mine	Reference
	Explorer	1864	"Calumet Ancient Pit"	Hulbert 1893, 1899
Hulbert, John, Jr.	"Explorer"	1864	Another pit 1200 feet SW of the "Calumet Ancient Pit"	Hulbert 1899:8-9
Hunter, J. W.	Captain	1850	Minesota	Brown 1850
Jacka, William	Captain	1879	Minong	Winchell 1881: 602-603
Jackson, Charles T.	U. S. Geologist	1848	Lake Superior	Jackson 1849:448
	U. S. Geologist	1848	North West	Jackson 1850:284
	U. S. Geologist	1848	North West	Jackson 1855:*342
Jennings, Edward	Captain (Cliff Mine)	1850	Aztec	Aztec Mining Company 1851:12
	Agent	1854	Arctic	Arctic Mining Company 1854:7
	Captain	1855	Rockland	Pontiac Jacksonian 1855:253
Jones, George C.	?	1864	Coulter	Coulter Mining Company 1864: 7, 13, 14
Knapp, Samuel O.	Agent	1848	Minesota	Hunt's Merchants' Magazine and Commercial Review 1848; Littell's Living Age 1848; Bagg and Harmon 1848; Foster and Whitney 1850: 159-160; Trowbridge 1881
Lewis, Theodore H.	Explorer	1888	Caledonia	Lewis 1889:295
	Explorer	?	Knowlton	Lewis 1890:180
Livingston, Robert B.	Superintendent	1852	Forest	Forest Mining Company 1852:7
Lord, Thomas H.	Agent (Shawmut Mines)	1855	Mass	Lord 1856; Mass Mining Company 1856:18-19

138

Reporters and Discoverers

Name	Occupation	Year	Mine	Reference
Mandlebaum, Simon	Superintendent	1852	Phoenix	Brown 1852; Phoenix Copper Company 1853:8
Martin, "Capt."	Lessee	1880	Caledonia	Lawton 1881:91
Mason, Milton	Superintendent	1853	Adventure	Adventure Mining Company 1854:5
	Superintendent	1857	Adventure	Adventure Mining Company 1857:11; Mason 1858a, 1858b
Mather, W. W.	Geologist	1848	Copper Falls	New-York Tribune 1848
	Geologist	1848	Eagle River	New-York Tribune 1848
McKenzie, H.	Newspaperman	1864	Ontonagon	McKenzie 1864a
	"	1864	Caledonia	McKenzie 1864a
	"	1864	Winona	McKenzie 1864a
	"	1864	Shelden and Columbian	McKenzie 1864b
	"	1864	South Pewabic	McKenzie 1864c
	"	1865	South Pewabic	McKenzie 1865
Mendenhall, Cyrus	?	1848	La Fayette	La Fayette Mining Company 1863:9
Merritt, Alfred	?	1873–74	?	Fox 1911:84
"Miskowabikokewi"	?	1863	Minesota	"Miskowabikokewi" 1863
Moss, Theodore F.	Mining engineer	1860	Bohemian	Bohemian Mining Company 860:7
Moyle, Richard	Agent (Nebraska Mine)	1855	Mass	Mass Mining Company 1856:4, 17
Newberry, J. S.	State geologist of Ohio	1862	?	La Fayette Mining Company 1863:8-9
	State geologist of Ohio	1862	Republic	Republic Mining Company 1864:5-6
Palmer, Charles H.	Agent	ca. 1855	Pewabic	Wood 1907:289
	?	1864	Adams	Forster 1879:166-167; Hollingsworth 1978:1-2

Name	Occupation	Year	Mine	Reference
Palmer, Walter W.		1850	Aztec	Aztec Mining Company 1851:12-13
Peck, S. F.	?	1866	Bohemian	Philadelphia Ledger 1883; Cresson 1884:55
Petherick, William	Agent	1859	Copper Falls	Copper Falls Mining Company 1859:8
	Agent	1860	Copper Falls	Copper Falls Mining Company 1860:8-9
"R., C."	?	1853	Lake Superior	"C. R." 1853
Reid, James D.	Agent	1860	Columbian	Columbian Mining Company 1862:4
Robingson, John	Member of Board of Directors	1854	Central	Central Mining Company 1855:4
Rudolph, Albert	Geologist	1851	Great Western	Great Western Copper Mining Company 1852:25
	Geologist	1852	National	Lawton 1881:81
"S. S."	?	1852	Phoenix	"S. S." 1852
Sales, Edward	?	1863	Penn	Coulter Mining Company 1864:13
Sanderson, C. M.	Agent	1865	Hilton	Hagar 1865:310
Scott, Amos H.	Foreman	1864	Another pit 1200 feet SW of the "Calumet Ancient Pit"	Hulbert 1899:8-9
	Foreman	1864	Calumet Ancient Pit	Wood 1907:291
Senter, M. D.	?	1854	Cliff	Whittlesey 1863:10
Shaw, Cornelius G.	Miner	1850	Lookout	Shaw 1847:24
	"	1850	Scoville	Foster and Whitney 1850:162
Shelden, Ransom	Merchant	1850?	Isle Royale	Clarke 1990:6

Reporters and Discoverers

Name	Occupation	Year	Mine	Reference
Shepherd, Forrest	Economic geologist	1863	Forrest Shepherd	Shepherd, F. 1864:13; Cox 2002:22-23
Slawson, John	Superintendent	1854	Central	Central Mining Company 1855:4; Tenney and Leeds 1855a:291; Whittlesey 1863:12
Sowden, James	Agent	1865	Norwich	Norwich Mining Company 1865:14
Spalding, William	Miner	1848	Minesota	Spalding 1901: 689-690
Stevens, William H.	Agent	1850	Forest	Foster and Whitney 1850:160-161
	Agent	1850	Forest	Forest Mining Company 1852:4
	Agent		Agate Harbor	Tenney and Partz 1854a:554-555
	Agent	1854	Agate Harbor	Stevens 1854
Tenney, William S.	Editor	1853	Copper Falls	Tenney 1853b:172
	"	1853	Waterbury	Tenney 1853b:173
	"	1853	Adventure	Tenney 1853b:174
	"	1856	Nebraska	Tenney 1856a: 117-118
	"	1857	Evergreen Bluff	Tenney 1857b:280
	"	1857	Superior	Tenney 1857e:383
	"	1857	Rockland	Tenney 1857f:572
Townsend, J. B.	Agent (Minesota Mine)	1857	Superior	Tenney 1857c:99
	Agent	1857	Rockland	Hobart and Wright 1883:512-513
	Agent	1858	Minesota	Townsend 1858a:334, 1858b:266
	?	1863	Hilton	Hilton Mining Company 1863:10
Tresider, Joel	Miner	1862	Bohemian	Emerson 1862a, 1862b
Trowbridge, S. V. R.	Assistant Agent, United States Mineral Lands	1849	Minesota	Trowbridge 1850a: 53-54, 1850:2

Appendix 1

Name	Occupation	Year	Mine	Reference
Uren, R.	Captain	1874	Minong	Minong Mining Company 1875:12 Lane 1898:16
West, George A.	Explorer	1875	"South Shore Group"	West 1929:43, 52
White, F. G.	?	1874	Minong	Minong Mining Company 1875:14
Whitney, Josiah D.	Geologist	1848	Minnesota	Whitney 1854: 293- 294
	"	1853	Native Copper	Whitney 1854:263
	"	1853	Aztec	Whitney 1854:292
	"	1853	Rockland	Whitney 1854:296
	"	1853	National	Whitney 1854:297
	"	1853	Forest	Whitney 1854:297
	"	1853	Isle Royale	Whitney 1854:302
Whittlesey, Charles	Superintendent	1853	Webster	Webster Mining Company 1853:9
Wilson, Daniel	Historian	1855	Minesota	Wilson 1856: 227-228
Wood, Alvinus B.	?	1855	Quincy	Wood 1907:288
	Agent	1857	Huron	Wood 1907:290-291
	?	1862	Mesnard	Wood 1907:290

Appendix 2: Lake Superior Copper Explorers 1819-1874

Appendix 2

John Bacon
*Daniel Beaser
Martin Beaser
Samuel Bell
Abel B. Bennet
James Bonner
Edward Breitung
Daniel D. Brockway
Samuel H. Broughton
William Brown
Dr. G. Brunshweiler
Robert Buckley
Jerome Butler
Augustus Coburn
*John Chynoweth
*Joseph Coulter
Jonathan Cox
James Crawford
Henry F. Q. D'Aligny
William Davey
*A. C. Davis
*C. C. Douglass
Edward Duncan
*Richard Edwards
*John H. Forster
*William B. Frue
Robert Gravreat
*John R. Grout
Philip Harrington
John Hays
Joseph Hempstead
*Samuel W. Hill
David Hodgson
John Hodgson
John P. Hodgson
Franklin Hopkins
Douglass Houghton
*Jacob Houghton, Jr.
Albert Hughes
*Edwin J. Hulbert
George Johnstone
William Keating
Benjamin R. Livermore
*Robert R. Livingston
William N. MacLeod
John McIntyre
Richard Menhenet
Algernon Merriweather
Peter Mitchell
*Richard Moyle

John Mullowney
Samuel Newcomb
James Paul
Joseph Paull
*William Petherick
Henry Pilgrim
"Capt." Plummer
*James D. Reid
*Edward Sales
Henry R. Schoolcraft
*Amos H. Scott
*Ransom Sheldon
Abner Sherman
*John Slawson
Dr. Edwin Smith
*William H. Stevens
John Stewart
*John Uren
Johnson Vivian
Luke Welch
Peter White
*Charles Whittlesey

*These names also appear in Appendix 1:
Reporters and Discoverers.

(Table based on Hulbert 1893:13–14.)

144

Appendix 3:
Historic Copper Mines
Founded on "Ancient Diggings"

Appendix 3

SITE	PITS	AREA	TOOLS AND FACILITIES	REFERENCE
KEWEENAW COUNTY				
Agate Harbor Mining Co. (20KE61)	"Numerous"	¼ mile in a continuous line	Unreported	Tenney and Partz 1854a:554-555
	"Numerous"	¼ mile in a continuous line	Unreported	Stevens 1854
	Many	Unspecified	Unreported	Whittlesey 1863:6
Central Mining Co. (20KE46) (Absorbed the Northwestern Mining Co. property)	"ancient excavation"	Unspecified	Unreported	Central Mining Company 1855:4
	Unspecified	Unspecified	Unreported	Tenney and Leeds 1855a:291-292; 1855d:172
	Many	Unspecified	Hammerstones; socketed copper	Whittlesey 1863:11-12
	Unspecified	Unreported	Unreported	Hagar 1865:310
	"an old Indian working"	Unspecified	Unreported	Bauerman 1866:457
	"three large masses were found uncovered"	Unspecified	Unreported	Egleston 1879:281
	"an ancient excavation"	Unspecified	Unreported	Lawton 1881:60
	"an ancient excavation"	Unspecified	Unreported	Kayner 1883b:326
	"an ancient excavation"	Unreported	Unreported	Pinkham 1888:11
	"a pit filled in with rubbish...5 feet deep and 30 long"	Unspecified	"Broken stone mauls"	Packard 1893a:181; 1893b:74
	"ancient excavation"	Unspecified	Unreported	Hubbard 1895:74-75
	Unspecified	Unreported	Unreported	Wood 1907:292

146

Historic Copper Mines Founded on "Ancient Diggings"

SITE	PITS	AREA	TOOLS AND FACILITIES	REFERENCE
"Cliff Mine" (20KE53) (legally known as the Pittsburgh and Boston)	"ancient excavations"	Unspecified	Unreported	Whitney 1854:276
	Unspecified	Unreported	Unreported	Pittsburgh and Boston Mining Company 1855:16-19
	"trenches"	Unspecified	Grooved stone hammer	Whittlesey 1863:10
	Unspecified	Unreported	Unreported	Wood 1907:292
Copper Falls Mining Co. (20KE63)	"an excavation"	"several feet in depth and several rods in length"	"stone axes of large size"	New York Tribune 1848
	"old works"	"ancient diggings can be traced for a mile and a half"	Unreported	Brown 1851
	"extensive ancient works"	Unspecified	Unreported	Hodge 1851a:338
	"a new vein with extensive ancient works upon it"	Unspecified	Unreported	Hodge 1851b:255
	"ancient works... are very extensive"	Unspecified	Unreported	New-York Tribune 1852
	"Deep ancient pits"	"a continuous chain...for more than one mile"	"stone hammers and copper arrow heads were frequently met with"	Copper Falls Mining Company 1852:16-18
	Unspecified	Unreported	Wooden shovels; "spearhead"; hammerstones	Whittlesey 1852a:28,29
	Unspecified	Unspecified	Unreported	Brown 1853a
	Unspecified	Unreported	Wooden shovels; "spearhead"; hammerstones	Whittlesey 1853:133,134
	Unspecified	"wrought extensively"	Unreported	Tenney 1853b:172

SITE	PITS	AREA	TOOLS AND FACILITIES	REFERENCE
	"deep ancient pits"	Unspecified	Unreported	Copper Falls Mining 1854:10
	"Extensive"	Unspecified	Unreported	Tenney and Partz 1854e:669
	"extensive ancient workings"	Unspecified	Unreported	Whitney 1854:265
	"considerable ancient workings"	Unspecified	Unreported	Emerson 1857d
	"ancient surface workings"	Unspecified	Unreported	Copper Falls Mining Company 1859:8
	"ancient diggings"	Unspecified	Unreported	Copper Falls Mining Company 1860:8-9
	"very extensive"	"total length of several miles"	"Broken stone mauls are common"; wooden shovels	Whittlesey 1863:9-10
	"extensive ancient workings"	"mark[ed] the surface outcrop for a great distance"	"stone hammers and copper arrow-heads"	Lawton 1881:38-39
	"extensively worked"	Unreported	"stone hammers, copper arrow-heads, and other evidences"	Kayner 1883b:330
Eagle Harbor Mining Co. (20KE64)	Unspecified	Unspecified	"ancient cedar shovel"	Whittlesey 1852a:30
	"ancient diggings; ancient excavations"	Unspecified	Unreported	Eagle Harbor and Waterbury Mining Companies 1858:6, 18
	"pits"	Unspecified	"broken stone mauls"	Whittlesey 1863:10
Eagle River Mining Co. (20KE81)	"ancient shafts"	Unreported	Unreported	*New-York Tribune* 1848
Fulton Mining Co. (20KE65)	One	Unreported	Unreported	Fulton Mining Co. 1853:3

Historic Copper Mines Founded on "Ancient Diggings"

SITE	PITS	AREA	TOOLS AND FACILITIES	REFERENCE
Garden City Mining Co. (20KE66)	"ancient mine work"	Unspecified	Unreported	Garden City Mining Co. 1856:6-7
"Humboldt Location" (20KE67)	Unspecified	Unreported	Hammerstones	Whittlesey 1863: 12-13
Iron City Mining Co. (20KE79)	"...that extensive chain of pits known to the explorers of the Northwest, Hartford, Mandan Bluff and **Iron City** mines as 'ancient diggings'." (This is the only known mention of "ancient diggings" at the Iron City Mining site.)			Moore 1915:11
Lake Superior Copper Co. (20KE68)	Unreported	Unreported	"Indian hammers in the soil"	Jackson 1849:448
Mandan Mining Co. (20KE78)	"...that extensive chain of pits known to the explorers of the Northwest, Hartford, **Mandan Bluff** and Iron City mines as 'ancient diggings'." (This is the only known mention of "ancient diggings" at the Mandan Mining Co. site.)			Moore 1915:11
Meadow Mining Co. (20KE69)	"transverse fissure veins... lined with ancient pits"	Unspecified	Unreported	Lawton 1881:55
	"ancient pits"	"abundant along the transverse fissure veins"	Unreported	Kayner 1883b:331
	"Where the Ashbed is crossed by transverse fissures, bunches of ore occur and such places usually show pits of prehistoric miners"	Unspecified	Unreported	Weed 1918:887
Natick Copper Co. ("Hulbert Tract") (20KE70)	"extensive line of ancient workings"	Unspecified	Unreported	Thomson & Co 1864:10
Native Copper Co. (20KE72)	"a line of ancient excavations"	Unspecified	Unreported	Whitney 1854:263
	"well developed"	Unspecified	Unreported	Whittlesey 1863:6

SITE	PITS	AREA	TOOLS AND FACILITIES	REFERENCE
North Cliff Mining Co. (20KE50)	"some ancient mine pits"	"a lode…has a channel of ancient mine work in it several hundred feet in length"	Unreported	North Cliff Mining Company 1858:20
	"ancient miners had excavated between the walls of this vein"	Unspecified	Unreported	Lathrop 1901a:5; 1901b:250; 1961:121
North West Mining Co. (20KE34) (This mining property was later known under many names.)	Unreported	Unreported	"elaborately fashioned stone hammers"	Greeley 1847
	Unreported	Unreported	"A great many Indian hammers were found in the soil"	Jackson 1849:445
	"remains of ancient pits"	Unreported	"stone hammers"	Hodge 1849a
	"a depression of six feet"	Unspecified	"Barrels full of hammers, much worn"	Jackson 1850:284
	Unspecified	Unreported	"Stone hammers are found in abundance"	Brown 1851
	"some"	Unspecified	Unreported	Unreported
	"Old Indian mine"	Unspecified	Hammerstones	Jackson 1855:*342
	"conspicuous"	Unspecified	Hammerstones	Whittlesey 1863:6
	"an ancient pit"	Unspecified	Unreported	Wood 1907:292
Northwestern Mining Co. (20KE82)	"A great many Indian hammers were found in the soil…"	Unspecified	Hammerstones	Jackson 1849:445; Clarke 1975:8

SITE	PITS	AREA	TOOLS AND FACILITIES	REFERENCE
Phoenix Mining Co. (20KE68) (Successor to the Lake Superior Copper Co. on the same location.)	"a line of ancient and extensive diggings equal to any in the country"	Unspecified	Unreported	Brown 1852
	Unspecified	Unspecified	Unreported	"S. S." 1852
	"ancient pits"	"can be traced several hundred feet"	Unreported	Phoenix Mining Co, 1853:10
	Unspecified	Unreported	Unreported	Whittlesey 1863:20
	"some"	Unspecified	Unreported	Lawton 1881:28
	"ancient pits"	Unspecified	Unreported	Kayner 1883b:334
Waterbury Mine (20KE73)	Unspecified	Unreported	wooden shovel	Whittlesey 1852a:27-28
	"old works"	Unspecified	Unreported	Brown 1853d
	Unspecified	Unreported	wooden shovel	Whittlesey 1853:133
	One	Unspecified	Unreported	Tenney 1853b:173
	One	"artificial cavern"	Cedar "shovels"; wooden scoop or bowl; cedar-bark gutter	Whittlesey 1863:7-8
HOUGHTON COUNTY				
Adams Mining Co. (20HO289)	"ancient works"	"several hundred feet [on the course of the vein]"	Stone hammers; "vein rock, the separation of the [copper] masses after had taken place, was piled up across the vein in a regular wall"	Forster 1879: 166-167
Arcadian Mining Co. (20HO290)	"extensive remains of workings"	Unspecified	Unreported	Forster 1879:158

Appendix 3

SITE	PITS	AREA	TOOLS AND FACILITIES	REFERENCE
Atlantic Mine (A consolidation of the Adams and South Pewabic mines)	"The extent of these workings was very great."	Unreported	Unreported	Smith 1915:142
Columbian Mining Co. (20HO291)	"ancient pits; Indian diggings"	"many hundred feet"	Unreported	Columbian Mining Company 1862: 4, 8, 11
	"Ancient Pit Vein"	Unspecified	Unreported	Emerson and McKenzie 1862
Hulbert Mine (20HO4) (Later Calumet and Hecla. Next to Knapp's revelation at Minesota Mine, this is probably the most famous of the early discoveries.)	"Calumet Ancient Pit"	Unreported		McKenzie 1865b; Forster 1879:168-169; Hulbert 1893, 1899; *Detroit Free Press* 1898; Lathrop 1901a:6-7; 1901b: 255-258; Smith 1915:142; Griffin 1966
Huron Copper Co. (20HO275)	"signs of ancient working"	Unspecified	"a leather string about 30 in. long"	Wood 1907: 290-291
Isle Royale Mining Co. (20HO298)	"numerous"	Unspecified	Unreported	Tenney 1853a:415
	Unspecified	"a row of ancient pits had been traced for a mile"	Unreported	Whitney 1854:302
	Many	1/3 mile in length	Hammerstones	Whittlesey 1863: 16-17
	Numerous	"Indian diggings, which extended along its outcrop across the greater section"	Unreported	Lawton 1881:116
	"number of Indian diggings"	"along its out-crop [Isle Royale lode] across the greater section"	Unreported	Kayner 1883a:263
	"Indian diggings"	Extended along the outcrop of the Isle Royale lode	Unreported	Beahan 1899:106

152

SITE	PITS	AREA	TOOLS AND FACILITIES	REFERENCE
	Unspecified	"a line of Indian pits for nearly a mile along the outcrop"	Unreported	Stevens 1900a:231
	"extensively worked"	Unreported	Unreported	Wood 1907:290
	"The extent of these workings was very great."	Unreported	Unreported	Smith 1915:142
	"line of prehistoric pits"	Unreported	Unreported	Clarke 1990:6
Mesnard Mining Co. (20HO276)	One	Unspecified	"half a dozen stone hammers"	Emerson 1862c
	One	Unspecified	Unreported	Lexington Mining Co. 1864:9
	One	Unspecified	Unreported	Swineford 1876:53
	One	Unspecified	"marks of the ancient miner's hammer"	Forster 1879:164
	One	Unspecified	"multitude"	Lawton 1881:140
	One	Unspecified	"multitude"	Kayner1883a:264
	One	Unspecified	"stone hammers and burnt wood"	Lawton 1886:62
	One	Unspecified	"multitude of stone hammers"	Pinkham 1888: 68-69
	One	Unspecified	"stone hammers, copper chisels and much charcoal about the mass"	Forster 1891: 21-22
	One	Unspecified	Unreported	Packard 1893a:182, 1893b:74-75
	One	Unspecified	"raised on skids"	Stevens 1900b:107
	One	Unspecified	Unreported	Lathrop 1901a:5, 1901b: 250, 1961:121-122
	One	Unspecified	"a cart-load of broken hammers"	Wood 1907:290

Appendix 3

SITE	PITS	AREA	TOOLS AND FACILITIES	REFERENCE
	One	Unspecified	Unspecified	Griffin 1961a: 73-74
Pewabic Mining Co. (20HO130)	"numerous"	Unspecified	Unreported	Tenney and Partz 1854c:74
	At least three	Unspecified	Unreported	Emerson 1857a
	One	Unspecified	Unreported	Wood 1907: 289-290
	"...line of some ancient pits..."	Unspecified	Unreported	Coon 1940:93
Pontiac Mining Co. (20HO299)	"extensive and show a regular system"	Unspecified	"Piled up around are heaps of stone hammers... being oblong pebbles of hard rock weighing from two to four pounds...These stone hammers are found piled into one large and two or three smaller heaps, containing a number of cart-loads, and every one of these hammers exhibits a fracture, generally at the small or cutting end, showing that the piles are of 'damaged tools' thrown aside...A copper chisel, stone hammers with bands hollowed out for the reception of handles of twisted twigs, and other relics of the miners"	J. Houghton 1862:1
Portage Mining Co. (20HO300)	"numerous"	Unspecified	Unreported	Tenney 1853a:415

SITE	PITS	AREA	TOOLS AND FACILITIES	REFERENCE
	Unspecified	Unspecified	Unreported	Brown 1853c
	One	Unspecified	"several stone hammers"	Emerson and McKenzie 1863
	"a large pit"	Unspecified	Unreported	Pope 1901a:11; 1901b:22
Quincy Mining Co. (20HO128)	"a line of pits"	Unspecified	Unreported	Emerson 1857a
	"an ancient working"	Unspecified	"rude chisels and the whetstones of ancient miners," "miner's skids"	*Detroit Free Press* 1858a, 1858b
	Many in the glacial drift and on bedrock veins.	Unspecified	Hammerstones	Whittlesey 1863:14-16
	"storage pits"	"between the eastern line of the Ripley location and Dollar-Bay"	Unreported	Hulbert 1893: 115-117
	"broad and deep pits in the gravel"	Unspecified	Unreported	Packard 1893a:181; 1893b:74
	One	Unspecified	Hammerstones; "scaffolding of white birch poles"	Wood 1907: 288-289
Ripley Mining Co. (20HO301)	"Indian pits... particularly numerous"	Unspecified	"great quantities of stone hammers were stored"	Pope 1901a:11, 1901b:22
Shelden and Columbian Co. (20HO291)	Unreported	Unreported	"rude Indian knife of copper"	McKenzie 1864a
South Pewabic Copper Co. (20HO302)	One	Unspecified	"several well worn stone hammers"	McKenzie 1864c
	"ancient diggings which are more extensive than any we have here- tofore seen"	Unspecified	Unreported	McKenzie 1865
	"A remarkably deep trench"	"several hundred feet in length"	Unreported	Packard 1893a:183; 1893b:75

SITE	PITS	AREA	TOOLS AND FACILITIES	REFERENCE
Winona Mining Co. (20HO303)	"several of these pits"	Unspecified	Unreported	McKenzie 1864b
	"ancient pits"	Unspecified	Unreported	Lawton 1881:105
	"ancient pits"	Extensive	Unreported	Beahan 1899:116
	"ancient pits"	Outcrops "worked extensively"	Unreported	Hobart and Wright 1883:529
	"a line of Indian pits along the outcrop"	Unspecified	Unreported	Stevens 1900a:276
	"…a line of old Indian pits along the outcrop…"	Unspecified	Unreported	Weed 1918:925
ONTONAGON COUNTY				
Adventure Mining Co. (20ON251)	"remarkable for its ancient workings"	Unreported	Unreported	Hodge 1849c:752
	Unreported	"…pits are dug to considerable depth."	Unreported	Anonymous 1850
	"frequent"	Unreported	Unreported	Whittlesey 1850:326
	"numerous caves, grottoes, holes, pits, trenches and cavities"	"scores of places"	Unreported	Brown 1853b
	Unspecified	"scores of places"	Unreported	Tenney 1853b:174
	"ancient diggings of very considerable dimensions"	Unspecified	Unreported	Adventure Mining Co. 1857:11; Mason 1858
	"the pits and diggings of the ancient miners are found in great numbers"	Unspecified	Unreported	Emerson 1857h

Historic Copper Mines Founded on "Ancient Diggings"

SITE	PITS	AREA	TOOLS AND FACILITIES	REFERENCE
	"From the ancient diggings near the summit of Adventure bluff... they are now taking a mass of almost pure copper, which... will weigh five tons"	Unspecified	Unreported	Emerson 1857j
	"an ancient digging of very considerable dimensions at the surface"	Unspecified	Unreported	Mason 1858a, 1858b; McElrath 1858
	Unspecified	Unreported	Unreported	Whittlesey 1863:20
	Unspecified	"A long line of pits"	Unreported	Stevens 1900a:281
	"a line of old pits"	Unspecified	Unreported	Stevens 1907:11
	"a line of ancient pits"	Unspecified	Unreported	Weed 1918:825
Arctic Mining Co. (20ON219)	Unspecified	"two lodes...can be traced for a great length, with 'Indian Diggins' on them"	Unreported	Arctic Mining Company 1854:7
Aztec Mining Co. (20ON271) (Later South Lake Mining Co.)	"very extensive ancient works"	Unreported	Unreported	Hodge 1849c:752
	"numerous ancient pits or workings"	Unspecified	Unreported	Aztec Mining Co. 1851:11-13
	Unspecified	Unreported	Wooden shovel	Whittlesey 1852a:27-28; 1852b:133
	"very extensively worked over by the ancient miners"	Unspecified	Unreported	Whitney 1854:292
	Multiple	Unspecified	Unreported	Tenney and Leeds 1855e:544

SITE	PITS	AREA	TOOLS AND FACILITIES	REFERENCE
	Unspecified	Unreported	Unreported	Whittlesey 1863:20
	"unusual amount of ancient mine work"	Unreported	Unreported	Lawton 1881: 101-102
	"ancient Indian diggings"	"one mile in length"	Stone hammer 11 ½ pounds	Draper 1882:19
	"an unusual number"	Unreported	"stone hammers, remains of fire, etc."	Lawton 1883: 169-170
	"ancient pits"	Unspecified	Unreported	Pinkham 1888: 41-42
	"ancient pits found on the surface of the lode"	Unreported	Unreported	Stevens 1900b:90
	"...the remains of extensive workings by a prehistoric race..."	Unreported	"A 16-lb. Stone hammer..."	Weed 1918:913
Bohemian Mining Co. (20ON272)	"Indian works"	Unreported	Unreported	Bohemian Mining Co. 1851:14
	"many excavations"	Unreported	Unreported	Bohemian Mining Co. 1860:7
	"ancient pits have been opened on a lode some 200 feet south of the one on which the mine is being sunk...large and quite numerous	Unreported	"besides the usual amount of stone hammers and beaten copper, some copper tools, probably used as sockets"	Emerson 1862a
	One pit	Unspecified	"pieces of charred wood and stone hammers in profusion"	Emerson 1862b
	Map location			Great Western Copper Mining Co. 1863a

Historic Copper Mines Founded on "Ancient Diggings"

SITE	PITS	AREA	TOOLS AND FACILITIES	REFERENCE
"Boston and Lake Superior Mining Company"	"Traces of ancient mining operations have been discovered at various points along this portion of the trap range, similar to those described as found at the Minnesota Company's location." (As there is no specific location given for the "ancient mining operations," no archaeological site number has been assigned.)			Jackson 1849:751
Caledonia Mining Co. (20ON255)	"a place where these old miners have burrowed into the rock in various directions for fifteen or twenty feet"	Unreported	Unreported	McKenzie 1864b
	"old Indian pit"	Unspecified	Unreported	Lawton 1881:91
	"ancient pits"	Unreported	"Many battered stone hammers lay around the mouth of the pit"	Johnson 1882:901
	"ancient mining pits"	Unreported	"132 grooved hammers…a broken stone axe and several large hand hammers"	Lewis 1889:295
Carp Lake Mining Co. (20ON189/190)	"ancient diggings"	"extensive and numerous"	Unreported	Carp Lake Mining Co. 1858:3-4
Devon Mining Co. (20ON262)	"Numerous indications of ancient or 'Indian diggings' "	Unspecified	Unreported	Devon Mining Company 1863:7
Evergreen Bluff Mining Co. (20ON265)	"a number of Indian diggings"	Unspecified	"a great many Indian hammers"	Evergreen Bluff Mining Co. 1854:7
	"ancient Indian diggings"	Unspecified	Unspecified	Tenney, Leeds and Partz 1854:429-430
	"filled with ancient pits"	Unspecified	"stone hammers… are found here in abundance"	*Pontiac Jacksonian* 1855: 252
	"very distinctly marked by ancient pits"	Unreported	Unreported	Emerson 1857g

Appendix 3

SITE	PITS	AREA	TOOLS AND FACILITIES	REFERENCE
	"A number of copper veins are discovered ... attention having been first directed to them by the ancient pits [6] in the trap rock"	Map location	Unreported	Evergreen Bluff Mining Company 1857, 1858:210
	"very distinctly marked by ancient pits"	Unspecified	Unreported	Tenney 1857b:280
	"many ancient diggings"	Unspecified	Unreported	Lawton 1881:94
Flint Steel River Mining Co. (20ON425)	"extensive 'ancient diggings' of considerable depth"	Unspecified	Unreported	McElrath 1858:150
	"Very extensive traces of aboriginal mining"	Unspecified	Unreported	Bauerman 1866:459
	"some old Indian pits"	Unspecified	Unreported	Lawton 1881:90, 91
	"some old Indian pits"	Unspecified	Unreported	Hobart and Wright 1883:519
	"old Indian pits"	Unspecified	Unreported	*Daily Mining Gazette* 1899
Forest Mining Co. (20ON290) (Earlier known as Cushin, later Victoria)	"ancient diggings"	"can be traced for half a mile"	Unreported	Forest Mining Company 1852a:4, 7, 14, 20
	"very extensive excavations"	Unspecified	Unreported	Whitney 1854:297
	"a row of ancient diggings"	Unspecified	Unreported	Tenney and Partz 1854d:431
	Unspecified	Unreported	Wooden bowl	Whittlesey 1863:20

SITE	PITS	AREA	TOOLS AND FACILITIES	REFERENCE
	"a series of pits, some of which were fourteen feet deep"	They were arranged in four lines, following the courses of four veins or feeders	Wooden bowl	Foster 1874:267
	Unspecified	Unspecified	Unreported	Lawton 1881:106
	Unspecified	Unspecified	Unreported	Hobart and Wright 1883:529
	Unspecified	"A line of Indian pits was traced along the outcrop"	Unreported	Stevens 1900a:309
	"…a line of prehistoric pits containing masses of native copper…"	Unspecified	Unreported	Weed 1918:918
Forest Shepherd Mining Co. (Insufficient information to assign a site number)	"depressions similar to those found to be ancient diggings"	Unspecified	Unreported	Shepherd, F. 1864:13
Great Western Copper Mining Co. (20ON272)	"a range of ancient diggings"	Unspecified	Unreported	Great Western Copper Mining Co. 1852:25
	Map location			Great Western Copper Mining Co. 1863a
	"numerous… ancient diggings"	Unspecified	Unreported	Great Western Copper Mining Co. 1863b: 9-11, 19
	"Indian pits"	Unspecified	Unreported	Lawton 1881:103
	"Indian pits"	Unspecified	Unreported	Hobart and Wright 1883:528
Hartford Mining Co. (20ON482)	"…that extensive chain of pits known to the explorers of the Northwest, **Hartford**, Mandan Bluff and Iron City mines as 'ancient diggings'." (This is the only known mention of "ancient diggings" at the Hartford Mining Co. site.)			Moore 1915:11

161

Appendix 3

SITE	PITS	AREA	TOOLS AND FACILITIES	REFERENCE
Hilton Mining Co. (20ON426)	"heavily marked at the surface by 'ancient diggings'"	Unspecified	Unreported	Hilton Mining Company 1863:10, 12
	"old pit"	Unspecified	"a leather bag... eleven inches long and seven inches wide"	Hagar 1865:310
Hudson Mining Co./ Eureka Mining Co. (20ON264)	"Traces of ancient mining..."	Unreported	Unreported	Jackson 1849:704-705
	"Indian digging vein..."	Unreported	Unreported	Hudson Mining Co. 1864:4
Knowlton Mining Co. (20ON270)	Unspecified	Unreported	Unreported	Knowlton Mining Co. 1863:1
	"ancient mining pit on the S. E. ¼ of section 1, township 50, of range 39..."	Unreported	copper "chisel"	Lewis 1890:180
LaFayette Mining Co. (20ON182)	"ancient diggings"	Unspecified	Unreported	LaFayette Mining Company 1863:4, 9-11, 16-17
Lake Superior Copper Co./ Rockland Mining Co. (20ON427)	Unspecified	Unspecified	Unreported	"C. R." 1853
Mass Mining Co. (20ON274/275)	"a line of ancient Indian pits"	"can be traced all through to the eastern boundary line"	Unreported	Lord 1856
	"ancient diggings"; "three noble ancient diggings"; ancient works"	Unspecified	"Many stone tools"; birch and poplar scaffolding	Mass Mining Company 1856:2, 4, 10, 11, 18-19
	"many ancient diggings"	Unspecified	Unreported	Mass Mining Company 1876:11
	"many ancient diggings"	Unspecified	Unreported	Lawton 1881:94

SITE	PITS	AREA	TOOLS AND FACILITIES	REFERENCE
	"old Indian diggings"	Unspecified	"five and one-half pound stone axe"	Draper 1882:19
Minesota Mining Co./ Michigan Copper Mining Co.) (20ON1)	One	Unspecified	Copper gad and narrow copper chisel; "more than a ton" of cobble hammerstones; skids of black oak; large wooden wedges	"B." 1848a-h, 1849; "E.C." 1849; "K." 1849; Page 1849; Bagg and Harmon 1848; Warburton 1850
	"sunken trench"	"extends for a mile"	Stone hammers, copper gad, copper chisel; large timbers used as props	Ingersoll 1848
	Unspecified	Unspecified	"tools"	"E. C." 1849
	"A series of open cuts"	Unspecified	"Large quantities of stone hammers"; Cob-work	Jackson 1849:745-746
	"lines of pits"	"more than two miles in length"	"stone hammers or picks"	Hodge 1849a, 1849b:545, 1849c
	Unspecified	Unspecified	Unspecified	"E. C." 1850
	"Three parallel rows of ancient diggings"	"extend for two or three miles... twenty to fifty yards apart"	Stone hammer	Trowbridge 1850a:54; 1850b
	Unspecified	Unspecified	tree-trunk ladder	Brown 1850a, 1850b
	Unspecified	2 ½ miles	10 cart loads of hammerstones; copper gad; copper chisel; cob-work supporting mass of copper	Foster and Whitney 1850: 159-160
	Unspecified	Unreported	"fifty cart loads of [hammers] might be collected"; "charred skids"	Hodge 1850b:307

SITE	PITS	AREA	TOOLS AND FACILITIES	REFERENCE
	One	Unreported	"thousands" of stone hammers; "cobwork" of timber	Schoolcraft 1851:96-97
			Copper chisel; copper gad; cob-work	Whittlesey 1852a:27, 28, 1853:132-133
	One	Unspecified	"large quantities of stone hammers and a few other tools"; cob-work	Whitney 1854:293-294
	"extensive trenches"	Unspecified	"stone hammers or mauls"; "artificial cradle of black-oak"	Wilson 1856: 227-228
	"an old Indian digging"	Unspecified	"Stone hammers, decayed timber, charcoal, etc."	Emerson 1857c
	"a channel or trench"	Unspecified	Unreported	Emerson 1857k
	One	Unspecified	Unreported	Emerson 1857m
	"In October last we cleared one of the large "ancient diggings" about 180 feet west of the Rockland line"	Unspecified	Unreported	Townsend 1858a:334; 1858b:266; Minesota Mining Company 1858b
	"They are now clearing out several deep pits or Indian diggings between shafts Nos. 6 and 9…They have not yet reached the bottom of the pits"	Unspecified	Unreported	Emerson 1858c
	Unspecified	Unreported	Hammerstones; copper chisel and a copper maul; "cob work" supporting 6-ton mass of copper; tree-trunk ladder	Whittlesey 1863:4, 11, 17-20

Historic Copper Mines Founded on "Ancient Diggings"

SITE	PITS	AREA	TOOLS AND FACILITIES	REFERENCE
	"The evidences of ancient mining on this tract, both on the middle and south bluffs were very distinct, perhaps more so than at any other point on the Lake"	Unspecified	Unreported	"Misk-wabikokewin" 1863
	"old working"	Unspecified	"skids"	McKenzie 1864a
	"indentations, which were plainly discernable along the surface outcrop of the vein"	Unspecified	"great numbers of stone hammers... bits of burnt wood, a copper chisel, with a socket for holding a handle, etc."; timbers supporting the copper mass	Lawton 1881: 68-69
	Unspecified	Unreported	"numerous stone hammers or mauls"; "skids"	Trowbridge 1881:122; Balch 1882
	Unspecified	Unreported	"numerous stone hammers"; "cob-work of logs"	MacLean 1887:76-77
	"indentations... plainly discernible along the surface outcrop of the vein..."	Unreported	Unreported	Pinkham 1888: 29-32
	One	Unreported	"raised on skids"	Stevens 1900a:298
	Two	Unspecified	Hammerstones; "oak skids"	Spalding 1901:689-690
	"A prehistoric copper mine ... near shaft A of the Michigan Copper Mining Company's mine..."	Unspecified	"large stone hammers, the weight of some being as much as 30 lbs"; "a pile of charred logs"	Baum 1903

165

SITE	PITS	AREA	TOOLS AND FACILITIES	REFERENCE
	One	Unreported	large number of stone hammers; copper chisels	Wood 1907:203
National Mining Co. (20ON279)	"Ancient mine work was discovered to have been carried on extensively"	Unspecified	"stulls or timbers forming a scaffolding across the shaft"	Whitney 1854:297
	Unspecified	Unreported	"scaffolding"	Tenney 1856b:309
	"ancient diggings"	Around Shafts 1 and 2	Unreported	National Mining Company 1857:foldout map
	"ancient diggings"	Between Shafts 6 and 7	Unreported	National Mining Company 1859:foldout map
	"ancient diggings"	Between Shafts 6 and 7	Unreported	National Mining Company
	"ancient diggings"	Between Shafts 6 and 7	Unreported	National Mining Company 1862:foldout map
	"The outcrop was plainly marked by ancient excavations"	Unspecified	"scaffolding"	Lawton 1881:81
	"The outcrop was plainly marked by ancient excavations"	Unspecified	Unreported	Beahan 1899:160
	"relics of ancient mining operations, including a shaft of 50' depth..."	Unspecified	"...timbered and and scaffolded down one side"	Weed 1918:892
Nebraska Mining Co. (20ON281)	Unspecified	Unspecified	Unreported	Tenney, Leeds, and Partz 1854:427, 429-430
	"frequent and very extensive"	Unreported	Unreported	Tenney and Leeds 1855b:189; 1855e:542
	Seven	Unspecified	Unreported	Tenney 1856a:117-118

Historic Copper Mines Founded on "Ancient Diggings"

SITE	PITS	AREA	TOOLS AND FACILITIES	REFERENCE
	Unspecified	"The top of the bluffs is very much marked by diggings and pits"	"quantities of their rude tools can be picked up"	Emerson 1857f
	"wonderful surface show of copper in ancient pits"	Unspecified	"Cartloads of hammers have been taken out"	Emerson 1857i:4
	"new openings in some of the ancient pits"	Unspecified	Unreported	Emerson1857k
	"ancient diggings"	Unspecified	Unreported	Nebraska Mining Company 1859:4
	"Numerous"	Unspecified	Unreported	Whittlesey 1863:20
Norwich Mining Co. (20ON41)	"Indian diggings"	Unspecified	Unreported	Norwich Mining Company 1865:14
Ogima Mining Co. (20ON283)	Unspecified	Unspecified	Unreported	Emerson 1857l
	"extensive ancient workings"	Unspecified	Unreported	Ogima Mining Company 1863:8
	"old Indian pits"	Unreported	Unreported	Lawton 1881:95
	"old Indian pits"	Unreported	Unreported	Hobart and Lawton 1883:526
	Unspecified	"a line of Indian pits"	Unreported	Stevens 1900a:289
Ohio Mining Co. (20ON426)	"numerous caves, grottoes, holes, pits, trenches and cavities of every shape"	"scores of places"	Unreported	Tenney 1853b:174
	Unspecified	Unreported	Unreported	Whittlesey 1863:20
Ohio Trap Rock Co. (20ON40)	"several"	Unspecified	Unreported	Tenney 1853c:635
	"numerous pits"	Unspecified	Unreported	Tenney 1853a:415
	Unspecified	Unreported	Unreported	Tenney 1857d:390

Appendix 3

SITE	PITS	AREA	TOOLS AND FACILITIES	REFERENCE
Ontonagon Copper Co. (20ON11/290)	"a small shaft was sunk ten or twelve [feet?] and drifts extended each way for quite a distance"	Unspecified	Unreported	McKenzie 1864a
Peninsula Mining Co. (20ON149)	"lines of ancient pits"	Unspecified	Unreported	Hodge 1851c
Penn Mining Co. (20ON428)	"very promising 'ancient pits'"	Unspecified	Unreported	Coulter Mining Company 1864:7, 13
Piscataqua Mining Co. (20ON272)	"Indian diggings"	Unspecified	Unreported	Piscataqua Mining Company 1850:17-19
	"extensive Indian Diggings"	"can be traced... for a long distance"	Unreported	*Lake Superior Journal* 1850:646
Republic Mining Co. (20ON430)	"Veins opened by the 'Ancient Miners'"	Unspecified	Unreported	Republic Mining Company 1864: 5-6
Ridge Copper Co. (20ON429)	"Indian works... well defined..."	Unreported	Unreported	Whittlesey 1850:326
	Unspecified	Unreported	Unreported	Whittlesey 1863:20
	"numerous Indian diggings"	Unspecified	Unreported	Lawton 1881:98
	Unspecified	Unreported	Unreported	Stevens 1900a:281
Rockland Mining Co./ Lake Superior Mining Co. (20ON427)	"numerous ancient excavations"	Unspecified	Unreported	Whitney 1854:296
	"ancient pits"	Unspecified	Unreported	*Pontiac Jacksonian* 1855:253-254
	Unspecified	Unreported	Unreported	Tenney and Leeds 1855c:404
	"ancient pits and diggings on the three principal parallel veins"	Unspecified	Unreported	Rockland Mining Company 1856:5, 1858:14, 1859:12

SITE	PITS	AREA	TOOLS AND FACILITIES	REFERENCE
	"recent openings on ancient pits have exposed a vein of large size and great regularity"	Unspecified	Unreported	Emerson 1857j
	Unspecified	Unreported	Unreported	Tenney 1857g:572
	"In 1857…a copper mass in an Indian pit… weighing… forty tons…"	Unreported	"large number of stone hammers lying about it…"; "…propped up on skids…"	Hobart and Wright 1883: 512-513
	"works of the 'ancient miners'"	Unspecified	Unreported	Rockland Mining Company 1858:14
	"ancient diggings"	Unspecified	Unreported	Rockland Mining Company 1859:12
	"Numerous"	Unspecified	Unreported	Whittlesey 1863:20
	Unspecified	Unspecified	Unreported	Pinkham 1888:36
	Unspecified	Unreported	Unreported	Lawton 1881:88
Superior Mining Co. (20ON288)	"a remarkable crossing, deeply marked with ancient pits"	Unspecified	Unreported	Emerson 1857e
	"well marked by the ancient diggings"	Unspecified	Unreported	Tenney 1857c:99
	"Several lines of ancient pits"	Unspecified	Unreported	Tenney 1857e:383

		ISLE ROYALE		

Isle Royale was the focus of twentieth-century archaeological surveys that located a number of ancient mining pits, most of which were not documented in early Euro-American mining documents. Sites located in the twentieth century are marked by an asterisk.

SITE	PITS	AREA	TOOLS AND FACILITIES	REFERENCE
*Dustin (20IR10)	"group of ancient pits"	At least 27 pits	Hammerstones	Dustin 1957:9-10; Clark 1995:46
Epidote Mine (20IR51)	Unspecified	Unreported	Unreported	Fox 1911: 84; Dustin 1957:7; Clark 1995:80-81

Appendix 3

SITE	PITS	AREA	TOOLS AND FACILITIES	REFERENCE
*Ferguson-Dustin (20IR9)	"…five trenches, penetrating to a depth of more than fifteen feet…"	Longest trench "…nearly five hundred feet in length…"	Hammerstones	Ferguson 1924: 451-452; West 1929: 26; Bastian 1963a:41; Clark 1995:46
*Island Mine (20IR11)	"sparse Indian workings"	Unreported	Unreported	Dustin 1957:10; Bastian 1963a: 69-70; Rakestraw 1965:13; Clark 1995:46
*Little Siskiwit or "Old Fort Diggings" (20IR6)	This appears to be a location where drift copper was mined from the gravel.	Unknown	Hammerstones	Gillman 1873:17; Ferguson 1923:158-162; 1924; Dustin 1957:7-9; Bastian 1963a:40-41; Clark 1995:44-45, 1996
Lookout Mine (20IR30)	"a place where someone had dug for a vein"	Unspecified	Unreported	Shaw 1847:24
	Unspecified	Unreported	"multitude of hammers"	Foster and Whitney 1850:162
	Unspecified	Unreported	"Hammers… abounded"	Foster 1874:268
	Five pits	Unreported	Hammerstones	Bastian 1963a:49-53, 1963b:9-10; Clark 1995:66
Minong Mining Co. (20IR24)	Unreported	"almost a continuous line for more than two miles"	Hammerstones "found by cart-loads"; wooden bowl	Gillman 1873:173
	"ancient mine-pits"	"two and a half miles"	"tons of stone and stone hammers"; "large granite boulders used to to hold up the hanging ground"	Davis 1875
	"ancient mining work"	Unspecified	Unreported	Minong Mining Company 1875: 5-15

Historic Copper Mines Founded on "Ancient Diggings"

SITE	PITS	AREA	TOOLS AND FACILITIES	REFERENCE
	Unspecified	1¾ miles by an average of 400 ft. averaging 20 ft in depth	"battered stone hammers… running into the millions"	Houghton 1876a:80
	"works exceed in extent all other works of the kind heretofore seen in the copper region put together"	"400 feet wide and 1½ miles long"	Unreported	Swineford 1876:52, 61
	"extensive ancient workings"	Unreported	Hammerstones	Forster 1879:183
	Unspecified	"cover an area …three to four hundred feet wide and more than a mile in length	"Their hammers, frequently to the number of several thousand, are found in heaps where they were evidently placed at the end of the season"	Johnson 1882:901
	Unspecified	"ancient pits extending over a length of nearly two miles, and covering an average width of 400 feet"	"large quantities of charred fire brands and numberless stone hammers"	Swineford 1884:9-10, 15-16
	Unspecified	"the works of the old miners mark their [the copper veins] course more than twelve hundred feet"	Hammerstones "collected by the cartload"; boulders used as supports; wooden drains	Lane 1898:16-18
	Unspecified	[Three pits] " distant one from the other nearly a mile"	Unreported	Lane 1898:18 (footnote quoting A.C. Davis)
	Unspecified	An area "at least half a mile"	A "plenitude of hammers"	Holmes 1901

SITE	PITS	AREA	TOOLS AND FACILITIES	REFERENCE
	Unreported	"one and three-quarter miles in length…an average width of nearly four hundred feet"	"dams to prevent the passage of water from one pit to another …a drain sixty feet long…"	Lathrop 1901a:5, 1901b:250-251, 1961:122
	Unreported	Unreported	Unreported	Wood 1907: 292-293
	"line of ancient pits…"	"largest 400' and about 1½ miles long"	Unreported	Weed 1918:889
	"pit"	Unspecified	Hammerstones	Christian 1932:33-35
	1500-2000 pits	1.6 miles by 300-400 feet	Hammerstones	Bastian 1963a: 41-49
*Outer Hill Island (20IR37)	"prehistoric mining remains"	Unspecified	Unspecified	Dustin 1957:6; Clark 1995:68
*Phelps (20IR81)	At least one pit	Unspecified	Hammerstones	Clark 1995:97-98; Bastian 1963a:70
*Ransom Mine (20IR43)	Unreported	Unreported	Hammerstone	Clark 1995:73
Scoville Mine (20IR32)	"evidences of mining"	"can be traced lengthwise for a mile"	"stone hammers and wedges"	Foster and Whitney 1850:162; Lane 1898:8-9; Dustin 1957:6; Bastian 1963a:74; Clark 1995:67
*Singer (20IR80)	"Three mining pits"	Unreported	Hammerstones	Clark 1995:93-97; Bastian 1963a:70
Siskowit (20IR41)	Unspecified	Unreported	Unreported	*Commercial Bulletin* 1849
	"it has been worked before by some one; they worked on the vein about fifteen feet deep"	Unreported	Unreported	Siskowit Mining Company 1850:18
	"ancient diggings"	Unreported	"the usual supply of stone hammers, &c."	Brown 1854

SITE	PITS	AREA	TOOLS AND FACILITIES	REFERENCE
	At least three pits	Unreported	Hammerstones	Dustin 1957:6-7; Bastian 1963a: 55-56; Clark 1995:69-72
*Stoll (20IR26)	Three pits	Unreported	Hammerstones	Bastian 1963a:54-55; Clark 1995:54
*Third Island (20IR84)	Several enlarged crevices	Unreported	Unreported	Bastian 1963a: 53-54, 70; Clark 1995:99
Triangle Island (20IR106)	"circular pits of the ancient miners"	Unreported	Unreported	Gillman 1874:387-389; Fox 1911:97-100; Clark 1995:100

Appendix 4: Profiles

Appendix 4

John Wells Foster (1815-1873)

Foster graduated from Wesleyan University in Connecticut graduating in 1834. He then moved to Ohio, where he studied law and was admitted to the bar at Zanesville. Geology had always attracted him, and he accepted a position as assistant in the Geological Survey of Ohio in 1837. In 1847 he made the fateful decision to join Charles T. Jackson in the survey of the "Lake Superior Land District." Ultimately, with Charles Jackson's dismissal, Foster and Josiah Whitney took over the completion and publication of the Lake Superior report. In later years he served as President of the American Association for the Advancement of Science and lecturer on physical geography at the University of Chicago. Toward the end of his life he became interested in North American archaeology and published *Pre-Historic Races of the United States.*

William Hickok

William Hickok is probably the most controversial and the most elusive of all the characters making up the Minesota Mine controversy. It seems that without his legal skills, Samuel Knapp never would have been able to obtain ownership of the Minesota. Hickok was frequently in Michigan after the "ancient diggings" discoveries, often in the company of Knapp, being interviewed by newspapermen, etc. Hickok also seems to have taken possession of the artifacts discovered in the mine, with Knapp showing little, if any, interest in them after the initial discovery. Hickok had offices ("William Hickok & Co.") in New York City in the 1840s and in Camptown, NJ, in the 1860s. As late as 1881, he was living in Tarrytown, NY. Correspondence with colleagues in New York has yet to produce significant information on Hickok. It is perhaps fitting that the years of his birth and death are unknown.

Samuel W. Hill (1815-1889)

Unlike most of the characters in this story of early copper mining, Samuel W. Hill was largely self-taught in the skills that would serve him well for many years. He was born in Starksboro, Vermont, and he attended township district schools until he was sixteen. After attending a local Friends' school for two years, where he paid his way by teaching in the winters, he graduated in engineering and surveying. He continued to teach in winters and work on his father's farm in the summer until he reached legal age. While teaching and farming, he also managed to procure surveying instruments and used them on small jobs in various parts of Vermont. In 1839 he did some surveying in New York, and then moved on to Milwaukee, Wisconsin, where he taught school in the winter of 1839-1840. Between 1840 and the spring of 1845, Hill was employed in U. S. public land surveys and the beginnings of the Great Lakes hydrographic survey. In 1845 he went to Lake Superior and worked with Douglass Houghton on the geological and lineal survey of the Upper Peninsula. Hill was placed in charge of a party of on the geological examination of the mining region. Over the next several years, he stayed in the Lake Superior region. He surveyed and investigated metalliferous deposits until 1848, when he joined up with Charles Jackson's Lake Superior Land District survey. After that, he became one of the most knowledgeable individuals available for hire in the entire copper-mining region.

James T. Hodge (1816-1871)

Hodge was born in Plymouth, Massachusetts, and graduated from Harvard in 1836. He was particularly interested in mineralogy and geology, and his knowledge and zeal caught the attention of professionals in those fields. As a result of this attention he was employed on the state geological survey of Maine under Charles T. Jackson and the state geological survey of Maine under Henry D. Rogers. At times, he also served on geological surveys in Ohio and New Hampshire. He published extensively, including numerous articles in the *New American Cyclopedia*. For several years he was assistant editor for mining and metallurgy in the *American Railroad Journal*, in which he offered useful views and opinions of early mining activities in the Keweenaw Peninsula. Returning from geological investigations in the Lake Superior region, he was lost in the sinking of the steamer *R. G. Coburn* in Lake Huron on October 15,1871.

Charles T. Jackson (1805-1880)

Dr. Jackson was probably the most controversial and colorful of the early mining figures. Trained in both geology and medicine (he held an M. D.), Jackson was appointed United States Geologist for the Lake Superior Land District. His leadership here was such a disaster that he was dismissed and completion of the survey was turned over to his assistants John Wells Foster and Josiah Dwight Whitney. This effectively removed him from the ancient diggings story. Jackson had severe mental problems (more on this later). Principal among them was a propensity for claiming prior discovery of, among other things, guncotton, the telegraph, the anesthetic effects of ether and the digestive action of the stomach. He spent the last seven years of his life in a mental asylum and died in 1880.

Appendix 4

Josiah D. Whitney (1819-1896)

Whitney was born in Northampton, Massachusetts. He was the oldest of 12 children. In his youth he was educated at schools in Northampton, Plainfield, Round Hill, New Haven, Connecticut, and Andover, Massachusetts. Entering Yale University in 1836, he studied astronomy, chemistry and mineralogy. After graduating from Yale in in 1839, he continued his studied of chemistry in Philadelphia. In 1840 he joined a geological survey of New Hampshire as an unpaid assistant to Charles T. Jackson. In 1841 he considered entering Harvard Law School, but was derailed when he heard a lecture on geology by Charles Lyell. He changed his career plans and in 1842 sailed to Europe to continue his scientific discoveries. Over the next five years he traveled throughout Europe, studying chemistry and geology in France and Germany. Upon returning home in 1847, he was hired, with Foster, to assist Charles Jackson in completing the survey of the Lake Superior Land District. Following the dismissal of Jackson, Foster and Whitney completed the monumental report. Whitney became a mining consultant and in 1854 wrote *The Metallic Wealth of the United States*, a tome that was a standard reference for many years.

Charles Whittlesey (1808-1886)

Without a doubt, outside of Samuel O. Knapp, Charles Whittlesey is the most important figure in the story of the "ancient diggings." Soldier, engineer, lawyer, geologist, artist, archaeologist, author, historian and museum founder, the range of his knowledge and experiences is unparalleled (he fought in the Blackhawk War and the Civil War). Without his *Ancient Mining on the Shores of Lake Superior* (published in 1863, but actually completed in 1856), our knowledge of ancient mining activities would be nothing short of pathetic. (See Baldwin 1887 and Tribble 2008.)

Appendix 5: Further Readings

Appendix 5

For readers interested in copper use during specific time periods, the author has provided this list of abbreviated citations. Sources are organized alphabetically within time period. Complete citations can be found in the bibliography.

Copper Usage in the Middle-Late Archaic Periods

Anonymous 1902; Anonymous 1954; Anonymous 1991:Pl. 2; Anonymous 1966a; 1966b; Anonymous 1990; Anonymous 1996; Alex 2000:57, 74–75; Atkeson 1959; Babcock 1928; Baerreis, Daifuku and Lundsted 1957; Beer 2014; Behm 1991, 1995; Benn and Thompson 2009:519–520; Bennet 1952:110; Berg 1945:7–10; Birmingham and Eisenberg 2000:75–77; Boyle 1888:54–56, 1889a:50, 1889b:39–40, 1889c:78–80, 1891: 59, 62, 1893:21, 1895:85–89, 1897b:53–54, 1898a:25, 1898b:53, 1900:32, 1904a, 1904b, 1905, 1908; Brose 1994a:218; Brower and Bushnell 1900:Plates IV, VI, XII, XIII; C. E. Brown 1901, 1902, 1904a, 1904b, 1905a, 1905b, 1916, 1917:7, 1923a:117–118, 1923b, 1924, 1926a, 1926b:12, 1927, 1929:88–90, 1930a:144, 1930b, 1931, 1938a, 1938b, 1940a, 1940b, 1942; C. E. Brown and R. H. Becker 1917:87, 91; J. Brown, 2004:677–679; T. Brown 1930; Bruhy and Wackman 1980:403,447; Bruseth 1991: 11, 15, 17; Bryan 2005:108–109; B. Butler 2009:626; J. Butler 1876a, 1876b, 1877, 1881, 1882a, 1882b, 1887; Caine 1969:39–44, 120–122, Plate 21; 1974:59–60, 63; Carlson 1997:12; *Central States Archaeological Journal* 1979:95, 1987:24, 1992:62, 2000:192, 2002:113, 2003:123, 2005:133, 2006:70, 84; J. Chapman 1982:18, Fig. 26, 1985:51; C. Clark 1996; W. Clarke 1884; Cleland and Wilmsen 1969; H. E. Cole 1922:92–93, 102, 106; Custer 1910:50; Dawson 1966, 1983:68; Douglas 1936; Draper 1876:37, 1879a:26, 1879b:47–49, 1879c:81, 1882a:18, 1882b:37; Draper et al. 1879; Eames 1922; Eberle 1997; Ellis et al. 1990:87–89, 2009:807, 808, 811; Finney et al. 1992:10, 16–17; A. K. Fisher 1932; J. Fisher 1924; Flaskerd 1940, 1943:17, 1945; Fogel 1963; Fregni 2009; Fryklund 1941; Fundaburk and Foreman 1957:Pl. 4; Galm 1981:140, 146, 238, 1984:212; *Genuine Indian Relic Society, Inc.* 1978:4–31, 37–49, 52–54; Gibbon 2012:77, 82–84, 122, 124; J. B. Griffin 1952b:355, 1961c:152, 1964:233–234, 1966b, 1972; Hedican 1981; Hedican and McGlade 1982, 1993; C. Hill 1982: 93, 94, 106, Fig. 15; H. H. Hill and Co. 1883:92; Hinsdale 1925:108–111, Pls. XXIX–XXXI; Hodge 1912a, 1912c, 1913; Holmquist 1948; Holsten 1938; Hoppa et al. 2005:240, 246, 251–253; Hruska 1967; Jefferies 1987:63–64, 80, 1996:54, 61, 62, 74, 75, 77, 2004:115, 2008: 236, 238, 297, 2009:643, 657; Jefferies and Butler 1982:632, 668, 1207, 1230; Jennings 1952:257; Jenson 1962; Jenson and Birch 1963:67–68; D. Johnson 2012; E. W. Johnson 1994, 1996, 1997; E. Johnson 1962:162, 1964, 1967; G. L. Johnson 1980; T. Johnson 1975; Johnston 1968a:8–9, Plate I; 1984:45–46; M. Jones 1980; Jury 1965, 1973; Kammerer 1964; Kellar 1993:34; Kennedy 1962:127–128, 1967, 1970:59–63; Kent 1991; Kenyon 1959a:13, Pl. V, 1959b; Kidd 1952:80; Kidder and Sassaman 2009:680, 684; D. R. King 1961: 49, Fig. 1, Fig. 4; M. Knight 1895:105;

Further Readings

V. J. Knight 2004:734–735; Korp 1990:16; Kruse and Soulen 1940; Kuhm 1928:64–67, 69–71; Laidlaw 1899, 1915:71–72; Lambert 1982:169, 173–174, Figs. 20, 89, 103, 116, 117; Landon 1940; Langevin et al. 1995; Lapham 1855:74–77, 88; Lawson 1902, Leechman and de Laguna 1949:30; Lemmer 1960; Leonard 1910:23–24; R. B. Lewis 1996a:61–62, 74, 75, 77; T. H. Lewis 1886:370, 1894:208, 1895:320; T. M. N. Lewis and M. Kneberg 1958:30–31, 1959:164, 169, Fig. 5:j, 172, Table 1 (177); T. M. N. Lewis and M. Kneberg Lewis 1961:Pl. 41, 103, 118, 1995:20,72–173; Lilly 1937b; W. D. Logan 1976:87; Long 1958; Lovis 2002:8–3, 2009:734, 740; Luedtke 1980; Lueger 1977:43–45, 98; Lurie et al. 2009:763; Maasch 1994; MacCurdy 1917:93–94; Mallam 1978; Marckel 2004; Marquardt and Watson 1983:333–334; P. S. Martin et al. 1947:299–303; S. R. Martin 1993, 1994, 1995, 2004, 2008, 2009; S. R. Martin and T. C. Pleger 1999; C. I. Mason and Ronald J. Mason 1961, 1967; Richard P. Mason 1989; Richard P. Mason and C. I. Mason 1994a, 1994b; Ronald J. Mason 1981:199, 212, 214–216, 218–220, 224, 226, 251, 258–259; McElrath et al. 2009:327, 339, 345, 347; McGee and Thomas 1905:358–364; McKern 1931b, 1942:167–169; McKusick 1964a:478, interior and exterior of back page, 1964b:75–77; K. David McLeod 1987:31–33; M. McLeod 1980: 23–26, 57, 78, 88–89, 96–97, 111; Miles 1951; Millar 1978:335–338; Miner 1941:50; Moffat and Speth 1999; Moll and Moll 1959:55–56; Montgomery 1913:6, Figs. 8–10; Moody 1883:546; Moorehead 1900:59, 62, 63, 182, 320–328, 335–336, 1910:161–234, Figs. 568–569, 575–606, 611, 613, 614; Morgan 1952:85; Mulholland and Pulford 2007; Mulholland et al. 2008:114–115, 124–127, 136–137; Muller 1986:67; Neuman 1984:80, 102–107, 309; Noakes 2007; Nolan and Fishel 2009:414, 445, 451; R. B. Orr 1912, 1913, 1917:25, 34, 1928; Osborn 1917; L. W. Patterson 2000:32, Table 5, Fig. 10, 38; Peet 1906b; Pelletier 1997:49–57; Penman 1977, 1978; Penney 1985a:25, 29, Pls. 13–15, 1985b:148; 1987:40–41; Perino 1961, 1968b; E . L. Perkins 1897; F. S. Perkins 1887; Perttula and Bruseth 1990:105; Platt 1984:20–23; Pleger 1992, 1998; Pleger and Stoltman 2009:704–712; Popham and Emerson 1954; du Pouget 1890:176; Pratt 1883b:153; Pulford 1999, 2000, 2009; Purtill 2009:589–591; Quimby 1952:100–101, 1954, 1957b, 1959a, 1960b:52–63, 1961, 1962; Quimby and Spaulding 1957; Rau 1885:138–139, 152–154, 181–182; Reynolds 1856; D. K. Riddle 1982:Cover page, 20, Fig. 10, g; W. Riddle 2005; Ritzenthaler 1957a, 1957b, 1958a, 1958b, 1960a, 1960b, 1963, 1967, 1969, 1970; Ritzenthaler and Wittry 1957; Ritzenthaler et al. 1957; Rolingson 1967:344–345; Rolingson and Schwartz 1966:90–92; Romano and Mulholland 2000; Ross 1980, 1982; Rothschild 1979:664; Ryan 1911:24; *St. Paul Dispatch*:1889; *St. Paul Pioneer Press* 1897; Salzer 1974:45; Scheidegger 1962; Schroeder and Ruhl 1968; S. A. Schultz 1913; J. P. Schumacher 1918: 133, 136; J. P. Schumacher and J. H. Glaser 1913:123–124; J. P. Schumacher and W. A. Titus 1913:78; Sciulli and Heilman 2004:51, Fig. 3; M. R. Sharp 1952; Shetrone 1930:69–70; Short 1879:98–99; Shriver 1987, 1988; Singer 1969; Slafter 1879; Cyril S. Smith 1968; H. I. Smith 1906; L. Smith 2002; S. D. Smith 1971:122–124, Fig. 8; Sohrweide 1932; K. Spaulding 1983; Spence 1986:86; Steinbring 1966, 1967, 1968, 1970a, 1970b, 1971a, 1971b, 1974:65, 68–73, 1975, 1980, 1991; Steinbring and

Sanders 1996; Steinbring and Whelan 1970:12, 33; Steponaitis 1986:375; Stoltman 1978:717, 1983:214, 215, 1997:127–133, 2005:75; Storck 1981:11; 14; Stothers and Abel 1993:75, 78, 81; A. B. Stout and H. L. Skavlem 1908:77, 78, 81, 93, 97; G. C. Stowe 1938, 1942; Tanner 1977; Thatcher 1971; Thayer 1940:8, 11, 1944; Theler and Boszhardt 2003:77–80; Thomas 1898:109; Thompson 1968; A. S. Tiffany 1876a; Titterington 1950:26–30; Titus 1914, 1916; Trevelyan 1987, 2003; Trottier 1973:62; Troyer 1955; Van Ert 2015; Van Tassel 1901:32, 49; Vastokas 1970:80–96, 103; Vehik and Baugh 1994:251, 252; Wagner 1995; Walbrandt 1979; Walley 2015; Walthall 1990: 92–93; Walthall and DeJarnette 1974:5, 6–8, 11, 17, 22, 28; Wang 1969; Ware 1917:39–41; G. D. Watson 1972, 1990:17, Fig. 7; Webb 1950:296, 300, 1951:432, 1974:167, 193, 232, 238, 279–281, 316; C. H. Webb 1982:12, 35, 36, 60, Table 14, 67; W. S. Webb 1974:167, 182, 183, 193, 232, 238, 279–281, 316; W. S. Webb and Haag 1947a:22, Fig. 6; West 1884, 1929, 1931, 1932, 1933, 1939; Wiant et al. 2009:262, 263, 265, 266, 268, 274–276; Wiegand 1955; Wilford 1960:48; C. L. Wilson and M. Sayer 1935; T. Wilson 1899:980–981; Winn 1924a:47, 49–50, 1924b, 1942, 1946, 1947a, 1947b; Wintemberg 1928:178, 184, 185–187, 1931:79–81, 110, Pl. X, 1935:231, 237–238; Winters 1968; Wittry 1954:116–117, 1957:214–216; Wittry and Ritzenthaler 1957; Woolworth 1968; Wormington 1964:150–152; Wormington and Forbis 1965:113; Worthing 1915; C. M. Wright 2000; G. F. Wright 1885; G. F. Wright and F. B. Wright:1906; J. V. Wright 1963:3, 1972a:27; Pl. I, Fig. 3, Pl. XIV, Figs. 20–21, Pl. XXV, Figs. 4, 5, Pl. XXXII, Figs. 7, 8, 1981b: Fig 3.

Copper Usage in the Red Ocher Period

Banks 2005; C. E. Brown 1916:73-75, 1824b:68, 73, 75, 76, 78-79, 80-81, 1940c; Cole and Deuel 1937:65-68; Conrad 1986:308; Esarey 1986; Farnsworth and Asch 1986:348- 349; Faulkner 1960; Griffin 1961c:153; Halsey 1972; Herrera 2012; Hothem 2000; G. L. Johnson 1980; Jury 1978; Justice 2006:40-42; M. E. King 1968:117,122; Knoblock 1939:72-73-, 201-210, 334-345; Krakker 1007:9-11; Mangold and Nesius 1983; Martin et al, 1947:291-292; Moorehead 1910: Fig.570; Morse 1959; Morse and Morse 1964; Niehoff 1959; Orr 1922; Penney 1987:43; Pleger and Stoltman 2009:715-718; Quimby 1957, 1960a; Ritzentaler 1965; 8Ritzentaler and Niehoff 1958; Ritzentaler and Quimby 1957, 1960a, 1972:249; Rueping 1944; Schupp 1944:30-31; Seeman 1992b:18-19; Shallenberger 1879; Snyder 1898:22-23, 1908:43, 1909a:53-55, Fig. 3; Spence and Fox 1986:11; Theler and Boszhardt 2003:89-91, 93-95; Wilson and Maginel 1976; Wintemberg 1928:176, 177; Wray 1937, 1952:153, 158.

Copper Usage in the Glacial Kame Period

Anonymous 1950; Anonymous 1977; Baby 1959; Britt 1985; D. Collins 1979:3-4; Converse 1972, 1980, 2003:107-143; Cunningham 1948; Dodds 1950a; Donaldson

182

and Wortner 1995; Dougherty 1956; Dunlap 1961; Ellis et al. 1990:115-119; Ellis et al. 2009:824, 825, 827; Greenman 1931; Griffin 1964:234; Hills 1898; Ireton 1982; Kellar 1993:32; Lilly 1942; Martin et al. 1947:262-263; Matson 1872; Moorehead 1909:138, 141-142; Penney 1987:43; Quimby 1952:101; Savage 1976; Sciulli et al. 1984:141, 143; Sciulli and Aument 1987:122; Sciulli et al. 1993: Table 1; Spence 1986:86; Spence and Fox 1986:11-14; Tharp 2004:49, 54; Vastokas 1970:96; Vietzen 1976:71; Wintemberg 1928:177; Young et al. 1961.

Copper Usage in the Early Woodland (Adena) Period

Anonymous 1991:Pl. 3; Andrews 1880; Baby 1949; Baby and Mays 1959; Bache and Satterthwaite 1930; Berner 1974; Black 1946; Brantner 1947:146; Bravard 1972; Brine 1894:59–61; Britt 1967; Brose 1985:53, 1994a:219–220; L. Carr 1877:15; Carskadden 1968, 1990; Carskadden and Fuller 1967; Carstens and Knudsen 1958; *Central States Archaeological Journal* 1965:94, 1972:38, 1981:144; Childs 1994; D. Collins1979:6, 7; Converse 2002; Dragoo 1951:16, 27–28, 32, 1963; Dresslar and Clay 1972:109; Fecht 1985:10; Fetzer and Mayer-Oakes 1951:12–13; Funkhouser and Webb 1935:96–97, Fig. 16, 1937:201; Greber 2005; Greenman 1932; J. B. Griffin 1952b:356, 1964:237; Hays 1994:84, 85, 86, 150, 272, 275, 277–278, 319, 324, 333, 337; Hemmings 1978:29, 31; Hermann 1948; Homsher 1884; Hooge 1999; Hranicky 2011:11, 20, 35, 40–41, 55, 87, 92; Holzapfel 2005; Jefferies 2004:117–118; Jolly 1976:54; Kellar 1993:38–39; Kellar and Swartz 1971:124–129; Kenyon 1959a:14, 16, Pl. VI; B. B. King n.d.:12–14, 1939:147–151; F. W. King n.d.:10; Knudsen and Radford 1957; T. M. N. Lewis and M. Kneberg 1957:15, Fig. 29a, 35, 41; Macpherson 1879:221; P. Martin et al 1947:265–266; Mayer-Oakes 1955; McBeth 1951; McConaughy 1990:5, 8; McGuire 1899:383; McManus 1952; McMichael 1956; McMichael and Mairs 1969; McKnight 1972; McNeal 1956; McPherson 1952; W. C. Mills 1900, 1902a, 1902b, 1907b, 1912:Fig. 12; Moats 2011; C. B. Moore 1895b; Morgan 1941a, 1952:86–87; Morse and Morse 1983:147, 155, 157; Mortine and Randles 1981; Murphy 1975: (Comprehensive listing in Index), 1984, 1986:63, 70; Norona 1957b, 1962:14, 2008; Norris 1985:130; Norris and Skinner 1985; Oberholtzer 1924; Peet 1884:133; Peter 1873; Pollack et al. 2005; Potter 1971:7; Prahl 1970:104, 106. 1991:77; Pulford 1999, 2009; Putnam 1876:18, 1882:93–94, 97, 100, 105, 108, 110, 111; Rafferty 2005; Railey 1996:95, 96; Rolingson 2004:535; Royer 1961; Schlarb 2005; Schoolcraft 1843, 1845:384, 397, 400–401, Plate II; Schroeder and Ruhl 1968; Setzler 1930; S. M. Skinner 1985:141, 142, 144–146, Figs. 6, 8; A. G. Smith 1955; W. E. Smith 1964:29; Solecki 1953; A. C. Spaulding 1952:263, 266; Starr 1960; E. T. Stevens 1870:370; Stoltman 1978:718; C. B. Stout and C. H. Stout 1970; Thomas 1888:51–53, 57, 1894; Tomlinson 1843; Trevelyan 1987;Van Blair 1981; Vehik and Baugh 1994:253; W. S. Webb 1940, 1941, 1942, 1943a:547, 1843b:591, 594; W. S. Webb and Baby 1957; W. S. Webb and Elliott 1942; W. S. Webb and Funkhouser 1940:232, 261–262; W. S. Webb and

Haag 1947b:73–85; W. S. Webb and Snow 1959, 1974; Weer 1945; R. F. Williamson 1980; J. Wood and J. Berner 1969; K. M. Wood 1992:9, Fig. 7; Young 1910:224–233.

Copper Usage in the Middle Woodland (Hopewellian) Period

Anonymous n.d.; Anonymous 1823; Anonymous 1868; Anonymous 1883c; Anonymous 1855a; Anonymous 1885b; Anonymous 1891; Anonymous 1925a; Anonymous 1925b; Anonymous 1925c; Anonymous 1948; Anonymous 1991:Pls. 13, 14, 40; W. R. Adams 1949; Alex 2000:100–101; Alex and Green 1995:51–62; Aument 1985:71, 73; 1990:213, 215, 218, 233, 234, 242, 392, 393–400; Baby 1961a; 1961b; Bacon 1986; Baker 1981; J. Baldwin 1980, 1982, 1984; Beaubien 1953a, 1953b; Beck 1995; C. E. Belknap 1922:42; Bennett 1944, 1952:115–117; Black 1933, 1934, 1936; Blazier et al 2005:107, 111; Blitz 1986; Blumer 1883; Bohannon 1972:65, Fig. 21; Bowman 2003; Boyle 1897a:55, 1897b:53, Brashler et al 2006:262; Braun 1979:70–72; Braun et al 1982; Brine 1894:70–73; Brookes 1976:6–8, 15, 25, 27; Brose 1974:36, 1979:141, 143–146, 1994a:222–227, 1985:66, 67, 70, 75, 77, 80, Pls. 37–40; Calvin S. Brown 1992:284–285; Charles E. Brown 1905a:219; Charles E. Brown and R. H. Becker 1917:22; J. Brown 1979:213–214, 2004:679; 2006:487; J. Brown and R. Baby 1966:78–80; R. Brown 1944:123; Bunnell 1897:87–90; Burks 2002, 2011:22–23, 27, 31–33; Burks and Pederson 2006:400; B. Butler 1979:151, 153; J. D. Butler 1895; Butts 1998; Calvert and Stephens 1991; Campbell 1816:1–2; Cantwell 1980; C. Carr 2005; C. Carr 2008 a-e; Carr, Goldstein, and Case 2008; Carr, Weeks, and Bahti 2008; Carroll 2001; Case, Carr, and Evans 2008; *Central States Archaeological Journal* 1968:162, 1970:170, 1971:7, 1977:194, 1988:47, 2003:213, Champion 1965: Fig. 47; C. C. Chapman & Co. 1881:126–129; Chapman 1952:142, 1980; Charles et al. 2004:60–62; *Cincinnati Gazette* 1870; Cochran and McCord 2001:13; Coffinberry 1885, 1962; Coffinberry and Strong 1876; G. Cole 1981; H. Cole 1922:99; D. Collins 1979:12, 22; H. Collins 1926:92, Fig. 93; Converse 2003:214–307; Conway 1980:42, 43; Cooper 1933:71, 1959:19, 22–23, 41, Plate IX, B, 1967:8, 9; Cowan 1996; Church 1983, 1987; Clifford 1820:99; Cole and Deuel 1937; Cotter and Corbett 1951:37–39; Cree 1992; Cumming 1958:54, Plate 18; Daniels 1904:14–16; Dawson 1980:66, 1983:73; DeBoer 2004; DeHart 1909:134; DeJarnette 1952:278–279; Delafield 1834:27; Demeter and Robinson 1999; Deuel 1952a, 1952b; Dockstader 1985:63, Pl. 8; Dodds 1950b; Douglas 1936; *Dowagiac Times* 1888; Dragoo 1964; Drake 1815:206, 207; Drooker 1992:197–200; Dun 1886:232; Elrod 1884:114–115; Emerson 1955:45; Espy 1871:10; Essenpreis and Moseley 1984:7; Everman 1971:138; Fairbanks et al. 1946:127; Farnsworth 2004; Farnsworth and Atwell 2015; Farquharson 1876a, 1876b; Faulkner 1968:17–22; Fecht 1961:84, 91, 92; Finlayson 1977; A. K. Fisher 1932; A. W. Fisher 1895; E. B. Fisher 1918:17, 20–21; Fitting 1970:96–99, 113, 1971; 1972:246, 248–249, 1979:110; Flanders 1968, 1969, 1979; Fleming 2004; Flotow 2006:48–49; Force 1879:49; B. Ford 2010; J. A. Ford 1963; J. A. Ford and Willey 1940:122–124;

Further Readings

Fortier 2006:331; Fortier et al. 1989:404; J. W. Foster 1852:375, 1869a:260–266, Pl. 25, 26, 1869b:244, Pl. 23,1874:143–144; Fowke 1888:65, 1892:73, 1896:506–507, 510, 512, 1900:197, 1902, 1928:447–448, Plate 78, 461; Fowke and Moorehead 1895; Fowler 1952a:164, 1957; Fundaburk and Foreman 1957:Pl. 5; Futato 1983:305–306, 429, 1988:34, 39, 45; Garland and DesJardins 2006:231; General Electric Company 1997: Photo Catalog; *Genuine Indian Relic Society* 1978:32–35; Gibson and Shenkel 1988:14, 16–17; Goad 1978, 1979; Greber 1979:32–34, 2005, 2006:91, 97, 99, 102, 2010:342–344; Greber and Ruhl 1989; J. B. Griffin 1941a, 1941b, 1952a:229, 1952b:360, 1964:239, 242, 244, 249, 1966a:58, 220–221, 1984; Griffin, Flanders, and Titterington 1970; Hall 1979:258, 2006; Haymond 1869:199; Heestand 1973; Heintzelman 1974; Hemmings 1984:44, 46; Henderson 1872:644–645, 1884:690–691, 695; Henriksen 1965; Henshaw 1890; Herold 1970, 1971; Hesselberth 1946:10; Hewes 1949:325; Hiestand 1951:18, 34; Hildreth 1837, 1842:342, 1843:248; G. Hill 1878:263, 266, 1880:14, 16, 17; N. N. Hill 1881:529; Hinkle 1984:22–24, 27–30, 31–36, 38, 42–43, 46, 48, 52, 54, 56–57, 59–60, 62, 65–68, 70, 72, 77, 81–86, 92; Hodge 1912a, 1912b, 1913:111–112; Hohol and Beck 1985; Holmes 1896;36–37; Hooge 1999; House 1996:142, 143; Hurley 1974:114–115, Fig. 29; Irwin 1900; Jefferies 1979:164, 166–168, 2004:121–123; 2006:161, 169, 172, 175; Jenkins and Krause 1986: 49, 58, 59, 70; Jenkins 1979:175, 178, 180; Jennings 1952:262–263; Jensen 1968:28–32; R. Jeske 1994:9, 2006:295; Jeter 1984:95–96; A . Johnson 1979:91–92; G. Johnson 1969; Johnston 1968a:16, 21, Pl. V, Plate VIII, 1968b:21, 25, 74; M. Jones 1980; Justin 1995a:Pl. 7–13, 1995b:Appendix G; Keel 1976:71, 149; Kellar 1960, 1979:102–103, 105, 1993:43–48; Kellar et al. 1962a:65–68, 71; Kellar et al. 1962b:Fig. 3, 344, 352–354; Kennedy 1970:64–70; Kent 1991; Kenyon 1986; H. F. Kett & Co. 1878:844; Kidd 1952:79; Kidder 2004:549–551; Kime 1985; B. B. King 1937:87, 88, 1939:147–150; V. J. Knight 2004:735–736; Korp 1990:20, 25; Krakker 2012, 2015:135; Lambert 1982:166; Langford 1927:178–179, Pl. XV; Larsen and O'Brien 1973; Laval 1923:137–138; Leeson 1881:126–129; Lepper 1989:125, Fig. 5; 1995, 2002; A. Lewis 2003; C. M. Lewis 1970:5, 7, Figs. 2, 6; R. B. Lewis 1996a:96, 98, 101, 103, 106, 109; T. H. Lewis 1896a:209, 1896b:317; T. M. N. Lewis and M. Kneberg 1958:65–66; Lilly 1937a:202–210; Lindley 1877; Lloyd 1998:6; B. Logan 1993:75, Fig. 7.10, 151–152, 2006:342; W. D. Logan 1971:38, Fig. 3, 40–41, 1976: (Comprehensive listing in Index); Lucas 1973:91–92; J. P. MacLean 1903:48, 57–58; MacNeish 1952:49; Magrath 1940, 1945, 1959:94; Mainfort 1986:65, 1988a:165, 1988b:43, 45, 1996a, 1996b:82; Mangold and Schurr 2006:211, 214, 223, 225; Markman 1988:271–272, Plates 11.2, 8–11; Marsh 1866:3; L. H. Marshall 1885; Martin et al: 1947:270, 274, 293, 294, 326, 352, 392, 410; O. T. Mason 1868, 1880:431, 432, 434, 1881:448; Matheny 1905:85, 93; Matson 1872:10, 12, 14; Maxwell 1952:185; McAdams 1882:58–64, 1884, 1895:258–267; McAllister 1932:14, 19, 26, 48, 53, 58–59, Plates 9, 27; McBeth 1944, 1954, 1960; McClurkan et al. 1966:7–8, Figs 9–10, 21, 23, 24; McClurkan et al. 1980; McConaughy 2011; McGee and Thomas 1905:358–364; McGregor 1952:66–67, 1958:58–59, 152–153; McGuire 1899:515, 519; McKendry

Appendix 5

1993; McKern 1929a, 1929b, 1931a, 1932; McKinley 1977:4, 6, 8, 15, 16; McKinney 1954:17; McKusick 1964b: 113, 115, 131; McMichael 1964; McNutt 1996a:212, 216, 1996b:171; Merriam 1923; Metz 1911; Meyer and Hamilton 1994:103; Milanich 1999, 2004:193, 194; Milanich and Fairbanks 1980:84, 86; T. B. Mills 1919:172, Figs. 4, 6; W. C. Mills 1899, 1900, 1907a, 1909a, 1909b, 1916, 1917a:402–404, 1917b, 1921:152, Figs. 30, 31, 1922; B. Mitchell 1880; E. C. Mitchell 1908:306–311; Moorehead 1887:298; 1889a:506–507, 1889b:835,836, 837, 838, 1890 a-d, 1892 a-f, 1894, 1895 a-c, 1896, 1897 a-h, 1898, 1899, 1903, 1910: Figs. 572, 573, 608–610, 612, 615–616, 619–621, 2000: (Comprehensive listing in Index); Morgan 1941b:386, 1946, 1952:89–91; R. W. Morris 2004, 2006, 2008; Morse 1963: 35–36, 65, Plate XI, Fig. 2; Morse and Morse 1962:156, 1965, 1983:162–163, 167–168; 1996:11; Mortine and Randles 1978; Morton and Carskadden 1983, 1987; Moses 1878:37, Pl. 2; Muller 1986:96, 115, 117–118; Murphy 1975: (Comprehensive listing in Index), 1978, 2001, 2008; Myron 1954; Nansel 1977:373; Neiburger 1989; Neuman 1984:142, 162, 209; Ochsner 1972:20; Osburn 1946:16, 19–20; J. T. Patterson 1937:47–48, Pl. 25; Payant 1929; Peet 1881:155, 1884, 1896, 1906a, 1906b, 1909; Peithman 1947:52, 53, 54; Penney 1985b:148, Figs. 24, 25, 32; Perino 1958:185–186, 1967:146, 149, 1968a, 1968b, 1969:51, 52, 1972, 1973:26, 1992; R. W. Perkins 1965; Perrin 1873; G. B. Phillips 1926:Figs. 1–5, 7; T. E. Pickett 1875:18, 36–37; Porter and McBeth 1960; du Pouget 1890:165, 167, 172, 174–176, 178; Power 2004:(Comprehensive listing in Index); Prahl 1970, 1991:87–91, 93, 106–107; Pratt 1877b:46, 1883a:90, 1883b:155, 1886:18, 19, 24; Priest 1835:181–183, 268–270; Prufer 1963:134, Tafel II., 1964; Pulford 1999, 2009; Putnam 1876:19,1881, 1882:95, 106, 1883c, 1886a, 1886b:508, 1886c, 1886d:495, 1888; Putnam and Willoughby 1896a, 1896b; Quimby 1941, 1944, 1952:102–103, 1959b; Railey 1990:251, 260, 275–276, 279, 324, 325, 1996:101, 103, 106, 109 ; Randall 1916:296, 300; Rau 1886:103–107; Rayner 1910:5; Read 1875:39, 1892:53–54; Read and Whittlesey 1877:107–112; Redmond 2007a:5, 12–13, Figs.4, 6, 17, 2007b:Fig. 5, 198, Fig. 7, 220–222, 225–226; Reeb and Carskadden 1982; H. A. Reid 1881:406–407; Reifel 1915:71; E. Richards and Shane 1974; J. F. Richmond 1907:448; M. D. Richmond and Kerr 2005; D. K. Riddle 1980:156, Fig. 85, b; 1981:225–226, 233, 262–264, Figs. 3, 16d, Figs 4, 20F; Riley 1979; Robertson 1875a:379, 1875b:380–381; Rolingson 2004:535, 537; Romain 2000; Rothschild 1979:666; Rowe 1984; Royer 1959; Ruby 1997:43, 337–338, 344–347, 354, 356–357, 2006:200; D. L. Ruhl 1981:198–201, Figs. 57–67; K. C. Ruhl 1992, 1996, 2005; K. C. Ruhl and Seeman 1998; Rummel 2006, 2009a, 2009b; Rummel and Balazs 2006; Rusnak 2009, 2010, 2011; Salzer 1969:Figs. 26, 35, 47, 66, 121–122, 165–166, 186, 222, 267–268, 290, 312–313, 346–348, 351-352, 1978:48–49; Sanford 1913; Sargent 1795, 1799; Schambach 1982:147, 1997:55, 57–58; Schoenbeck 1941, 1947, 1949:45; Schroeder and Ruhl 1968; C. Schultz 1810:155; Schurr 1997:128, 131, 136; Scoville 1878; Sears 1962:6, 7, 8, 1964:262, 265, 267, 268; Seeman 1979:311–325, 1985:22, 23, 1986:570; 1992a:12, 107, 1992b:14, 21, 28–29, 1995; Seeman and Cramer 1982; Seeman and Soday 1980; F. C. Sharp 1954; M. R. Sharp 1952; H. A. Shepherd 1887:69–70, 71–72, 75–76, 79, 83, 92, 94–97, 115–116,

118; Shetrone 1924:349, 1926, 1930, 1938; Shetrone and Greenman 1931; Shippee 1967:83, Fig. 51,c; Short 1879:82, 85, 87, 88; Shriver 1982, 1983; Sibley and Jakes 1986; Sieg 2005; Sieg and Burks 2010:57–59, 69; Skinner 1987; B. A. Smith 1979:182, 184, 187; G. P. Smith 1996:108; H. I. Smith 1893; R. Smith 1966:8; W. Smith 1961; W. E. Smith 1964:37, 45–47; Smucker 1875:103, 104, 116, 1881:270; Snyder 1893:183, 1895a:79–81, 1895b:109–110, 1898:17–19; Sohrweide 1932; Spence 1967; Spence et al. 1979:117–121; Spence et al. 1990; Spielmann 2009; Squier 1850; Squier and Davis 1848; Stafford and Sant 1985; F. Starr 1887, 1888, 1893, 1897; S. F. Starr 1960:23, 35, 44, 46, 49, 55; Steinen 2006:180, 183; Steponaitis 1986: 381; Stevens 1870:358, 359, 364, 366; Stevenson 1879:94, 95, 96–97; Stoltman 1973:17, 30, 1978:721, 1979:125–126, 128–130, 2006:326; Story et al. 1990:280, 284, 287, 288, 290, 292; Stothers 1974:24; Struever 1964; Struever and Houart 1972; Syms 1977:79, 80–81, 84; Tankersley 2008:282–283; Theler and Boszhardt 2003:114–116, 119; Thomas 1884c, 1886, 1888:15, 25, 31, 35, 46, 48–53, 57, 1894 (Comprehensive listing in Index), 1898:110; Thornberry-Ehrlich 2010:13; Thunen and Ashley 1995:5; Tichenor 1910:34; A. S. Tiffany 1876b; Till 1977:193, 196, 200, 201, 203; Tomak 1983:74, 75, 1994; Tomak and Burkett 1996; orbenson et al. 1994:439, 1996:76–77; Toth 1979:194, 1988:51–56; Van Gorder 1916:260; Van Tassel 1901:32, 37, 49, 54, 60–62; Vehik and Baugh 1994:254, 255; Vickery 1979:62; Vietzen 1965: Fig. 128, 1976:59, 60; Walthall 1979:200, 202–203, 205–208, 1985, 1990:118, 125–127, 131, 145, 146, 152, 154, 158, 159, 171, 182; Waring 1945; Warner, Beers & Co. 1883:227; Wayman et al. 1992; W. S. Webb 1943a:528–531; W. S. Webb and D. DeJarnette 1942:27, 29, 30–32, 93, 154, 155, 157–158, 162–163, 166, 169, 172, 217, 227–229, 297–298, 304–305; W. S. Webb and C. G. Wilder 1951; Wedel 1938:104, 1940:308, 311, 1986:91; Weinland 1910; Weisman 1995; Whitman 1977; Wied 1906:148; Wilford 1950a:165, 1950b:232, 1955:130, 1960:50, 52; Williams Brothers 1880:28, 30, 396; H. Z. Williams and Bro. 1881:16; R. Williams 1968:92; Williamson 1887; Willoughby 1903, 1917, 1935, 1938; Willoughby and Hooton 1922; R. L. Wilson 1965:19–20, Pl. II, P, Q; T. Wilson 1893:138, 1898:499–503; S. B. Wimberly and H. A. Tourtelot 1941:5, 8, Fig. 10, d, e; V. S. Wimberly 2004; Wintemberg 1928:175, 176, 185; Winters 1981; Wolf 1916; E. F. Wood 1936:219; J. A. Wood and J. C. Allman 1961; Woods 1855:665–666; Wray 1952: 154–156, 159; A. P. Wright 2014:286; J. V. Wright 1963:4, 1981b:90, Fig. 4; Wyman 1869:12–15, 1872:21; Wymer 2004; B. Young 1910:224–233; G. Young 1970, 1976; Zakucia 1992.

Copper Usage in the Late Woodland, Upper Mississippian, Fort Ancient and Oneota Periods

Alex 2000:201; Benn and Green 2000:439, 481; Bennett 1952:114, 118, 123; Bluhm and Liss 1961:Fig. 66; Bluhm and Wenner 1956; Boyle 1888:56, 1889a:49; Brashler et al. 2000:558–559, 566, 568–569; Brose 1985:85, 1994a:230; C. Brown and Skavlem 1914:66, 72–73; J. Brown et al. 1967:35; Bruhy et al. 1998:Fig. 12, 26, 1999:36, Fig.

18; Bullock 1942:38; Caldwell and Jensen 1969:68; Clark and Martin 2005; Conway 1977:55–58; Coté 1993:32; H. A. Davis 1958:2–3; DeVisscher 1957; Dirst 1985:99–102; Dorwin 1971:239; Faulkner 1972:138; Fowler 1952b:56; Gibbon 1973:8; J. B. Griffin 1966b:67, 89, 107, 128, 150, 189, 201, 282, 166–167; R. Hall 1962:141–146, Plates 79–81; Halsey 1981; Halsey and Brashler 2013; Harvey 1979:67, 91–92, 123, 137, 138, 140, 156, 177; Herold et al. 1990; Hill and Neuman 1966; Holtz 1992; Hurley 1975:287–288, Plate 44; J. A. Jeske 1922:24, 33; 1927:179–180; E. Johnson 1973:13, 40, Pl. 25; Langford 1927:Pl. XV; Lawshe 1947:81–82; A. Lewis 2003; R. B. Lewis 1996b:58; Lindeman 1967:86; Lovis 1973:188–190, Fig. 47 (192–193); J. A. MacLean 1931:113, 114, 116, 124, 125, Plates 23, 37; Martin et al. 1947:279, 280, 283; McKern 1927:49, 1928:Plate L, 1945:137–138; McKusick 1964b:131, Fig. 8.8; McPherron 1967; Meyer and Hamilton 1994:113–114; Miller 1922:23; Mills 1917a:402–404; W. D. Moore 1973:17; Moorehead 1910: Fig. 571; Morgan and Ellis 1943:28; Muller 1986:123; Nassaney 2000:716, 720; O'Brien and Wood 1998:224, 283; O'Gorman 1989:26, 1994:83–87, 1995:167–169, 191–192; Overstreet 1976:126, 138–140; Pasco and McKern 1947; Pasco and Ritzenthaler 1949; Pauketat 1988; Perino 1971c:211–212; Prahl 1991:110; Pulford 1999, 2009; Putnam 1882:87, 89, 91, 96, 112–117, 126; Quimby 1952:104; Redmond and McCullough 2000: 652, 657, 660; C. S. Reid 1975:33; D. K. Riddle 1980:177, 178, Fig. 78, f; Rowe 1956:57–58; Salzer 1974:49–50; Santure et al. 1990:103; Scheffers 2004:17–18; Seeman 1985:24, 1992b:29; Seeman and Dancey 2000: 599, 601; Steponaitis 1986:384; Stoltman and Christiansen 2000:500, 503, 504, 511–512; Stothers and Bechtel 2000:7; Syms 1977:99, 113, 123; Theler and Boszhardt 2003:160–163, 169–170, 178; Titus 1915:20–21; Vastokas 1970:98–101; Vehik and Baugh 1994:256; Vickers 1945:91; M. M. Wedel 1959:63–75; W. R. Wedel 1961:118, 1986:112–113, 215; Wickersham 1885:835; Wilford 1960:58, 60; T. Wilson 1896:Pl. 12, 13, 889–891; Wittry 1959:110–111, Fig. 7b; Wolbach 1877; W. R. Wood and S. L. Brock 1984:18, 33, 41, 51, 66, 101, 112, 2000:19, 33; Wray 1952:159; J. V. Wright 1967:69, 87, 89–90, 1969:19, 29–30, 43; Pl. V, 18, 19, Pl. XII, 17–22, Pl. XIV, 11, 1981b:Fig. 6, 95.

Copper Usage in the Middle Mississippian/Caddoan Periods

Anonymous 1881; Anonymous 1956; Adam and Duncan 1996; Adams 1972; Alex 2000:167; Ashley 2005:269–270; Ball et al. 1976:31–33; Barrett 1933:344–346, 348–350, Pl. 69; L. Belknap 2008; Bell 1947:182, 1972; Bennett 1952:121; Blitz 1993; Bohannon 1972, 1973; Brain and Phillips 1996:362–375; Brannon 1931; Brooks 1945; Brose 1994a:231; C. Brown 1982; C. E. Brown 1932; H. Brown 1980; J. A. Brown 1966, 1983:131, 142, 152, 1985:99, Figs. 12, 21, Pl. 101,1989:186–188, 200–202, 2004:679, 2007a:77, 78, 79, 82–83, 2007b:202, 221, 226–229, 237–240, 24; J. A. Brown and D. H. Dye 2007; J. A. Brown and H. W. Hamilton 1965:44–45, Pls. 10, 11, 12, 23, 24, 25, 26; J. A. Brown and J. E. Kelly 2000; J. A. Brown, R. A. Kerber,

Further Readings

and H. D. Winters 1990; J. A. Brown et al. 1967:35, 1996; Burnett 1945; Butkus 1973; Byers 1962, 1964; Cadle 1902; Caldwell and McCann 1941:16; *Central States Archaeological Journal* 1968:116, 1970:18, 1981:41, 1982:112, 1989:46, 2004:163, 2006:65, 2008:127; C. H. Chapman 1980; J. Chapman 1982:34, Fig. 59, 69, 1985:80; Chastain 2009; Chastain et al. 2011; Cherry 1877; W. M. Clark 1878:272–273; C. Cobb 1991:62–64; D. Cobb 1999:11–12; Coe 1995:236–238; Conant 1877:41–42; Conrad 1972:116–119,1989:102, 104, 108,110, 113, 1991:128, 132, 140, 141, 143–149, 153, 155; Cook and Pearson 1989:150–152, 155; Cotter 1952; Cox 2014; Crumpler 1969:14–16; DeJarnette 1952:281–282; DePratter 1991:153; DiBlasi and Sudhoff 1978:36; Dickens 1982: Exhibits 100, 109, 110, 111; Dickson 1956:92; Dille 1867:360; Douglas 1936; Douglass 1912:11–12; Drooker 1992:76,9 3–94, 167, 2001:177–178; Duncan 1967; Duncan and Diaz-Granados 2000:2–8; Durham and Davis 1974:14, 17–18, 28, 42,-43, 50–51, 54, 59, 79; Dye 2006:110, 116–117, 127, 131–132, 2007:152; Early et al. 1988:9, 129–130; T. Emerson 1997:34, 111 , 147, 168, 181, 226, 240, T. Emerson and J. Brown 1992:88, 95, 99; Emerson and Hargrave 2000:8–9; Emerson et al. 2010:54, 56; Emerson and Hughes 2001:150, 156; Emerson and Pauketat 2002:115; Esch 1984; Evans 1977:114; Fairbanks 1952:291, 295, 1956:31–32, 58, 70; Feagins 1988:45; Featherstonhaugh 1899:7; Finkelstein 1940:6, 9; Finney 2013:154–155; Fowler et al 1999:136–137; Fraser 2013; Fugle 1957:276; Fundaburk and Foreman 1957:Pls. 9, 29, 30, 31, 107–110; Galm 1981:146, 230, 232, 233; Gardner 1980; *Genuine Indian Relic Society* 1978:50–51, 55; Goggin 1947, 1949, 1952:123–124, 146, Pl. 9; Goodman 1984; J. B. Griffin 1952a:232, 235, 1964: 279; J. B. Griffin and D. F. Morse 1961; J. W. Griffin 1952:325; R. Hall 1989:241, 256; Hamilton 1952; Hamilton et al. 1974; Harrington 1920:223–226; Hatch 1974:97, 130–132, 134, 150, 152, 156–157, 160–161, 163–167, 171, 173–174, 176–177, 182, 187–188, 192–194, 201, 207, 220, 221, 224, 231–232, 250, 276, 1976a: 82–84, 97–98, 1976b:152, 169, 176; Hatchcock 1976; Hess 2014; Heye et al. 1918:Pls. VII, VIII, 16, 22–24, 96–97; F. W. Hodge 1912a, 1912b; Hohol and Beck 1985; Houck 1908:395–404; Howard 1968:64–75; Howland 1877; Hudson 1990:226–227; Jenkins and Krause 1986: 86, 87, 88, 94; Jennings 1941:203; B. C. Jones 1994:128–136; C. C. Jones 1873; J. Jones 1880:8, 45, 59, 137; W. B. Jones and D. L. DeJarnette n.d.:5, 6; J. E. Kelley et al. 2007:78–79; J. E. Kelley et al. 2008:300, 313, 314; A. R. Kelly 1970:30, 68–69; A. R. Kelly and L. H. Larson, Jr. 1956:43–44, 46–47; A. R. Kelly and R. S. Neitzel 1961:30, 52–53, Pls. IV-A, VIII-A; J. E. Kelly et al. 1990:188, 224, Pl. 4.6, d; J. E. Kelly 1991a:64, 65, 72, 73, 76, 1991b:69, 71, 1994:24, 27; A. King 2003:68–69, 75, 126, 2004:160–165; Kneberg 1952:191–192, 196–198; V. J. Knight, Jr. 2004:737; V. J. Knight. Jr. and V. Steponaitis 1998:17, 2011:203, 219–223, 227–228; Koldehoff and Kinsella 1995; Kowaleski 1996:31–32; Krieger 1945, 1946:177, 183, 184; Ladassor 1950:35, 37; Lankford 2007a:119, 121–122, 2007b:28–29; Larson 1954, 1957, 1958a:17, 1958b, 1959, 1971, 1989:139–141, 1993:171, 174–178; Latchford 1985; Lemley 1936:53, 54; R. B. Lewis 1996a:148; T. M. N. Lewis 1950; 1951:64, 1958:89; T. M. N. Lewis and M. Kneberg 1956, 1958:106–109; 1961, 1970:131; T. M. N. Lewis

189

and M. Kneberg Lewis 1995:20, 172–173; Ludwickson et al. 1993:163; Marshall 1992:61–64; Martin et al 1947:356, 360, 363, 366, 382, 424; Martoglio et al. 1992; Maxwell 1952:189; McAdams 1881:715, 1882:58–64, 1895:258–267; McDonald 1950; McGee and Thomas 1905:358–364; D. H. McKenzie 1965:168, 170, 1966:21–28; McKinley 1977:6, 8; McKusick 1964b:142; McNutt 1996a:230, 245, 256; Melbye 1963:13, Photo 18; Merriam and Merriam 2004:Figs. 30–54, 2005; Milanich 1994 (Comprehensive listing in Index), 2004:198–199; Milanich and Fairbanks 1980:164, 198, 199; Milner 1984:475, 480, 482, 1985:203, 1990:13, 1998:131, 135, 151; Mitchem and Hutchinson 1987:32–35; Montgomery 1906:644, 645, 646, Pl. XXXIV, 649, 1908:36, 1910:51, 52, 53; Moore 1894a, 1894b, 1895a, 1895b, 1896a, 1896b, 1896c, 1897a, 1897b, 1899, 1902, 1903a, 1903b, 1905a, 1905b, 1905c, 1907, 1909, 1910, 1911, 1912, 1915; Moorehead 1910:Fig. 617–618, 1924, 1925a, 1925b, 1932, 1939, 2000: (Comprehensive listing in Index); Morse 1986:74–76, 89; Morse and Morse 1961:126, 128, 1983:(Comprehensive listing in Index), 1989:43; Neuman 1984:226–232, 276; M. J. O'Brien 1996:50–51; M. J. O'Brien and Holland 1994:242–243; M. J. O'Brien and Marshall 1994:168, Fig. 5.22; M. J. O'Brien and Wood 1998:287–288, 309, 337, 341, 352; P. J. O'Brien 1972:191, 1988:34–35, 1992:401; O'Connor 1995:(Comprehensive listing in Index); K. G. Orr 1941, 1946:231, 236, 238, Pl. XXIX, 1952:249; Pauketat 1994:83–84, 95–96, 155, 1997:4, 10, 2004: (Comprehensive listing in Index); Payne 1994:306–307, 310–311, 313, 2002:202–203; Peebles 1974:68,77, 93, 95–96, 120, 121, 130, 131, 133, 139, 145, 159, 161, 162, 163, 172, 175, 177, 178, 180, 182, 183, 185–189, 1983:189, 191, 192; Penney 1985b:148–149, Pl. 116–120; Pepper 1917; Perino 1940:11, 1959:132, 138, 1966a, 1966b, 1968b, 1971a:130–132, 1971b:158–164, 1971d:83–86, Figs. 1 and 2, 1976, 1985:48, 1986a, 1986b, 2002; Perttulla 1996:311, 312; G. B. Phillips 1923, 1925a:Plate 1, 1925b:Figs. 1–4, 1926:Fig. 6; Phillips and Brown 1978:185–208; Poehls 1944; Polhemus 1987:Figs. 6.9, 6.12, 6.15, 6.17, pp. 822–825; Power 2004: (Comprehensive listing in Index); Pulford 1999; Putnam 1882:85, 98, 99, 102, 121, 122; Reilly and Garber 2007:2, 3, 4, 5, 8; J. D. Richards et al. 2012:98, Fig. 22, 101; T. T. Richards 1871; Robb 2010; Rodell 1991:275, 1997:415–416, 488; Roedl and Howard 1957:72, Fig. 19:13; Rolingson 1961:53, 1990:35, 2004:540, 543; Rudolph and Halley 1985:314–315, 318, 470–471, 533; Rust 1877:534; Salo 1969:47, 51; Sampson 1991; Sampson and Esarey 1993; Sank and Sampson 1994:6; C. M. Scarry 1998:79, 84; J. F. Scarry 1990:181–182, 1992:178–179, 1999; 2007; Schambach 1982:152, 1997:65; Scharf 1883:97; Schoolcraft 1819:198; Schreffler 1988; Schroeder and Ruhl 1968; Sears 1951:3, 4, 1953a:11–13, 29–22, 85–86, 87–91, 1953b:224, 225, 228,1956:28, 1959:9,1964:278, 280; Seeman 1985:25; Shaeffer 1957:95, 98; Shippee 1956:8, 1972:10; Shiras 1965; Short 1879:94; Sibley and Jakes 1986, 1989, 1994; Sibley et al. 1986; Sibley et al. 1996; Simpson 1934, 1937, 1952:72; Skinner et al. 1969:101–102, 106; Sly 1958:Fig. 50, 91; H. G. Smith 1951:19; K. E. Smith 1992:183–187, 1994:97–98, 107; K. E. Smith and M. C. Moore 1999:107; M. T. Smith 2000: 5, 15, 26, 29, 44–45, 90, 94, 112, Figs. 1:A, 2:A, C; Snyder 1908:40, 1909b:82; Springer 2007; Stack 1946; Stelle 1871:410, 413, 414;

Further Readings

Stephenson 1871; Steponaitis 1986:389, 391, 1991:208, 209, 221, 222; Stirling 1935:379, 393, Plate 7; Stoltman 1978:727; N. R. Stowe 1989:128; Strezewski 2003:57, 58–59, 78, 208–212, 236–238, 242–244, 250–252, 285,334, 352, 359, 360, 377, 403, 417, 441; Strong 1989:212, 214–215, 224–226, 230, 232–233; Struever 1968:22, 209, 246–247, 280–281, 287–288, 310–313, 349–352, 359–360, 367–368, 373–375; Suhm and Krieger 1954:165, 174, 181, 202, 208, 214; Syms 1977:120; Thoburn 1926:146–147, 1929:216, 225, 226, 1931:58, 68; C. Thomas 1884a, 1884b, 1885, 1888:64–71, 75, 93–95, 98–106, 1894:153, 324, 1898:110–113; Throop 1928; Thruston 1888:390, 1890:79, 80, 169, 171, 298–304, 342–346, 1892, 1897:99, 1904:137, 139; Thunen and Ashley 1995:6–7; J. A. Tiffany 1991:187, 189; Titterington 1938:11, Figs. 29, 48, 1940:21; Trevelyan 1987, 2003; Trocolli 2002:171–172, 176–177; Vavak 1967; Vietzen 1976:142–143, 153; Walthall 1990:190, 200, 209, 217, 228, 230, 232, 233, 240, 241, 242, 244, 245; Wardle 1906; Waring1968:35–36, 39, 40–43, 47, 50, 59; Waring and Holder 1945; V. D. Watson 1950; C. H. Webb and M. Dodd, Jr. 1939; C. H. Webb and H. F. Gregory 1986:5–9; C. H. Webb and R. R. McKinney 1975:52, 53, 54, 64, 103, 104; W. S. Webb 1938:109, 111; W. S. Webb and DeJarnette 1942:51, 93, 115, 116, 118, 122–123; Welch 1990:208, 214, 218, Whiteford 1952:211, 216–218, 220; Wilkins 2001; S. Williams 1980: Fig. 2, 109; S. Williams and J. P. Brain 1983:267; S. Williams and J. M. Goggin 1856; L. A. Wilson 1981; T. Wilson 1896:886–887, 1898:499–503; Winters 1974:42; 1981; W. R. Wood 1963:4, 6, 8, Fig. 3, 1967:13, 114; Wray 1952:157, 161; Wyckoff 1974:51, 55, 65, 70, 82, 83, 94, 115, 1980:345, 350, 354, Table 62, Fig. 46, 373, Fig. 49, 393, Table 69, 397, 398, 411–412, 2001.

Copper Usage in the Late Prehistoric/Early Historic Periods

Author's note: Some of the metal objects cited here may be brass rather than native copper. At the time of these publications, it was unclear to many of the authors how to distinguish the metals.

Anonymous 1883d; Anonymous 1885a:15–16; Abel and Burke 2014; Anselmi et al. 1997; Battles 1969; Battles and Battles 1972; Boyle 1891:60–62, 1892:51; Brannon 1931; Brose 1970b; 1994b:118; I. Brown 1985:86; Buchanan 1986:287, 291,292–293, 295, 297, 298–299,302, 305, 307, 308, 309, 311–314, 322–323; Channen and Clarke 1965:16–17; Cornett 1976; Crowder 1988:57; Curren 1984:23, 170; DePratter 1991:153; Drooker 1995:5, 8, 10, 12, 1997:153–170; Eastman 2001:63, 65, 2002:50–53; Ehrhardt 2013; Ehrhardt et al. 2000; Feldser 1982; Finlayson et al. 1998: 204, 215, 275, 226, 275, 284, 331, 344–345, 360, 385; Finlayson and Pihl 1980:9; Fitzgerald 1990; Foster 1902; Fox 1991, 2004; Fox et al. 1995; Goggin 1954; Greer 1966; Hamilton et al. 1974:168–170; Herrick 1957; Hoffman 1977; Hooton and Willoughby 1920:69–71; Howard 1953:134; Jolly 1969, 1975; Kannenberg and

Stowe 1938:50; Kidd 1953; V. J. Knight, Jr. 1982; Kuhm 1928; Lehmer 1954:37–38, 40, 73, 114; Little and Curren 1981:129; Loughridge 1855; Low 1880:56; Manson and MacCord 1941, 1944; McCharles 1887:72, 74; Miner 1941:59, 69; Mitchem 1989, 1996; Mitchem and Hutchinson 1987:55–58; Mitchem and Leader 1988; P. A. Morse 1981:Fig. 10; Moxley 1988:40; Munson and Hancock 2001; Myers 1964; Noble 1971:46, 2004:184; Norona 1957a; O'Brien and Williams 1994:273, Fig. 7.11; R. B. Orr 1911:10–11, 1912, 1913; Overton 1929, 1931 (Age/ages of these sites are uncertain); A. C. Pickett 1851:85; Pollack 1998:209, 266, 268, 279, 285, 296, 407, 416, 417, 2004: (Comprehensive listing in Index); du Pouget 1890:177; Pulford 1999, 2003, 2009; Putnam 1878:307–308, 343–344; Quimby 1963, 1966:36–44; Radin 1911:520, 1923:97; Schoolcraft 1853; Seeman 1985:26, 27; W. E. Sharp 1990:485, 489, 494; A. Skinner 1920; C. S. Smith and R. T. Grange 1958:110–111; H. G. Smith 1956:13, 16, 17, 26, 32; M. T. Smith 1987, 1989; Sohrweide 1932; F. Starr 1887:362, 1897:99; Stothers and Abel 1991; W. D. Strong 1935:114; Swanton 1911:286, 328, 345, 1928a:685, 709, 1928b:503–510; Syms 1979:293; Trevelyan 1987; Trigger 1976; Vehik 2002:47, 48, 55; Vehik and Baugh 1994: 257–260, 263, 264; Vickers and Bird 1949; Vietzen 1974:21, 62, 80, 81, 88, 91, 92, 116; Wedel 1959:63–75; Willey 1998:124, 145, 235, 240, 241, 243, 266, 270, 271, 277, 281, 292, 298–299, 308, 318–319, 326, 334, 336, 344, 394, 467, 487; M. H. Williamson 2003: (Comprehensive listing in Index); Wintemberg 1900:85, 1926:41, 1948:5; J. V. Wright 1974:111, 132, 144, 157, 168, 178–179, 198–199, 206, 220, 246, Pl. III, 10, Pl. VI, 11, 12, Pl. IX, 20, Pl. X, 8, Pl. XI, 19, 21, Pl. XV, 15–17, 1981a:51, Fig. 2:25, 27.

Bibliography

Bibliography

Author's note: It was a fairly common practice in early nineteenth-century newspapers to publish letters or articles with the author using only initials, sometimes three, sometimes two and occasionally only one. In a few cases, for a variety of reasons, we can make an educated guess as to who lies behind the initials, e.g., C. W. is Charles Whittlesey. However, it is impossible at this time to assign most initials (or nicknames) to real people. To standardize this situation, if an individual chose to be identified as A. B. C., he would be alphabetized under C.

Anonymous

n.d.   *A Description of the Exhibit from M. C. Hopewell's Farm, Ross County, Ohio.* (No authors or place of publication given. It was probably Warren K. Moorehead).

1823   Indian Antiquities in Ohio. *Collections, Historical and Miscellaneous: and Monthly Literary Journal* 2:47–48.

1850   From Lake Superior. *Detroit Free Press* [Detroit, Michigan], July 2, p. 2.

1868   Indian Relics. *Proceedings of the American Antiquarian Society, at the Semi-Annual Meeting, held at the Hall of the American Academy of Arts and Sciences, in Boston, April 29, 1868*, pp. 45–50.

1881   Relics of the Mound Builders Near Joliet, Illinois. *The American Antiquarian and Oriental Journal* 3(2):155.

1883a   An Extensive Find of Copper Implements. *The American Antiquarian and Oriental Journal* 5(1):83.

1883b   One of the Old Homesteads of Jackson. *Michigan Pioneer and Historical Collections* 4:281–283.

1883c   Mound Explorations in the Little Miami Valley, Ohio. *Science* 1(17):496–497.

1883d   Report of the Peabody Museum. *Science* 1(11):308–309.

1885a   American Shell-Heaps and Aboriginal Mounds. *Appletons' Annual Cyclopaedia and Register of Important Events of the Year 1884*, pp. 14–16.

1885b   Archaeological Researches in Rush County Ind. *The Hoosier Mineralogist and Archaeologist* 1(10).

1890   Description of Some Copper Relics of the Collection of T. H. Lewis in the Macalester Museum of History and Archaeology. *Macalester College Contributions: Department of History, Literature, and Political Science* 6:175–181.

1891   Untitled. *Science* 18(460):298–299.

1892   Prehistoric Copper Implements. *Popular Science Monthly* 42:137.

1896   Bay View Sketches: How the Place Was Discovered. *The Bay View Magazine* 3(4):159–160.

1897a   Old Copper Mines: Seven of Them Discovered on the North Shore: Still Rich in Ore and Near By Is a Mound That May Contain Valuable Relics. *Toronto Daily Mail*, January 2, p. 10.

1897b   Ancient Mines. *The Mineral Collector* 3(12):178–180.

1902   A Copper Pickax Found. *The American Antiquarian and Oriental Journal* 24(5):421.

1925a   Disappearance of a Famous Indian Mound. *Indiana History Bulletin* 2(9):213–214.

1925b   An Archaeological Find. *Indiana History Bulletin* 3(1):16.

1925c   Pearl Burial Unearthed in Pricer Mound. *The Museum News* 3(8):1.

1948   Exciting Discoveries at Frankfort Mound. *Museum Echoes* 21(7):54–55.

1950   A Red Ochre Site in Peoria County? *Journal of the Illinois State Archaeological Society* 1(4):118.

1954   Ed Borg Finds Collection of Ancient Hardened Copper Tools on Farm at Grayling. *Aitkin Independent Age* [Aitkin, Minnesota] 72(10), June 20.

1966a   They Came for Copper. *Earth Science* 19(3):104–105.

1966b   Here's Proof That Hammer Stones Can Still Be Found. *Earth Science* 19(6):266–267.

1977   Figure 69. *The Redskin* 12(3):85.

1990   Finds of the Year. *Iowa Archaeological Society Newsletter* 40(2):4.

1991   *Recovering a Heritage: Prehistoric Art of West Central Illinois*. Western Illinois
        University Art Gallery, Macomb, Illinois. (October 21–November 13).
1996   *"Old Copper" in Manitoba*. Manitoba Culture, Heritage and Citizenship.

Abbott, Charles C.
1881   *Primitive Industry: or Illustrations of the Handiwork, in Stone, Bone and Clay,
        of the Native Races of the North Atlantic Seaboard of America*. George A. Bates,
        Salem, Massachusetts.
1885   The Use of Copper by the Delaware Indians. *The American Naturalist* 19(8):774–777.

Abel, Timothy J. and Adrian L. Burke
2014   The Protohistoric Time Period in Northwest Ohio: Perspectives from the XRF
        Analysis of Metallic Trade Materials. *Midcontinental Journal of Archaeology*
        39(2):179–199.

Adam, Charles and Jim Duncan
1996   The "Meppen Long-Nosed God Maskettes," a New Description and Preliminary
        Analysis. *Central States Archaeological Journal* 43(4):206–208.
2004   The "Meppen Long-Nosed God Maskettes," a New Description and Preliminary
        Analysis. *Central States Archaeological Journal* 51(4):22–23.

Adams, Chuck
1972   The Scott's Site Copper Eagle. *Central States Archaeological Journal* 19(4):170–172.

Adams, William R.
1949   *Archaeological Notes on Posey County, Indiana*. Indiana Historical Bureau,
        Indianapolis.

Adventure Mining Company
1857   Report of the Adventure Mining Co. of Michigan. Robert M. Riddle, Pittsburgh,
        Pennsylvania.
1858   Report of the Adventure Mining Company. *The Mining and Statistic Magazine*
        10(1):82–86.

Alex, Lynn M.
2000   *Iowa's Archaeological Past*. University of Iowa Press, Iowa City, Iowa.

Alex, Lynn M. and William Green
1995   Toolesboro Mounds National Historic Landmark Archaeological Analysis and
        Report. *Research Papers* Vol. 20, No. 4, Office of the State Archaeologist,
        University of Iowa, Iowa City, Iowa.

Allen, Emory A.
1885   *The Prehistoric World: or, Vanished Races*. Central Publishing House, Cincinnati,
        Ohio.

Altuna, Linda M.
1984   *A Re-examination of the Assumption of Reciprocity among Middle Woodland Cultural
        Manifestations Participating in Hopewellian Exchange in the Midwestern-Riverine
        and the Western and Upper Great Lakes Region*. PhD dissertation, University of
        Toronto, Toronto, Ontario, Canada.

Bibliography

Andrews, E. B.
    1880   Report of Explorations of Mounds in Southeastern Ohio. In *Reports of the
                *Peabody Museum of American Archaeology and Ethnology in Connection with
                Harvard University (1876–79)*, 2(1): 51–74.

Anselmi, L. M., M. A. Latta, and R. G. V. Hancock
    1997   Instrumental Neutron Activation Analysis of Copper and Brass from the Auger
                Site (BdGw-3), Simcoe County, Ontario. *Northeast Anthropology* 53:47–59.

Anttila, Oliver N.
    2000   Not One…Not Two…But Three...Caches!!! *Indian Artifact Magazine* 19(3):50–
                51, 72–73.
    2002   Gone Fishin' the Old Copper Complex Way. *Indian Artifact Magazine* 21(4):44–47, 76.
    2006   Another Copper Cache! *Central States Archaeological Journal* 53(3):152–153.
    2009   Ancient Copper Crosses Borders: Copper Toolmakers Migrate Out of Minnesota
                and into Northwestern Ontario. *The Minnesota Archaeologist* 68:107–119.

Anttila, Oliver and William Reardon
    2006   Proving the Age of Old Copper. *Central States Archaeological Journal* 53(4):189.

Arctic Mining Company
    1854   Prospectus of the Arctic Mining Company. Published by the Arctic Mining
                Company, no location given.

Ashland Press [Ashland, Wisconsin]
    1873   Who First Mined Copper on Lake Superior. January 25.

Ashley, Keith H.
    2005   Archaeological Overview of Mt. Royal. *The Florida Anthropologist* 58(3–4):265–286.

Atkeson, T.
    1959   A Burial by the River. *Journal of Alabama Archaeology* 5:13–15.

Atwater, Caleb
    1820   Description of the Antiquities Discovered in the State of Ohio and Other Western
                States. *Transactions and Collections of the American Antiquarian Society* 1:105–267.

Aument, Bruce W.
    1985   Results of the Boyd County Mounds [project and the Preliminary Interpretation
                of Prehistoric Mortuary Variability. In *Woodland Period Research in Kentucky*,
                edited by David Pollack, Thomas N. Sanders, Charles D. Hockensmith, pp.
                63–85. Kentucky Heritage Council, Frankfort, Kentucky.
    1990   *Mortuary Variability in the Middle Big Darby Drainage of Central Ohio between
                300 B.C. and 300 A.D.* PhD dissertation, Ohio State University, Columbus, Ohio.

Aztec Mining Company
    1851 *Charter and By-Laws of the Aztec Mining Company of Michigan, with Letters
                Descriptive of the Location, and the Value and Character of the Mine, Extracts
                from the Minutes, &c.* George Parkin & Company, Pittsburgh, Pennsylvania.

"B."
    1848a   The Copper Region. *Littell's Living Age* 18(223):375.

1848b The Copper Region. *Niles' National Register* 74(1910):159–160.

1848c The Copper Region—Singular Discovery. *Lake Superior News and Mining Journal* [Sault Ste. Marie, Michigan], August 18, p. 2.

1848d The Copper Region. —Singular Discovery. *Friends' Review: A Religious, Literary and Miscellaneous Journal* 1(49):782.

1848e The Copper Region. —Singular Discovery. *Latter Day Saints' Millennial Star* 10(22):351.

1848f Singular Discovery. *The Times* [London], September 15, p. 3.

1848g Singular Discovery in the Copper Region. *Hunt's Merchants' Magazine and Commercial Review* 19(3):340.

1848h The Copper Region. —Singular Discovery. *Massachusetts Eagle* [Lenox, Massachusetts], August 11.

1849 Singular Discovery. *The Moreton Bay Courier* [Moreton Bay, Australia], March 24, p. 4. http://ndpbeta.nla.gov.au/ndp/de1/article/3715525 <13 January 2009>

Babcock, Willoghby M.
  1928 A Minnesota Copper Pike. *The Wisconsin Archeologist* 7(4):218–219.

Baby, Raymond S.
  1949 Cowan Creek Mound Exploration. *Museum Echoes* 22(7):54–55.
  1959 The Clifford M. Williams Site. *Ohio Archaeologist* 9(3):79.
  1961a A Hopewell Human Bone Whistle. *American Antiquity* 27(1):108–110.
  1961b A Unique Hopewellian Breastplate. *Ohio Archaeologist* 11(1):13–15.
  1963 Prehistoric Hand Prints. *Ohio Archaeologist* 13(1):10–11.

Baby, Raymond S. and Asa Mays, Jr.
  1959 Exploration of the William H. Davis Mound. *Museum Echoes* 32:95–96.

Bache, Charles and Linton Satterthwaite, Jr.
  1930 Excavation of an Indian Mound at Beech Bottom. *The Museum Journal* 21(3–4):133–163.

Bacon, Willard S.
  1986 Middle Woodland Panpipes. *Tennessee Anthropologist* 11(2):73–99.

Baerreis, David A., Hiroshi Daifuku, and James E. Lundsted
  1957 The Burial Complex of the Reigh Site, Winnebago County, Wisconsin. *The Wisconsin Archeologist* 38(4):244–278.

Bagg, A. S. and J. H. Harmon
  1848a A New and Important Discovery in the Copper Region—American Antiquity. *Detroit Free Press* [Detroit, Michigan], August 2, p. 2.
  1848b The Copper Country. *Detroit Free Press* [Detroit, Michigan], December 14, p. 2

Baker, Stanley W.
  1981 The Copper Head from the Hopewell Mound Group. *Ohio Archaeologist* 31(2):15–17.

Baldwin, Charles C.
  1887 *Memorial of Colonel Charles Whittlesey, Late President of the Western Reserve Historical Society.* Western Reserve Historical Society, Tract No. 68. Williams' Book Publishing House, Cleveland.

Bibliography

Baldwin, John
> 1980 Ohio Hopewell Mound Offerings. *Ohio Archaeologist* 30(4):8–14.
> 1982 Pileated Woodpecker Effigy Axe. *Central States Archaeological Journal* 29(4):172–173.
> 1984 Spoonville—Classic Michigan Hopewell. *Ohio Archaeologist* 34(4):32–39.

Baldwin, John D.
> 1872 *Ancient America, in Notes on American Archaeology*. Harper & Brothers, New York.

Ball, Donald B., Victor P. Hood, and E. Raymond Evans
> 1976 The Long Island Mounds, Marion County, Tennessee – Jackson County, Alabama. *Tennessee Anthropologist* 1(1):13–47.

Banks, Alan
> 2005 Some Information on the Etley. *Central States Archaeological Journal* 52(2):90–98.

Barnett, LeRoy
> 2004 "What the Sam Hill!" *Michigan History Magazine* 88(3):14–19.

Barrett, S. A.
> 1926 Aboriginal Copper Mines at McCargoe's Cove, Isle Royal. *Yearbook of the Public Museum of the City of Milwaukee* 4:20–36.
> 1933 Ancient Aztalan. *Bulletin of the Public Museum of the City of Milwaukee* 13:1–602.

Bartlett, John
> 1980 *Familiar Quotations*. 15th ed. Edited by Emily Morison Beck. Little, Brown and Company, Boston.

Bass, William A. and Carol J. Loveland
> 1983 Biological Anthropology of the George Preston, Smith, and Sam Sites. *Bulletin of the Texas Archeological Society* 53:83–100.

Bastian, Tyler J.
> 1961 Trace Element and Metallographic Studies of Prehistoric Copper Artifacts in North America: A Review. In *Lake Superior Copper and the Indians: Miscellaneous Studies of Great Lakes Prehistory*, edited by James B. Griffin, pp. 151–175. Anthropological Papers 17. Museum of Anthropology, University of Michigan, Ann Arbor, Michigan.
> 1962 Prehistoric Copper Mining in the Lake Superior Region. *American Antiquity* 27(4): 598–599.
> 1963a *Archaeological Survey of Isle Royale National Park, Michigan, 1960–1962*. United States Department of the Interior, National Park Service.
> 1963b *Prehistoric Copper Mining in Isle Royale National Park, Michigan*. Master's thesis, Department of Anthropology, University of Utah, Salt Lake City, Utah.

Battles, Mrs. Richard E.
> 1969 One Foot in the Grave. *Journal of Alabama Archaeology* 15(1):35–38.

Battles, Richard and Juanita Battles
> 1972 Copper and Lithic Artifacts. *Journal of Alabama Archaeology* 18(1):32–35.

Bauerman, Hilary
    1866   Remarks on the Copper-Mines of the State of Michigan. *The Quarterly Journal of the Geological Society of London* 22:448–463.

Baum, Henry M.
    1903   North America: United States. *Records of the Past* 2(7):223.

Bayliss, Joseph E.
    1946   John N. Ingersoll: Pioneer Upper Peninsula Editor. *Michigan History Magazine* 30(4):663–674.

Bay View Magazine
    1910   Bay View, Yesterday and To-Day, by J. M. H. 17(8):541, May 1910. HathiTrust https://hdl.handle.net/2027/mdp.39015071503240, accessed March 20, 2018.

Beahan, S. J.
    1899   Producing and Prospective Copper Mines. In *First Annual Review of the Copper Mining Industry of Lake Superior*, by James Russell and Albert Hornstein, pp. 75–163. The Mining Journal Co., Limited, Marquette, Michigan.

Beatty, George
    1849   Copper Mines on Lake Superior. *Niles' National Register* 75(26) [June 27]:400.

Beaubien, Paul L.
    1953a   Cultural Variation within Two Woodland Mound Groups of Northeastern Iowa. *American Antiquity* 19(1):56–66.
    1953b   Some Hopewellian Mounds at the Effigy Mounds National Monument, Iowa. *The Wisconsin Archeologist* 34(2):125–138.

Beauchamp, William M.
    1884   New York Copper Implements. *The American Antiquarian* 6(1):32–34.
    1902   *Metallic Implements of the New York Indians.* Bulletin 55. New York State Museum, University of the Museum of the State of New York, Albany, New York.

Beck, Lane A.
    1995   Regional Cults and Ethnic Boundaries in "Southern Hopewell". In *Regional Approaches to Mortuary Analysis*, edited by Lane Anderson Beck, pp. 167–187. Plenum Press, New York.

Beer, James R.
    2014   Wisconsin's Monster Copper Spear. *Central States Archaeological Journal* 61(4):303.

Behm, Jeffery A.
    1991   Description of an Old Copper Monolithic Knife. *Fox Valley Archeology* 17:33–35.
    1995   The Jergenson Artifact Collection. *Fox Valley Archeology* 24:7–33.

Belknap, Charles E.
    1922   *The Yesterdays of Grand Rapids.* The Dean-Hicks Company, Grand Rapids, Michigan.

Belknap, Lori
    2008   Copper Technology of Native North America. *Cahokian* (Spring 2008):3–5.

Bibliography

Bell, Robert E.
    1947   Trade Materials at Spiro Mound as Indicated by Artifacts. *American Antiquity* 12(3):181–184.
    1972   *The Harlan Site, Ck-6, a Prehistoric Mound Center in Cherokee County, Eastern Oklahoma.* Memoir No. 2. Oklahoma Anthropological Society, Norman, Oklahoma.

Bell, Robert E. and Robert J. Block
    1972   A Copper Spearhead from Western Oklahoma. *Plains Anthropologist* 17(55):65–67.

Benn, David W. and William Green
    2000   Late Woodland Cultures in Iowa. In *Late Woodland Societies: Tradition and Transformation across the Midcontinent*, edited by Thomas E. Emerson, Dale L. McElrath, and Andrew C. Fortier, pp. 429–496. University of Nebraska Press, Lincoln, Nebraska.

Benn, David W. and Joe B. Thompson
    2009   Archaic Periods in Eastern Iowa. In *Archaic Societies: Diversity and Complexity across the Midcontinent*, edited by Thomas E. Emerson, Dale L. McElrath, and Andrew C. Fortier, pp. 491–561. State University Press of New York, Albany, New York.

Bennett, John W.
    1944   Hopewellian in Minnesota. *American Antiquity* 9(3):336.
    1952   The Prehistory of the Northern Mississippi Valley. In *Archeology of Eastern United States*, edited by James B. Griffin, pp. 108–123. University of Chicago Press, Chicago.

Berg, Ernest L.
    1945   Stone and Metallic Materials Used by Ancient Indians of Minnesota in the Making of Artifacts. *The Minnesota Archaeologist* 11(1):4–17.

Berlin, A. F.
    1885   Relics of Copper from Eastern Pennsylvania. *The American Antiquarian* 7(1):42–44.

Bernardini, Wesley and Christopher Carr
    2005   Hopewellian Copper Celts from Eastern North America. In *Gathering Hopewell: Society, Ritual, and Ritual Interaction*, edited by Christopher Carr and D. Troy Case, pp. 624–647. Kluwer Academic/Plenum Publishers, New York.

Berner, John F.
    1974   The James T. Robinson Mound. *Artifacts* 4(4):18–21.

Beukens, R. P., L. A. Pavlish, R. G. V. Hancock, R. M. Farquhar, G. C. Wilson, P. J. Julig, and William Ross
    1992   Radiocarbon Dating of Copper-Preserved Organics. *Radiocarbon* 34(3):890–897.

Binford, Lewis R.
    1961   The Haltiner Copper Cache. *The Michigan Archaeologist* 7(2):7–10.
    1963   The Hodges Site: A Late Archaic Burial Station. In *Miscellaneous Studies in Typology and Classification*, by Anta M. White, Lewis R. Binford, and Mark L. Papworth, pp. 124–148. Anthropological Papers No. 19. Museum of Anthropology, University of Michigan, Ann Arbor, Michigan.

Birmingham, Robert A. and Leslie E. Eisenberg
    2000   *Indian Mounds of Wisconsin.* University of Wisconsin Press, Madison, Wisconsin.

Bishop, Ronald L. and Veletta Canouts
    1993   Archaeometry. In *The Development of Southeastern Archaeology*, edited by Jay
                K. Johnson, pp. 160–183. University of Alabama Press, Tuscaloosa, Alabama.

Black, Glenn A.
    1933   The Archaeology of Greene County. *Indiana History Bulletin* 10(5):181–346.
    1934   Archaeological Survey of Dearborn and Ohio Counties. *Indiana History Bulletin*
                11(7):171–260.
    1936   Excavation of the Nowlin Mound, Dearborn County Site 7, 1934–1935. *Indiana*
                History Bulletin 13(7):187–342.
    1946   The Cato Site—Pike County, Indiana. *Proceedings of the Indiana Academy of*
                Science 55:18–22.
    1967   *Angel Site: An Archaeological, Historical, and Ethnological Study* (Two
                Volumes). Indiana Historical Society, Indianapolis.

Blake, William P.
    1875   The Mass Copper of the Lake Superior Mines, and the Method of Mining It.
                Transactions of the American Institute of Mining Engineers 4:110–112.

Blazier, Jeremy, AnnCorinne Freter, and Elliot M. Abrams
    2005   Woodland Ceremonialism in the Hocking Valley. In *The Emergence of the*
                Moundbuilders: The Archaeology of Tribal Societies in Southeastern Ohio, edited
                by Elliot M. Abrams and AnnCorinne Freter, pp. 98–114. Ohio University Press,
                Athens, Ohio.

Bleed, Peter
    1969   *The Archaeology of Petaga Point: The Preceramic Component.* Minnesota
                Historical Society, St. Paul, Minnesota.

Blitz, John H.
    1986   The McRae Mound: A Middle Woodland Site in Southeastern Mississippi.
                Mississippi Archaeology 21(2):11–40.
    1993   *Ancient Chiefdoms of the Tombigbee.* University of Alabama Press, Tuscaloosa,
                Alabama.

Bluhm, Elaine E. and Allen Liss
    1961   The Anker Site. In *Chicago Area Archaeology*, edited by Elaine H. Bluhm, pp.
                89–137. Bulletin No. 3. Illinois Archaeological Survey, Urbana, Illinois.

Bluhm, Elaine E. and David J. Wenner, Jr.
    1956   Prehistoric Culture of Chicago Area Uncovered. *Chicago Natural History*
                Museum Bulletin 27(2):5–6.

Blumer, A.
    1883   Exploration of Mounds in Louisa County, Iowa. *Proceedings of the Davenport*
                Academy 3:132–133.

Bohannon, Charles F.
    1972   *Excavations at the Pharr Mounds, Prentiss and Itawamba Counties, Mississippi*

and Excavations at the Bear Creek Site, Tishomingo County, Mississippi. U. S. Department of the Interior, National Park Service, Office of Archeology and Historic Preservation, Division of Archeology and Anthropology, Washington, DC.

1973   *Excavations at the Mineral Springs Site, Howard County, Arkansas.* Arkansas Archeological Survey, Publications on Archeology, Research Series No. 5.

Bohemian Mining Company

1851   *Charter and By-Laws of the Bohemian Mining Co, with Letters from the Mines.* Grattan & McLean, Philadelphia.

1860   *Report of the Directors to the Stockholders.* Inquirer Printing Office, Philadelphia.

Bostwick, Todd W.

2008   *Beneath the Runways: Archaeology of Sky Harbor International Airport.* Pueblo Grande Museum, Parks and Recreation Department and Aviation Department, Phoenix.

Boucher, Pierre

1883   True and Genuine Description of New France, Commonly Called Canada, and of the Manners and Customs and Productions of that Country. In *Canada in the Seventeenth Century. From the French of Pierre Boucher*, by Edward Louis Montizambert, pp. 5–85. George E. Desbarats & Co., Montreal, Quebec, Canada.

Bowman, David J.

2003   Grassy Lake and the Mounds of Roxana, Illinois. *Illinois Antiquity* 38(2):3–7.

Boyle, David

1888   Copper. In *Annual Report of the Canadian Institute, Session 1886–87. Being Part of Appendix to the Report of the Minister of Education, Ontario*, pp. 54–56.

1889a   Copper. In *Annual Report of the Canadian Institute, Session 1887–8. Being Part of Appendix to the Report of the Minister of Education, Ontario*, pp. 48–50.

1889b   Copper. In *Annual Report of the Canadian Institute, Session 1888–9. Being Part of Appendix to the Report of the Minister of Education, Ontario*, pp. 39–40.

1889c   Copper and Hematite. In *Annual Report of the Canadian Institute, Session 1888–9. Being Part of Appendix to the Report of the Minister of Education, Ontario*, pp. 78–80.

1891   Copper. In *Fourth Annual Report of the Canadian Institute, (Session 1891). Being an Appendix to the Report of the Minister of Education, Ontario*, pp. 59–63.

1892   Copper. In *Fifth Annual Report of the Canadian Institute, (Session of 1892–3.) Being an Appendix to the Report of the Minister of Education, Ontario*, p. 21.

1893   Copper. In *Sixth Annual Report of the Canadian Institute, (Session 1890–91.) Being an Appendix to the Report of the Minister of Education, Ontario*, p. 51.

1895   *Notes on Primitive Man in Ontario.* Warwick Bros. & Rutter, Printers, Toronto, Ontario, Canada.

1897a   Beads. *Annual Archaeological Report 1896–97, Being Part of Appendix to the Report of the Minister of Education, Ontario*, p. 55.

1897b   Copper. *Annual Archaeological Report 1896–97, Being Part of Appendix to the Report of the Minister of Education, Ontario*, pp. 53–55.

1897c   Sugar Island. *Annual Archaeological Report 1896–97, Being Part of Appendix to the Report of the Minister of Education, Ontario*, pp. 33–35.

1898a   Copper. *Annual Archaeological Report 1897–98, Being Part of Appendix to the Report of the Minister of Education, Ontario*, p. 25.

1898b   Copper Tools. *Archaeological Report 1898, Being Part of Appendix to the Report of the Minister of Education, Ontario*, p. 53.

1900   Pelee Island. *Archaeological Report 1899, Being Part of Appendix to the Report of the Minister of Education, Ontario*, pp. 30–34.

1904a   A Few Copper Tools. *Annual Archaeological Report 1903, Being Part of Appendix to the Report of the Minister of Education, Ontario*, pp. 88–91.

1904b   The Working of Native Copper. *Annual Archaeological Report 1903, Being Part of Appendix to the Report of the Minister of Education, Ontario*, pp. 36–43.

1905   Copper. *Annual Archaeological Report 1904, Being Part of Appendix to the Report of the Minister of Education, Ontario*, pp. 49–50.

1908   Copper. *Annual Archaeological Report 1907, Being Part of Appendix to the Report of the Minister of Education, Ontario*, p. 35.

Brain, Jeffrey P. and Philip Phillips
1996   *Shell Gorgets: Styles of the Late Prehistoric and Protohistoric Southeast.* Peabody Museum Press, Cambridge, Massachusetts.

Branch, Jonathan J.
2013   *A Study of Moundville Copper Gorgets.* Bachelor's with honors, Department of Anthropology, University of North Carolina at Chapel Hill, Chapel Hill, North Carolina.

Brannon, Peter A.
1931   Sacred Creek Relics Found in Alabama. *Alabama Anthropological Society* 19(1–2):3.

Brantner, J. H.
1947   *Historical Collections of Moundsville, West Virginia.* Marshall County Historical Society, Moundsville, West Virginia.

Brashler, Janet G., Michael J. Hambacher, Terrance J. Martin, Kathryn E. Parker, and James A Robertson
2006   Middle Woodland Occupation in the Grand River Basin of Michigan. In *Recreating Hopewell*, edited by Douglas K. Charles and Jane E. Buikstra, pp. 261–284. University Press of Florida, Gainesville, Florida.

Brashler, Janet G., Elizabeth B. Garland, Margaret B. Holman, William A. Lovis, and Susan R. Martin
2000   Adaptive Strategies and Socioeconomic Systems in Northern Great Lakes Riverine Environments: The Late Woodland of Michigan. In *Late Woodland Societies: Tradition and Transformation across the Midcontinent*, edited by Thomas E. Emerson, Dale L. McElrath, and Andrew C. Fortier, pp. 543–579. University of Nebraska Press, Lincoln, Nebraska.

Braun, David P.
1979   Illinois Hopewell Burial Practices and Social Organization: A Reexamination of the Klunk-Gibson Mound Group. In *Hopewell Archaeology: The Chillicothe Conference*, edited by David S. Brose and N'omi Greber, pp. 66–79. MCJA Special Paper No. 3. Kent State University Press, Kent, Ohio.

Braun, David P., James B. Griffin and Paul F. Titterington
1982   *The Snyders Mounds and Five Other Mound Groups in Calhoun County, Illinois.* Technical Report 13, Research Reports in Archaeology Contribution 8. Museum of Anthropology, University of Michigan, Ann Arbor, Michigan.

Bravard, Dudley C.
1972   The Schaefer Mound Revisited. *Artifacts* 2(2):4–5.

Bibliography

Brine, Lindesay
    1894   *Travels amongst American Indians*. Sampson Low, Marston & Company, Limited, London, England.

Brink, Jack
    1988   The Highwood River Site: A Pelican Lake Phase Burial from the Alberta Plains. *Canadian Journal of Archaeology* 12:109–136.

Britt, Claude
    1967   The Abbott Site: An Archaic-Adena Site in Shelby County, Ohio. *Ohio Archaeologist* 17(3):111–114.
    1985   Glacial Kame Sites Not Reported by Converse. *Ohio Archaeologist* 35(1):12.

Britton, Linda A.
    1967   *Copper and the Indians of New Jersey*. Master's thesis, Hunter College, University of New York, New York.

Brockett, Bernard O.
    1974   Four Thousand Years of Copper Mining. *Lapidary Journal*, July, pp. 712–721.

Brookes, Samuel O.
    1976   *The Grand Gulf Mound (22-Cb-522): Salvage Excavation of an Early Marksville Burial Mound*. Edited by Samuel O. McGahey and Priscilla M. Lowrey. Archaeological Report No. 1. Mississippi Department of Archives and History, Jackson, Mississippi.

Brooks, Mary Jane
    1945   A Copper Comb. *Tennessee Archaeologist* 1(2):18–19.

Brose, David S.
    1970a   *The Archaeology of Summer Island: Changing Settlement Systems in Northern Lake Michigan*. Anthropological Papers No. 41. Museum of Anthropology, University of Michigan, Ann Arbor, Michigan.
    1970b   Summer Island III: An Early Historic Site in the Upper Great Lakes. *Historical Archaeology* 4:3–33.
    1971   The Girdled Road Site, an Early Woodland Hunting Station in Lake County, Ohio. *Kirtlandia* 13:1–19.
    1974   The Everett Knoll: A Late Hopewellian Site in Northeastern Ohio. *The Ohio Journal of Science* 74(1):36–46.
    1979   An Interpretation of the Hopewellian Traits in Florida. In *Hopewell Archaeology: The Chillicothe Conference*, edited by David S. Brose and N'omi Greber, pp. 141–149. MCJA Special Paper No. 3. Kent State University Press, Kent, Ohio.
    1985   The Woodland Period. In *Ancient Art of the American Woodland Indians*, edited by Andrea P. A. Belloli, pp. 42–91. Harry N. Abrams, Inc., Publishers, New York, in Association with the Detroit Institute of Arts.
    1994a   Trade and Exchange in the Midwestern United States. In *Prehistoric Exchange Systems in North America*, edited by Timothy G. Baugh and Jonathon E. Ericson, pp. 215–240. Plenum Press, New York.
    1994b   *The South Park Village Site and the Late Prehistoric Whittlesey Tradition of Northeast Ohio*. Monographs in World Archaeology No. 20. Prehistory Press, Madison, Wisconsin.

Brower, J. V. and D. I. Bushnell
    1900   *Memoirs of Explorations in the Basin of the Mississippi: Mille Lac.* Vol. 3. St. Paul, Minnesota.

Brown, Calvin S.
    1992   *Archeology of Mississippi.* Introduction by Janet Ford. University Press of Mississippi, Jackson, Mississippi.

Brown, Catherine
    1982   On the Gender of the Winged Being on Mississippi Period Copper Plates. *Tennessee Anthropologist* 7(1):1–8.

Brown, Charles E.
    1901   The Milwaukee Public Museum Collection of Copper Implements. *The Wisconsin Archeologist* 1(1):11–13.
    1902   Archeological Notes. *The Wisconsin Archeologist* 1(4):102–107.
    1904a   The Native Copper Implements of Wisconsin. *The Wisconsin Archeologist* 3(2):49–84, Plates 1–12.
    1904b   The Native Copper Ornaments of Wisconsin. *The Wisconsin Archeologist* 3(3):101–121, Plates 13–17.
    1905a   The State Fair Exhibit of the Wisconsin Archeological Exhibit. *The Wisconsin Archeologist* 5(1):200–221.
    1905b   Wisconsin Caches. *Records of the Past* 4(3):82–95.
    1906   A Record of Wisconsin Antiquities. *The Wisconsin Archeologist* 5(3–4):293–429.
    1907   The Implement Caches of the Wisconsin Indians. *The Wisconsin Archeologist* 6(2):47–70.
    1916   Archaeological History of Milwaukee County. *The Wisconsin Archeologist* 15(2):23–105.
    1917   The Antiquities of Green Lake. *The Wisconsin Archeologist* 16(1):1–55.
    1923a   Waukesha County: Southern Townships. *The Wisconsin Archeologist* 2(2):69–119.
    1923b   Copper Implements in Northern Wisconsin. *The Wisconsin Archeologist* 2(4):178–179.
    1924a   Additional Notes on Vilas and Oneida Counties. *The Wisconsin Archeologist* 3(2):52–57.
    1924b   Indian Gravel Pit Burials in Wisconsin. *The Wisconsin Archeologist* 3(3):65–82.
    1926a   A Copper Pike. *The Wisconsin Archeologist* 5(2):66–67.
    1926b   Delavan Lake. *The Wisconsin Archeologist* 6(1):7–31.
    1927   Native Copper Harpoon Points. *The Wisconsin Archeologist* 7(1):50–55.
    1929   Checklist of Wisconsin Indian Implements. *The Wisconsin Archeologist* 8(3):81–94.
    1930a   The Kohler Museum. *The Wisconsin Archeologist* 9(3):143–145.
    1930b   The Largest Copper Knives. *The Wisconsin Archeologist* 9(3):145–146.
    1931   Large Copper Implements. *The Wisconsin Archeologist* 10(2):72–73.
    1932   A Copper Bird Effigy Ornament. *The Wisconsin Archeologist* 11(3):104–107.
    1938a   A Toothed Shank Copper Spearpoint. *The Wisconsin Archeologist* 18(3):77–78.
    1938b   An Ornamented Copper Knife. *The Wisconsin Archeologist* 19(1):16.
    1939   Myths, Legends and Superstitions about Copper. *The Wisconsin Archeologist* 20(2):35–40.
    1940a   Cache of Copper Implements. *The Wisconsin Archeologist* 21(2):34.
    1940b   A Fluted Copper Spud. *The Wisconsin Archeologist* 21(4):64–68.
    1940c   Red Paint with Wisconsin Burials. *The Wisconsin Archeologist* 21(4):74–76.
    1942   A Copper Adze. *The Wisconsin Archeologist* 23(1):17–18.

Bibliography

Brown, Charles E. and Robert H. Becker
    1917   The Chetek and Rice Lakes. *The Wisconsin Archeologist* 16(3):83–114.

Brown, Charles E. and H. L. Skavlem
    1914   Notes on Some Archeological Features of Eau Claire, Chippewa, Ruck and Dunn Counties. *The Wisconsin Archeologist* 13(1):60–79.

Brown, Helen C.
    1980   *A Catalogue of Mississippian Copper Plates.* Master's thesis, University of Georgia, Athens, Georgia.

Brown, Ian W.
    1985   *Natchez Indian Archaeology: Culture Change and Stability in the Lower Mississippi Valley.* Archaeological Report No. 15, Mississippi Department of Archives and History, Jackson, Mississippi.

Brown, J. Venen
    1850a   Copper from Ontonagon. *Lake Superior Journal* [Sault Ste. Marie, Michigan], July 10, p. 2.
    1850b   Discoveries in the Ancient Copper Diggings. *Lake Superior Journal* [Sault Ste. Marie, Michigan], September 25, p. 2.
    1850c   Discoveries in the Ancient Copper Diggings. *Detroit Free Press* [Detroit, Michigan], October 1.
    1851   Eagle Harbor Mines. *Lake Superior Journal* [Sault Ste. Marie, Michigan], August 13, p. 2.
    1852   Phoenix Mine—New Discovery. *Lake Superior Journal* [Sault Ste. Marie, Michigan], September 15, p. 2.
    1853a   Copper Falls Mine. *Lake Superior Journal* [Sault Ste. Marie, Michigan], June 18, p. 2.
    1853b   Ontonagon Region-Toltec Consolidated Mine-Adventure-Aztec and Ohio Mining Companies. *Lake Superior Journal* [Sault Ste. Marie, Michigan], June 25, p. 2.
    1853c   Portage Lake Mines. *Lake Superior Journal* [Sault Ste. Marie, Michigan], November 5, p. 2.
    1853d   Summary of the Mining News. *Lake Superior Journal* [Sault Ste. Marie, Michigan], May 28, p. 2.
    1854   Isle Royale. *Lake Superior Journal* [Sault Ste. Marie, Michigan], May 17, p. 2.

Brown, James A.
    1966   *Spiro Studies, Vol. 2: The Graves and Their Contents.* University of Oklahoma Research Institute, Norman, Oklahoma.
    1979   Charnel Houses and Mortuary Crypts: Disposal of the Dead in the Middle Woodland Period. In *Hopewell Archaeology: The Chillicothe Conference*, edited by David S. Brose and N'omi Greber, pp. 211–219. MCJA Special Paper No. 3. Kent State University Press, Kent, Ohio.
    1983   Spiro Exchange Connections Revealed by Sources of Imported Raw Materials. In *Southeastern Natives and Their Pasts: A Collection of Papers Honoring Dr. Robert E. Bell*, edited by Don G. Wyckoff and Jack L. Hofman, pp. 129–162. Oklahoma Archeological Survey Studies in Oklahoma's Past No. 11, Cross Timbers Heritage Association Contribution No. 2, Norman, Oklahoma.
    1985   The Mississippian Period. In *Ancient Art of the American Woodland Indians*, edited by Andrea P. A. Belloli, pp. 92–145. Harry N. Abrams, Inc., Publishers, New York, in Association with the Detroit Institute of Arts.
    1989   On Style Divisions of the Southeastern Ceremonial Complex: A Revisionist

Perspective. In *The Southeastern Ceremonial Complex: Artifacts and Analysis*, edited
by Patricia Galloway, pp. 183–204. University of Nebraska Press, Lincoln, Nebraska.

2004   Exchange and Interaction until 1500. In *Handbook of North American Indians,
Volume 14: Southeast*, edited by Raymond Fogelson and William Sturtevant, pp.
677–685. Smithsonian Institution, Washington, DC.

2006   The Shamanic Element in Hopewellian Period Ritual. In *Recreating Hopewell*,
edited by Douglas K. Charles and Jane E. Buikstra, pp. 475–488. University Press
of Florida, Gainesville, Florida.

2007   On the Identity of the Birdman within Mississippian Period Art and Iconography.
In *Ancient Objects and Sacred Realms: Interpretations of Mississippian
Iconography*, edited by F. Kent Reilly III and James F. Garber, pp. 56–106.
University of Texas Press, Austin, Texas.

2007   Sequencing the Braden Style within Mississippian Period Art and Iconography.
In *Ancient Objects and Sacred Realms: Interpretations of Mississippian
Iconography*, edited by F. Kent Reilly III and James F. Garber, pp. 213–245.
University of Texas Press, Austin, Texas.

Brown, James A. and Raymond S. Baby
1966   *Mound City Revisited*. Ohio Historical Society, Columbus, Ohio.

Brown, James A., Alice M. Brues, Lyle W. Konigsberg, Paul W. Parmalee and David H. Stansbery
1996   *The Spiro Ceremonial Center: The Archaeology of Arkansas Valley Caddoan
Culture in Eastern Oklahoma* (Vol. 2). Memoirs No. 29. Museum of
Anthropology, University of Michigan, Ann Arbor, Michigan.

Brown, James A. and David H. Dye
2007   Severed Heads and Sacred Scalplocks: Mississippian Iconographic Trophies. In
The Taking and Displaying of Human Body Parts as Trophies by Amerindians,
edited by Richard J. Chacon and David H. Dye, pp. 278–298. Springer Science
+Business Media, LLC, New York.

Brown, James A. and Henry W. Hamilton
1965   The Cultural and Artistic World of Spiro and Mississippian Culture. In *Spiro
and Mississippian Antiquities from the McDannald Collection*, pp. 11–52. The
Museum of Fine Arts, Houston, Texas.

Brown, James A. and John E. Kelly
2000   Cahokia and the Southeastern Ceremonial Complex. In *Mounds, Modoc, and
Mesoamerica*, edited by Steven R. Ahler, pp. 469–510. Illinois State Museum
Scientific Papers Vol. 28, Illinois State Museum, Springfield, Illinois.

Brown, James A., Richard A. Kerber, and Howard D. Winters
1990   Trade and the Evolution of Exchange Relations at the Beginning of the
Mississippian Period. In *The Mississippian Emergence*, edited by Bruce D. Smith,
pp. 251–280. Smithsonian Institution Press, Washington, DC.

Brown, James A. and J. Daniel Rogers
1989   Linking Spiro's Artistic Styles: the Copper Connection. *Southeastern Archaeology*
8(1):1–8.

Brown, James A., Roger W. Willis, Mary A. Barth, and Georg K. Neumann
1967   *The Gentleman Farm Site, LaSalle County, Illinois*. Report of Investigations No.
12, Illinois State Museum, Springfield, Illinois.

Bibliography

Brown, Ralph D.
  1944   The Harvey Rock Shelter on the St. Croix River (Washington County, Minnesota). *Minnesota Archaeologist* 10(4):118–123.

Brown, Theodore T.
  1930   The Largest Copper Knives. *The Wisconsin Archeologist* 9(3):145–146.

Bruhy, Mark E., Kathryn C. Egan-Bruhy, and Kim L. Potaracke
  1998   *The Butternut Lake Inlet Site (47 FR-137): Exploring Seasonality and Diet at a Woodland Tradition Settlement, the 1997 Field Season.* Chequamegon-Nicolet National Forest, Northern State Regional Archaeology Center, Report of Investigation No. 15, USDA Forest Service, Rhinelander, Wisconsin.
  1999   *Summary Report of Excavations at the Butternut Lake Inlet Site (47 FR-137): Exploring Seasonality and Diet at a Multi-Component Site in Forest County, Wisconsin.* Chequamegon-Nicolet National Forest, Northern State Regional Archaeology Center, Report of Investigation No. 19, USDA Forest Service, Rhinelander, Wisconsin.

Bruhy, Mark E. and John F. Wackman
  1980   Test Excavations at the Butternut Lake Site (47-FR-122), Forest County, Wisconsin. *The Wisconsin Archeologist* 61(4):389–451.

Bruseth, James E.
  1991   Poverty Point Development as Seen at the Cedarland and Claiborne Sites, Southern Mississippi. In *The Poverty Point Culture: Local Manifestations, Subsistence Practices and Trade Networks*, edited by Kathleen M. Byrd, pp.7–25. Geoscience and Man Vol. 29. Geoscience Publications, Department of Geography and Anthropology, Louisiana State University, Baton Rouge, Louisiana.

Bryan, Liz
  2005   *The Buffalo People: Pre-contact Archaeology on the Canadian Plains.* Heritage House Publishing Company Ltd., Surrey, BC, Canada.

Bryant, Charles S., ed.
  1881   William W. Spalding. In *History of the Upper Mississippi Valley*, p. 697. Minnesota Historical Company, Minneapolis.

Bryce, George
  1887   *The Souris Country: Its Monuments, Forts and Rivers.* Transaction No. 24. The Historical & Scientific Society of Manitoba, Winnipeg. Manitoba Free Press Print, Winnipeg, Manitoba, Canada.
  1890   The Winnipeg Mound Region: Being the Most Northerly District Where Mounds Have Been Examined on the American Continent. *Proceedings of the American Association for the Advancement of Science, Thirty-eighth Meeting, Held at Toronto, Ontario, August, 1889*, pp. 344–345.
  1904   *Among the Mound Builders Remains.* Transaction No. 66. The Historical & Scientific Society of Manitoba, Winnipeg.

Buchanan, William T.
  1986   *The Trigg Site, City of Radford, Virginia.* Special Publication No. 14. Archaeological Society of Virginia, Charles City County, Virginia.

Buchner, A. P.
  1979   *The 1978 Caribou Lake Project, Including a Summary of the Prehistory of East-Central Manitoba.* Final Report No. 8. Papers in Archaeology, Department of Cultural Affairs and Historical Resources, Historic Resources Branch, Winnipeg, Manitoba, Canada.
  1980   The Cemetery Point Site (EaKv-1): Report on the 1879 Field Season. In *Studies in Eastern Manitoba Archaeology*, pp. 1–36. Miscellaneous Papers No. 10. Papers in Archaeology, Department of Cultural Affairs and Historical Resources, Historic Resources Branch, Winnipeg, Manitoba, Canada.

Buckmaster, Marla M.
  2001   The Reindle Site. *The Michigan Archaeologist* 47(3–4):79–136.

Buell, John L.
  1906   Menominee Range. *Proceedings of the Lake Superior Mining Institute* 11:38–49.

Bullock, Harold R.
  1942   Lasley Point Mound Excavations. *The Wisconsin Archeologist* 23(2):37–44.

Bunnell, Lafayette H.
  1897   *Winona (We-No-Nah) and Its Environs on the Mississippi in Ancient and Modern Days.* Jones & Kroeger, Printers and Publishers, Winona, Minnesota.

Burks, Jarrod
  2002   Hopewell Copper in Context at Mound City Group. Ohio Archaeological Council, Columbus, Ohio.
  2011   *Prehistoric Native American Earthwork and Mound Sites in the Area of the Department of Energy Portsmouth Gaseous Diffusion Plant, Pike County, Ohio.* OVAI Contract Report #2009-22-2. Ohio Valley Archaeology, Inc., Columbus, Ohio.

Burks, Jarrod and Jennifer Pederson
  2006   The Place of Nonmound Debris at Hopewell Mound Group (33RO27), Ross County, Ohio. In *Recreating Hopewell*, edited by Douglas K. Charles and Jane E. Buikstra, pp. 376–401. University Press of Florida, Gainesville, Florida.

Burnett, E. K.
  1945   The Spiro Mound Collection in the Museum. *Contributions from the Museum of the American Indian, Heye Foundation* 14:1–47.

Butkus, Edmund
  1973   Mississippian Trade at the Anker Site. *Central States Archaeological Journal* 20(4):172–176.
  2002   Documentation of Chicago Area Upper Mississippian Artifacts. *Central States Archaeological Journal* 49(4):164–168.

Butler, Brian D.
  1979   Hopewellian Contacts in Southern Middle Tennessee. In *Hopewell Archaeology: The Chillicothe Conference*, edited by David S. Brose and N'omi Greber, pp. 150–156. MCJA Special Paper No. 3. Kent State University Press, Kent, Ohio.
  2009   Land between the Rivers: The Archaic Period of Southernmost Illinois. In *Archaic Societies: Diversity and Complexity across the Midcontinent*, edited by Thomas E.

Bibliography

Emerson, Dale L. McElrath, and Andrew C. Fortier, pp. 607–634. State University Press of New York, Albany, New York.

Butler, James D.
    1876a   Copper Tools Found in the State of Wisconsin. *Transactions of the Wisconsin Academy of Sciences, Arts, and Letters* 3:99–104.
    1876b   Prehistoric Wisconsin. *Report and Collections of the State Historical Society of Wisconsin, for the Years 1873, 1874, 1875 and 1876*, 7:80–101.
    1877   The Copper Age in Wisconsin. *Proceedings of the American Antiquarian Society, at the Semi-Annual Meeting, Held in Boston, April 25, 1877*, pp. 57–63.
    1881   Wisconsin Copper Finds and Lake Dwellings. *The American Antiquarian and Oriental Journal* 3(2):141.
    1882a   Recent Accessions of Historic Copper. *Report and Collections of the State Historical Society of Wisconsin for the Years 1880, 1881, and 1882*, 9:97–99.
    1882b   A Crucial Copper. *The American Antiquarian and Oriental Journal* 4(3):231–232.
    1887   Mr. Wyman's Copper Implements. *The American Antiquarian and Oriental Journal* 9(6):370–371.
    1895   Another Huge Deposit of Stone Relics. *The American Antiquarian and Oriental Journal* 17(4):249–250.

Butts, Bruce
    1998   Copena Culture. *Central States Archaeological Journal* 45(2):98.

Byers, Douglas S.
    1962   The Restoration and Preservation of Some Objects from Etowah. *American Antiquity* 28(2):206–216.
    1964   Two Textile Fragments and Some Copper Objects from Etowah, Georgia. In *XXXV Congreso Internacional de Americanistas, Mexico, 1962*. Actas y Memorias 1:591–598.

"E. C."
    1849   Mining Operations—Ancient Workings on Lake Superior, &c. &c. *Lake Superior News and Mining Journal* [Sault Ste. Michigan], July 13, p. 2.
    1850   Mines of the Ontonagon—Its Scenery, &c. *Lake Superior Journal* [Sault Ste. Marie, Michigan], July 17, p. 2.

Cadle, Cornelius
    1902   A Remarkable Prehistoric Ceremonial Pipe. *Records of the Past* 1:218–220.

Caine, Christy A. H.
    1969   *The Archaeology of the Snake River Valley*. Master's thesis, University of Minnesota, Minneapolis.
    1974   The Archaeology of the Snake River Region in Minnesota. In *Aspects of Upper Great Lakes Anthropology: Papers in Honor of Lloyd A. Wilford*, edited by Elden Johnson, pp. 55–63. Minnesota Prehistoric Archaeology Series No. 11. Minnesota Historical Society, St. Paul, Minnesota.

Caldwell, Joseph and Catherine McCann
    1941   *Irene Mound Site, Chatham County, Georgia*. University of Georgia Press, Athens, Georgia.

Caldwell, Warren W. and Richard E. Jensen
    1969   *The Grand Detour Phase*. Publications in Salvage Archaeology No. 13. River Basin Surveys, Museum of Natural History, Smithsonian Institution, Lincoln, Nebraska.

Callender, John A.
    1854   The Lake Superior Copper Mines. The Mining Magazine 2(3):249–253.

Calvert, Dan and Lynn Stephens
    1991   A Panpipe from Crawford County, Illinois. *Illinois Antiquity* 27(3):6–8.

Campbell, John P.
    1816   Of the Aborigines of the Western Country. *The Port Folio* 2(1):1–8.

Cantwell, Anne-Marie
    1980   *Dickson Camp and Pond: Two Early Havana Tradition Sites in the Central Illinois Valley*. Report of Investigations No. 36. Dickson Mounds Museum Anthropological Studies, Illinois State Museum, Springfield, Illinois.

Capes, Katherine H.
    1963   *The W. B. Nickerson Survey and Excavations, 1912–15, of the Southern Manitoba Mounds Region*. Anthropology Papers No. 4. National Museum of Canada, Department of Northern Affairs and National Resources, Ottawa, Ontario, Canada.

Carlson, Gayle F.
    1997   A Preliminary Survey of Marine Shell Artifacts from Prehistoric Archeological Sites in Nebraska. *Central Plains Archeology* 5(1):11–47.

Carp Lake Mining Co.
    1858   *Articles of Association, By-Laws, &c. of Carp Lake Mining Co.* Steam Press of Pinkerton and Nevins, Cleveland.

Carr, Christopher
    2005   *Final Report: Development of High-Resolution, Digital, Color and Infrared Photographic Methods for Preserving Imagery on Hopewellian Copper Artifacts*. Grant No. MT-2210-0-NC-12. The National Center for Preservation Technologies and Training, NSUBox 5682, Natchitoches, Louisiana 71497.
    2008a   Coming to Know Ohio Hopewell Peoples Better: Topics for Future Research, Masters' Theses, and Doctoral Dissertations. In *The Scioto Hopewell and Their Neighbors: Bioarchaeological Documentation and Cultural Understanding*, edited by D. Troy Case and Christopher Carr, pp. 603–690. Springer Science + Business Media, LLC, New York.
    2008b   Environmental Setting, Natural Symbols, and Subsistence. In *The Scioto Hopewell and Their Neighbors: Bioarchaeological Documentation and Cultural Understanding*, edited by D. Troy Case and Christopher Carr, pp. 41–100. Springer Science + Business Media, LLC, New York.
    2008c   Settlement and Communities. In *The Scioto Hopewell and Their Neighbors: Bioarchaeological Documentation and Cultural Understanding*, edited by D. Troy Case and Christopher Carr, pp. 101–150. Springer Science + Business Media, LLC, New York.
    2008d   Social and Ritual Organization. In *The Scioto Hopewell and Their Neighbors:*

Bioarchaeological Documentation and Cultural Understanding, edited by D. Troy Case and Christopher Carr, pp. 151–288. Springer Science + Business Media, LLC, New York.

2008e   World View and the Dynamics of Change: The Beginning and the End of Scioto Hopewell Culture and Lifeways. In *The Scioto Hopewell and Their Neighbors: Bioarchaeological Documentation and Cultural Understanding*, edited by D. Troy Case and Christopher Carr, pp. 289–328. Springer Science + Business Media, LLC, New York.

Carr, Christopher, Beau J. Goldstein, and D. Troy Case

2008   Contextualizing Preanalyses of the Ohio Hopewell Mortuary Data, I: Age, Sex, Burial–Deosit, and Intraburial Artifact Count Distributions. In *The Scioto Hopewell and Their Neighbors: Bioarchaeological Documentation and Cultural Understanding*, edited by D. Troy Case and Christopher Carr, pp. 523–574. Springer Science + Business Media, LLC, New York.

Carr, Christopher, Rex Weeks, and Mark Bahti

2008   The Functions and Meanings of Ohio Hopewell Ceremonial Artifacts in Ethnohistorical Perspective. In *The Scioto Hopewell and Their Neighbors: Bioarchaeological Documentation and Cultural Understanding*, edited by D. Troy Case and Christopher Carr, pp. 501–521. Springer Science + Business Media, LLC, New York.

Carr, Lucien

1877   Report on the Additions to the Museum and Library for the Year 1976. *Tenth Annual Report of the Trustees of the Peabody Museum of American Archaeology and Ethnology*, pp.13–29.

Carroll, Charles E.

2001   A Hopewell Pottery Vessel from the Blennerhassett Island. *Ohio Archaeologist* 51(4):18.

Carskadden, Jeff

1968   An Adena Stone Mound, Flint Ridge, Ohio. *Ohio Archaeologist* 18(3):80–81.
1989   Excavation of Mound D at the Philo Mound Group, Muskingum County, Ohio. *Ohio Archaeologist* 39(1):4–8.
1990   Bradley Atchinson and the High Hill Mounds. *Ohio Archaeologist* 40(1):4–7.

Carskadden, Jeff and Donna Fuller

1967   The Hazlett Mound Group. *Ohio Archaeologist* 17(4):139–143.

Carstens, C. J. and J. P. Knudsen

1958   An Archaeological Survey of the Rockhouse Ledge Area, Part II: The Source of the Copper from Rockhouse Spring Cave. *Journal of Alabama Archaeology* 4(2):13–17. (See also Knudsen and Radford 1957.)

Case, D. Troy, Christopher Carr, and Ashley E. Evans

2008   Definition of Variables and Variable States. In *The Scioto Hopewell and Their Neighbors: Bioarchaeological Documentation and Cultural Understanding*, edited by D. Troy Case and Christopher Carr, pp. 419–463. Springer Science + Business Media, LLC, New York.

Central Mining Company
1855   *First Annual Report of the Directors of the Central Mining Company together with the By-Laws.* Advertiser Print, Detroit.

Central States Archaeological Journal
1956   Figure 109 and Figure 110. 2(4):154-155.
1965   Figure 72. 12(3):94.
1968   Figure 86. 15(3):116.
1968   Figure 123. 15(4):162.
1970   Figure 8. 17(1):18.
1970   Figure 115. 17(4):170.
1971   Figure 3. 18(1):7.
1972   Figure 20. 19(1):38.
1977   Figure 123. 24(4):194.
1979   Untitled. 26(2):95.
1981   Untitled. 28(1):41.
1981   Untitled. 28(3):144.
1982   Untitled. 29(2):112.
1983   Untitled. 30(3):159.
1986   Untitled. 33(4):285.
1987   Untitled. 34(1):24.
1988   Untitled. 35(1):47.
1989   Untitled. 36(1):46.
1992   Untitled. 39(2):62.
2000   Untitled. 47(4):192.
2002   Untitled. 49(3):113.
2003   Untitled. 50(3):123.
2003   Untitled. 50(4):213.
2004   Untitled. 51(4):163.
2005   11⅜" Old Copper Knife Discovered. 52(3):133.
2006   Spiro Human Effigy Copper on Wood Face. 53(2):65.
2006   Untitled. 53(2):70, 84.
2008   Untitled. 55(3):127.
2015   Untitled. 62(3):143.

Champion, Don
1965   Crib Mound, Spencer County, Indiana. *Central States Archaeological Journal* 12(2):51–61.

Channen, E. R. and N. D. Clarke
1965   *The Copeland Site: A Precontact Huron Site in Simcoe County, Ontario.* Anthropology Papers No. 8. National Museum of Canada, Ottawa, Ontario, Canada.

Chapman & Co., C. C.
1881   *History of Kent County, Michigan.* C. C. Chapman & Co., Chicago.

Chapman Brothers
1890   Samuel O. Knapp. In *Portrait and Biographical Album of Jackson County, Michigan*, pp. 200–203. Chapman Bros., Chicago.

Chapman, Carl H.
1952   Culture Sequence in the Lower Missouri Valley. In *Archeology of Eastern United*

Bibliography

States, edited by James B. Griffin, pp. 139–151. University of Chicago Press, Chicago.

1980   *The Archaeology of Missouri, II*. University of Missouri Press, Columbia, Missouri.

Chapman, Jefferson
1982   *The American Indian in Tennessee: An Archaeological Perspective*. The Frank H. McClung Museum, University of Tennessee, Knoxville, Tennessee.
1985   *Tellico Archaeology: 12,000 Years of Native American History*. Report of Investigations No. 43, Department of Anthropology, University of Tennessee, Knoxville, Tennessee. Occasional Paper No. 5, Frank H. McClung Museum, University of Tennessee, Knoxville, Tennessee. Publications in Anthropology No. 41, Tennessee Valley Authority.

Chaput, Donald
1969   Michipicoten Island: Ghosts, Copper and Bad Luck. *Ontario History* 61(4):217–223.
1971   *The Cliff: America's First Great Copper Mine*. Sequoia Press, Kalamazoo, Michigan.

Charles, Douglas K., Julieann Van Nest and Jane E. Buikstra
2004   Minerals and Meaning in the Hopewellian World. In *Soils, Stones and Symbols: Cultural Perceptions of the Mineral World*, edited by Nicole Boivin and Mary Ann Owoc, pp. 43–70. UCL Press, London, England.

Chastain, Matthew
2009   *Prehistoric Copper Metallurgy at Cahokia, Illinois*. Master's thesis, Materials Science and Engineering, Northwestern University, Evanston, Illinois.

Chastain, Matthew L., Alix C. Deymier-Black, John E. Kelly, James A. Brown, and David C. Dunand
2011   Metallurgical Analysis of Copper Artifacts from Cahokia. *Journal of Archaeological Science* 38:1727–1736.

Cherry, Peter P.
1877   *The Grave-Creek Mound: Its History, and Its Inscribed Stone with Its Vindication*. Privately published, Steam Printing House, Wadsworth, Ohio.

Chicago Daily Tribune
1856   Ancient Copper Diggings of Lake Superior. February 22, p. 1.

Childs, S. Terry
1994   Native Copper Technology and Society in Eastern North America. In *Archaeometry of Pre-Columbian Sites and Artifacts: Proceedings of a Symposium Organized by the UCLA Institute of Archaeology and the Getty Conservation Institute, Los Angeles, California, March 23–27, 1992*, edited by David A. Scott and Pieter Meyers, pp. 229–253. Getty Conservation Institute, Los Angeles.

Christian, Sarah Barr
1932   *Winter on Isle Royale: A Narrative of Life on Isle Royale during the Years of 1874 and 1875*. Privately published.

Church, Flora
1983   An Analysis of Textile Fragments from Three Ohio Hopewell Sites. *Ohio Archaeologist* 33(1):10–15.

1987   Current Research on Ohio's Prehistoric Textiles. In *Ethnicity and Culture: Proceedings of the Eighteenth Annual Conference of the Archaeological Association of the University of Calgary*, edited by Réginald Auger, Margaret F. Glass, Scott MacEachern, and Peter H. McCartney, pp. 155–161. Archaeological Association, Department of Archaeology, University of Calgary, Calgary, Alberta, Canada.

Cincinnati Gazette
1870   The Mound Builders of Ohio. *The Fire Lands Pioneer* 10:75–76

Clark, Caven P.
1988   *Survey and Testing at Isle Royale National Park, 1987 Season*. Midwest Archeological Center, National Park Service, Lincoln, Nebraska.
1990   *Archeological Survey and Testing at Isle Royale National Park, 1986–1990 Seasons*. Midwest Archeological Center, National Park Service, Lincoln, Nebraska.
1991   *Group Composition and the Role of Unique Raw Materials in the Terminal Wood and Substage of the Lake Superior Basin*. PhD dissertation, Department of Anthropology, Michigan State University, East Lansing, Michigan.
1995   *Archaeological Survey and Testing at Isle Royale National Park, 1987–1990 Seasons*. Occasional Studies in Anthropology No. 32. Midwest Archeological Center, Lincoln, Nebraska.
1996   Old Fort and Old Copper: The Search for the Archaic Stage on Isle Royale. In *Investigating the Archaeological Record of the Great Lakes State*, edited by Margaret B. Holman, Janet G. Brashler, and Kathryn E. Parker, pp. 101–138. New Issues Press, Western Michigan University, Kalamazoo, Michigan.

Clark, Caven P. and Susan R. Martin
2005   A Risky Business: Late Woodland Copper Mining on Lake Superior. In *The Cultural Landscape of Prehistoric Mines*, edited by Peter Topping and Mark Lynott, pp. 110–122. Oxbow Books, Oxford, England.

Clark, D. E. and B. A. Purdy
1982   Early Metallurgy in North America. In *Early Pyrotechnology: The Evolution of the First Fire-using Industries*, edited by Theodore A. Wertime and Steven F. Wertime, pp. 45–58. Smithsonian Institution Press, Washington, DC.

Clark, W. M.
1878   Antiquities of Tennessee. In *Annual Report of the Board of Regents of the Smithsonian Institution*, 1877, pp. 269–276.

Clarke, Don H.
1975a   *Central Mining Company*. Copper Mines of Keweenaw No. 7. Privately published by the author.
1975b   *Northwest Copper Mining Association*. Copper Mines of Keweenaw No. 9. Privately published by the author.
1978   *Minesota Mining Company*. Copper Mines of Keweenaw No. 11. Privately published by the author.
1990   *Isle Royale Mine*. Copper Mines of Keweenaw No. 18. Privately published by the author.

Clarke, Robert E.
1853   Notes from the Copper Region. *Harper's New Monthly Magazine*. 6(35):577–588.

Bibliography

Clarke, W. P.
    1884   Letter to the Editor. *The American Antiquarian and Oriental Journal* 6(2):108.

Cleland, Charles E.
    1966   *The Prehistoric Animal Ecology and Ethnozoology of the Upper Great Lakes Region*. Anthropological Papers No 29. Museum of Anthropology, University of Michigan, Ann Arbor, Michigan.

Cleland, Charles E. and Edwin N. Wilmsen
    1969   Three Unusual Copper Implements from Houghton County, Michigan. *The Wisconsin Archeologist* 50(1):26–32.
    1966   *The Prehistoric Animal Ecology and Ethnozoology of the Upper Great Lakes Region*. Anthropological Papers No 29. Museum of Anthropology, University of Michigan, Ann Arbor, Michigan.

Clermont, Norman et Claude Chapdelaine
    1998   *Ile Morrison: Lieu Sacré et Atelier de l'Archaïque dans l'Outaouais*. Paléo-No. Quebec No. 28, Musée Canadien des Civilisations, Recherches Amérindiennes au Québec, Montréal, QC.

Clermont, Norman, Claude Chapdelaine, et Jacques Cinq-Mars
    2003   *L'île aux Allumettes: L'Archaïque Supèrieur dans l'Outaouais*. Paléo-Québec No. 30, Musée Canadien des Civilisations, Recherches Amérindiennes au Québec, Montréal, QC.

Clifford, John D.
    1820   Indian Antiquities. *The Western Review and Miscellaneous Magazine* 1(2):96–100.

Cobb, Charles R.
    1991   One Hundred Years of Investigations at the Linn Site in Southern Illinois. *Illinois Archaeology* 3(1):56–76.

Cobb, Dawn E.
    1999   Relocation and Reanalysis of the Rose Mound Group. *Illinois Antiquity* 34(4):9–12.

Cochran, Donald R. and Beth K. McCord
    2001   *The Archaeology of Anderson Mounds, Mounds State Park, Anderson, Indiana*. Reports of Investigation 61. Archaeological Resources Management Service, Ball State University, Muncie, Indiana.

Coco, Emily
    2014   Spatial Analysis of Debris from the Mound 34 Copper Workshop. Unpublished research paper under the direction of John Kelly. Washington University, St. Louis.

Coe, Joffre Lanning
    1995   *Town Creek Indian Mound: A Native American Legacy*. University of North Carolina Press, Chapel Hill, North Carolina.

Coffinberry, Wright L.
    1885   That Archaeological Find. *Grand Rapids Eagle*, June 4, p. 3.
    1962   Coffinberry Manuscript – Article IV (Cont'd). *The Coffinberry News Bulletin* 9(9):106–107.

Coffinberry, Wright L. and E. A. Strong
    1876   Notes upon Some Explorations of Ancient Mounds in the Vicinity of Grand Rapids, Mich. *Proceedings of the American Association for the Advancement of Science, Twenty-fourth Meeting, Held at Detroit, Michigan, August, 1875*, pp. 293–297.

Cole, Fay-Cooper and Thorne Deuel
    1937   *Rediscovering Illinois: Archaeological Explorations In and Around Fulton County*. University of Chicago Press, Chicago.

Cole, Gloria G.
    1981   *The Murphy Hill Site (1Msd300): The Structural Study of a Copena Mound and Comparative Review of the Copena Mortuary Complex*. Research Series No. 3. Office of Archaeological Research, University of Alabama, Tuscaloosa, Alabama and Publication No. 31, Tennessee Valley Authority, Chattanooga, Tennessee.

Cole, H. E.
    1922   Summary of the Archeology of Western Sauk County. *The Wisconsin Archeologist* 1(3):81–111.

Collins, David R.
    1979   *Archaeology of Clark County*. Clark County Historical Society, Springfield, Ohio.

Collins, Henry B.
    1926   Archaeological and Anthropometrical Work in Mississippi. In *Explorations and Field-Work of the Smithsonian Institution in 1925*. Smithsonian Miscellaneous Collections 78(1):89–95.

Columbian Mining Company
    1862   *Report of the Directors of the Columbian Mining Company, also, Reports of J. H. Forster, Esq., and Capt. W. B. Frue, of the Pewabic and Franklin Mines, together with Statement from the Books of the Treasurer*. Wm. D. Roe & Co., Stationers and Printers, New York.

Commercial Bulletin
    1849   Antiquity of Mining Operations on Lake Superior. *Lake Superior News and Mining Journal* [Sault Ste. Marie, Michigan], July 13, p. 1.

Conant, A. J.
    1877   The Mounds and Their Builders-or-Traces of Pre-Historic Man in Missouri. In *The Commonwealth of Missouri; a Centennial Record*, edited by C. R. Barns, pp. 1–122. Bryan, Brand & Co., St. Louis.

Conrad, Lawrence A.
    1972   *1966 Excavations at the Dickson Mound: A Sepo-Spoon River Burial Mound in the Central Illinois River Valley*. Master's thesis, University of Wisconsin, Madison, Wisconsin.
    1986   The Late Archaic/Early Woodland Transition in the Interior of West-Central Illinois. In *Early Woodland Archeology*, edited by Kenneth B. Farnsworth and Thomas E. Emerson, pp. 301–325. Kampsville Seminars in Archeology, Vol. 2. Center for American Archeology, Center for American Archeology Press, Kampsville, Illinois.
    1989   The Southeastern Ceremonial Complex on the Northern Mississippian Frontier:

Bibliography

Late Prehistoric Politico-religious Systems in the Central Illinois River Valley. In *The Southeastern Ceremonial Complex: Artifacts and Analysis*, edited by Patricia Galloway, pp. 93–113. University of Nebraska Press, Lincoln, Nebraska.
1991   The Middle Mississippian Cultures of the Central Illinois Valley. In *Cahokia and the Hinterlands*, edited by Thomas E. Emerson and R. Barry Lewis, pp. 119–156. University of Illinois Press, Urbana and Chicago.

Converse, Robert N.
1972   Glacial Kame Artifacts. *Ohio Archaeologist* 22(2):43–44.
1980   *The Glacial Kame Indians*. The Archaeological Society of Ohio.
2001   A Hopewell Shell Container and Ear Spools from Campbell Island. *Ohio Archaeologist* 51(3):22.
2002   The 1876 Excavation of the Baldwin Mound, Clark County, Ohio. *Ohio Archaeologist* 52(4):16–17.
2003   *The Archaeology of Ohio*. The Archaeological Society of Ohio.

Conway, Thor A.
1977   *Whitefish Island – A Remarkable Archaeological Site at Sault Ste. Marie, Ontario*. Data Box 310 (Research Manuscript Series). Ontario Ministry of Culture and Recreation, Historical Planning and Research Branch, Toronto, Ontario, Canada.
1980   Point aux Pins Archaeology: Woodland and Historic Components. In *Collected Archaeological Papers*, edited by David Skene Melvin, pp. 29–63. Archaeological Research Report 13. Ontario Ministry of Culture and Recreation, Historical Planning and Research Branch, Toronto, Ontario, Canada.
1981   *Archaeology in Northeastern Ontario: Searching for Our Past*. Ontario Ministry of Culture and Recreation, Historical Planning and Research Branch, Ottawa, Toronto, Canada.

Cook, Fred C. and Charles E. Pearson
1989   The Southeastern Ceremonial Complex on the Georgia Coast. In *The Southeastern Ceremonial Complex: Artifacts and Analysis*, edited by Patricia Galloway, pp. 147–165. University of Nebraska Press, Lincoln, Nebraska.

Cooley, Thomas M.
1860   *James H. Titus v. The Minnesota Mining Company*. In *Michigan Reports. Reports of Cases Heard and Decided in the Supreme Court of Michigan from January 4 to October 13, 1860*. 4:183–260. Published by the Reporter, Ann Arbor. F. Raymond & Co. and S. D. Ellwood, Detroit.

Coon, David S.
1940   The Quincy Mine. *Michigan History* 24(1):91–103.

Cooper, Dennis Glen
1939   Geographic Influences on the History and Development of Isle Royale, Michigan. *Papers of the Michigan Academy of Science, Arts and Letters* 24(3):1–8.

Cooper, H. Kory.
2007   *The Anthropology of Native Copper Technology and Social Complexity in Alaska and the Yukon Territory: An Analysis Using Archaeology, Archaeometry, and Ethnohistory*. PhD dissertation, University of Alberta, Edmonton, Alberta, Canada.

Cooper, James B.
  1901   Historical Sketch of Smelting and Refining Lake Copper. *Proceedings of the Lake Superior Mining Institute* 7:23–26.
  1903   Treatment of Lake Copper. *Mines and Minerals* 23(10):463–464.

Cooper, Leland R.
  1933   The Red Cedar Variant of the Wisconsin Hopewell Culture. *Bulletin of the Public Museum of the City of Milwaukee* 16(2), pp. 47–108.
  1959   Indian Mounds Park Archaeological Site, Rice Lake, Wisconsin. *Science Bulletin No. 6*. The Science Museum of the St. Paul Institute, St. Paul, Minnesota.
  1967   *Preliminary Report: Archaeological Survey and Excavation of the Stumme and Vach Sites, Pine County, Minnesota, 1966.* (Issued by the Minnesota Historical Society as part of the Minnesota Natural Resources program on Prehistoric Archaeology directed by the State Archaeologist, Elden Johnson.)

copperagates.com
  2016   Native Copper Artifacts. http://www.copperagates.com/home/native-copper-artifacts.

Copper Falls Mining Company
  1852   *Third Annual Report of the Directors of the Copper Falls Mining Company, with the Report of Samuel W. Hill, Superintendent of the Mine, and the Charter and By-Laws of the Company.* Damrell & Moore, Boston.
  1854   *Reports of the Directors of the Copper Falls Mining Company.* Evans and Plummer, Boston.
  1859   *Annual Reports to the Stockholders of the Copper Falls Mining Company, to April 30, 1859.* Geo. C. Rand & Avery, Boston.
  1860   *Report of the Copper Falls Mining Company for One Year Ending April 30, 1860.* Geo C. Rand & Avery, Boston.

Cordier, Pierre Louis Antoine
  1849   Native Copper in America. *Hunt's Merchants' Magazine and Commercial Review* 20(6):679–680.

Cornett, B. Kenneth
  1976   Excavations at Tallassee (40Bt8): An Historic Cherokee Village Site in East Tennessee. *Tennessee Archaeologist* 32(1–2):11–19.

Côté, Marc
  1993   Le Site DaGt-1: Un Établissement Algonquin du Sylvicole Supérieur en Abitibi-Témiscamingue. In *Traces du passé, Images du présent: Anthropologie amérindienne du Moyen-nord québécois,* edited by Marc Côté and Gaétan L. Lessard, pp. 5–59. Cégep-Éditeur, Cégep de l'Abitibi-Temiscamingue, Rouyn-Noranda, Québec.

Cotter, John L.
  1952   The Mangum Plate. *American Antiquity* 18(1):65–68.

Cotter, John L. and John M. Corbett
  1951   *Archeology of the Bynum Mounds, Mississippi.* Archeological Research Series No. 1. National Park Service, U. S. Department of the Interior, Washington, DC.

Bibliography

Coulter Mining Company
  1864   *Prospectus and Reports of the Coulter Mining Company of Michigan.* J. B.
           Chandler, Printer, Philadelphia.

Cowan, C. Wesley
  1996   Social Implications of Ohio Hopewell Art. In *A View from the Core: A Synthesis
           of Ohio Hopewell Archaeology*, edited by Paul J. Pacheco, pp. 128–148. Ohio
           Archaeological Council, Columbus, Ohio.

Cox, Jim E.
  2014   A Spiro Copper Hand-and-Eye Motif and Its Interpretations. *Central States
           Archaeological Journal* 61(4):264–270.

Crane, H.R. and James B. Griffin
  1964   "University of Michigan Radicarbon Dates IX." *Radiocarbon* 6: 1–27.
  1965   "University of Michigan Radicarbon Dates X." *Radiocarbon* 7: 123–152.

Cree, Beth
  1992   Hopewell Panpipes: A Recent Discovery in Indiana. *Midcontinental Journal of
           Archaeology* 17(1):3–15.

Cresson, H. T.
  1884   Notes on Prehistoric Copper Implements. *Proceedings of the Academy of Natural
           Sciences of Philadelphia*, 1883, 35:55–56.

Crowder, Lisa E.
  1988   Cultural Affiliations of the King Site. In *The King Site: Continuity and Contact in
           Sixteenth-Century Georgia*, edited by Robert L. Blakely, pp. 47–59. University of
           Georgia Press, Athens, Georgia.

Crucefix, Lanna
  2001   *Copper Use in the Old Copper Complex: A Comparative Analysis of Wittry
           VI-C Copper Axes and Three-Quarter Grooved Stones Axes.* Master's thesis,
           Department of Archaeology, Simon Fraser University, Burnaby, British Columbia,
           Canada._

Crumpler, Gus H.
  1969   The Wilkerson Indian Graveyard. *Central States Archaeological Journal* 16(1):12–21.

Cultural Resource Analysts, Inc.
  2009   Copper Earspools. http://crai-prehistory.blogspot.com/2008/07/copper-earspools.html.

Cumming, Robert B., Jr.
  1958   *Archaeological Investigations at the Tuttle Creek Dam, Kansas.* Bulletin 169. Bureau
           of American Ethnology, Smithsonian Institution, Washington, DC.

Cunningham, Wilbur B.
  1948   A Study of the Glacial Kame Culture in Michigan, Ohio, and Indiana. *Occasional
           Contributions from the Museum of Anthropology of the University of Michigan*
           No. 12. University of Michigan Press, Ann Arbor, Michigan.

Curren, Caleb
    1984   *The Protohistoric Period in Central Alabama*. Alabama Tombigbee Regional Commission, Camden, Alabama.

Curry, Dennis C.
    2002   The Old Copper Culture in Maryland? *Maryland Archeology* 38(2):33–34.

Cushing, Frank Hamilton
    1894a   Primitive Copper Working: An Experimental Study. *American Anthropologist* 7(1):93–117.
    1894b   Primitive Copper Working: An Experimental Study. *The Archaeologist* 2(4): 97–105; 2(5):129–141.

Custer, Milo
    1910   Archaeology of McLean County, Illinois. *The Archaeological Bulletin* 1(2):49–51.

Daily Mining Gazette [Houghton, Michigan]
    1899   May Be Explored Next Summer: The Flint Steel River Copper Property in Ontonagon County: Something about Its History: First Mined in the Early Fifties and Always Worked in a Desultory Manner. October 27, p. 6.

Daniels, E. D.
    1904   *A Twentieth Century and Biographical Record of LaPorte County, Indiana*. Lewis Publishing Company, Chicago.

Davis, A. C.
    1875   Antiquities of Isle Royale, Lake Superior. *Annual Report of the Board of Regents of the Smithsonian Institution*, 1874, pp. 369–370.

Davis, Hester A.
    1958   Weekend Excavations during the Fall of 1958. *Iowa Archeological Society Newsletter* 28:2–13.

Davis, Watson
    1924   *The Story of Copper*. The Century Co., New York.

Dawson, K. C. A.
    1966   Isolated Copper Artifacts from Northwestern Ontario. *Ontario Archaeology* 9:63–67.
    1980   The MacGillivray Site: A Laurel Tradition Site in Northwestern Ontario. *Ontario Archaeology* 34:45–68.
    1983   Prehistory of the Interior Forest of Northern Ontario. In *Boreal Forest Adaptations: The Northern Algonkians*, edited by A. Theodore Steegman, Jr., pp. 55–84. Plenum Press, New York.
    1987   The Martin-Bird Site. *Ontario Archaeology* 47:33–57.

Day, Grant L.
    1996   *Copper Mines and Mining in Ontonagon County, Michigan*. (2 volumes). Department of Social Sciences, Michigan Technological University, Houghton, Michigan.

Bibliography

DeBoer, Warren R.
>   2004   Little Bighorn on the Scioto: The Rocky Mountain Connection to Ohio Hopewell.
>   *American Antiquity* 69(1):85–107.

DeHart, R. P.
>   1909   *Past and Present of Tippecanoe County, Indiana* (Vol. 1). B. F. Bowen &
>   Company, Publishers, Indianapolis.

DeJarnette, David L.
>   1952   Alabama Archeology: A Summary. In *Archeology of Eastern United States*, edited
>   by James B. Griffin, pp. 272–284. University of Chicago Press, Chicago.

Delafield, J., Jr.
>   1834   *A Brief Topographical Description of the County of Washington, in the State of
>   Ohio.* J. M. Elliott, 6 Old Slip, New York.

Demeter, C. Stephan and Elaine H. Robinson
>   1999   *Phase I Archaeological Literature Review and Above-Ground Assessment of the
>   US-131 Curve Crossing of the Grand River in the City of Grand Rapids. Kent
>   County, Michigan.* Commonwealth Cultural Resources Group, Inc., Jackson,
>   Michigan.

DePratter, Chester B.
>   1991   *Late Prehistoric and Early Historic Chiefdoms in the Southeastern United States.*
>   Garland Publishing, Inc., New York.

Detroit Free Press [Detroit, Michigan]
>   1850   Copper from Ontonagon. *Lake Superior Journal*, July 18.
>   1855a   A Curiosity. *New York Journal of Commerce,* September 18.
>   1855b   Fatal Accident. *Lake Superior Journal*, October 4.
>   1855c   Another Fatal Accident. *Lake Superior Miner*, October 4.
>   1858a   Antiquities in Lake Superior Mines. *The Historical Magazine* 2(10):299.
>   1858b   Antiquities in the Lake Superior Mines. *Lake Superior Miner* [Ontonagon,
>   Michigan], August 21, p. 393.
>   1893   Michigan at the World's Fair. April 30, pp. 25–27.
>   1898   The Calumet & Hecla: An Interesting Story of the Discovery of This Greatest of
>   Copper Mines. November 28, p. 3.

Deuel, Thorne
>   1952a   The Hopewellian Community. In *Hopewellian Communities in Illinois*, edited
>   by Thorne Deuel, pp. 249–265. Scientific Papers 5(6). Illinois State Museum,
>   Springfield, Illinois.
>   1952b   Hopewellian Dress in Illinois. In *Archeology of Eastern United States*, edited by
>   James B. Griffin, pp. 165–175. University of Chicago Press, Chicago.

DeVisscher, Jerry
>   1957   Three Macomb County "Copper and Mica" Burials. *The Totem Pole* 40(4).

Devon Mining Company
>   1863   *Statement Concerning the Devon Mines of Lake Superior.* T. R. Marvin & Son,
>   Boston.

DiBlasi, Philip J. and Bobbie K. Sudhoff
    1978   *An Archaeological Reconnaissance of the Kentucky Side of the Smithland Pool Project on the Ohio River*. October 1–November 2. Prepared under Supervision of and with a Foreword by J. E. Granger, Ph. D., Principal Investigator. Contract No. DACW 27-77-C-0133. Submitted to United States Army Corps of Engineers, Louisville District, Louisville, Kentucky.

Dickens, Roy S., Jr.,
    1982   *Of Sky and Earth: Art of the Early Southeastern Indians*. October 1 – November 28, 1982. The High Museum of Art, Atlanta, Georgia. National Endowment for the Arts, the Georgia Council for the Arts and Humanities, and the National Endowment for the Humanities through the Georgia Endowment for the Humanities.

Dickson, Don F.
    1956   The Liverpool Mounds. *Central States Archaeological Journal* 2(3):85–92.

Dille, I.
    1867   Sketch of Ancient Earthworks. In *Annual Report of the Board of Regents of the Smithsonian Institution*, 1866, pp. 359–362.

Dirst, Victoria
    1985   *Three Classic Oneota Sites in East Central Wisconsin*. State Historical Society of Wisconsin, Madison, Wisconsin.

Dockstader, Frederick
    1985   The Thruston Collection Re-viewed. In *Arts and Artisans of Prehistoric Middle Tennessee: The Gates P. Thruston Collection of Vanderbilt University Held in Trust by the Tennessee State Museum*, pp. 37–58. Tennessee State Museum, Nashville, Tennessee.

Dodds, Gilbert F.
    1950a   Ancient Burial Mound Opened in Delaware County: 1949. *Ohio Indian Relic Collectors Society Bulletin* 23:6, 8.

    1950b   Hopewell Culture Mound Explored by Ross County Historical Society. *Ohio Indian Relic Collectors Society Bulletin* 23:7, 9.

Donaldson, William S. and Stanley Wortner
    1995   The Hind Site and the Glacial Kame Burial Complex in Ontario. *Ontario Archaeology* 59:5–108.

Dorsey, George A.
    1903   Remarks by George A. Dorsey Concerning "Sheet-Copper from the Mounds is Not Necessarily of European Origin." *American Anthropologist* 5(1):49.

Dorwin, John T.
    1971   The Bowen Site: An Archaeological Study of Culture Process in the Late Prehistory of Central Indiana. *Prehistory Research Series* 4(4):192–411. Indiana Historical Society, Indianapolis.

Dougherty, Irvin S.
    1956   Gravel Kame Artifacts. *Ohio Archaeologist* 6(3):105–107.

Bibliography

Douglas, Frederic H.
> 1936 Copper and the Indian. Denver Art Museum Leaflet 75–76. Denver, CO.

Douglass, Robert S.
> 1912 *History of Southeast Missouri* (Vol. 1). Lewis Publishing Company, Chicago and New York.

Dowagiac Times (Dowagiac, Michigan)
> 1888 The Mound Builders. Prof. Crane's Investigation of Their Burial Places in Sumnerville. October 25, p. 5.

Dragoo, Don W.
> 1951 *Archaeological Survey of Shelby County, Indiana.* Indiana Historical Bureau, Indianapolis.
> 1963 Mounds for the Dead: An Analysis of the Adena Culture. *Annals of Carnegie Museum*, Vol. 37. Pittsburgh, Pennsylvania.
> 1964 The Development of Adena Culture and Its Role in the Formation of Ohio Hopewell. In *Hopewellian Studies*, edited by Joseph R. Caldwell and Robert L. Hall, pp. 1–34. Scientific Papers 12(1). Illinois State Museum, Springfield, Illinois.

Drake, Daniel
> 1815 Natural and Statistical View; or Picture of Cincinnati and the Miami Country, Illustrated by Maps. Printed by Looker and Wallace, Cincinnati, Ohio.

Draper, Lyman C.
> 1876 Additions to the Cabinet. Twentieth Annual Report. In *Report and Collections of the State Historical Society of Wisconsin, For the Years 1873, 1874, 1875, and 1876*, Vol. 7, pp. 27–44. E. B. Bolens, State Printer, Madison, Wisconsin.
> 1879a Additions to the Cabinet. Twenty-third Annual Report. In *Report and Collections of the State Historical Society of Wisconsin, For the Years 1877, 1878, and 1879*, Vol. 8, pp. 13–32. David Atwood, State Printer, Madison, Wisconsin.
> 1879b Additions to the Cabinet. Twenty-fourth Annual Report. In *Report and Collections of the State Historical Society of Wisconsin, For the Years 1877, 1878, and 1879*, Vol. 8, pp. 32–60. David Atwood, State Printer, Madison, Wisconsin.
> 1879c Additions to the Cabinet. Twenty-fifth Annual Report. In *Report and Collections of the State Historical Society of Wisconsin, For the Years 1877, 1878, and 1879*, Vol. 8, pp. 60–85. David Atwood, State Printer, Madison, Wisconsin.
> 1882a Additions to the Cabinet. Twenty-sixth Annual Report. In *Report and Collections of the State Historical Society of Wisconsin, For the Years 1877, 1878, and 1879*, Vol. 9, pp. 13–21. David Atwood, State Printer, Madison, Wisconsin.
> 1882b Additions to the Cabinet. Twenty-fifth Annual Report. In *Report and Collections of the State Historical Society of Wisconsin, For the Years 1877, 1878, and 1879*, Vol. 9, pp. 60–85. David Atwood, State Printer, Madison, Wisconsin.

Draper, Lyman C., Fred S. Perkins, Charles Whittlesey, and Philo R. Hoy
> 1879 Fabrication of Ancient Copper Implements. In *Report and Collections of the State Historical Society of Wisconsin, For the Years 1877, 1878, and 1879*, Vol. 8, pp. 166–173. David Atwood, State Printer, Madison, Wisconsin.

Dresslar, Jim and Bob Clay
> 1972 The Porter Mound in Jennings County, Indiana. *The Redskin* 7(3):103–110.

Drier, Roy W.
> 1961   The Michigan College of Mining and Technology Isle Royale Excavations, 1953–
> 1954. In *Lake Superior Copper and the Indians: Miscellaneous Studies of Great
> Lakes Prehistory*, edited by James B. Griffin, pp. 1–7. Anthropological Papers No.
> 17. Museum of Anthropology, University of Michigan, Ann Arbor, Michigan.

Drier, Roy W. and Octave J. Du Temple
> 1961   *Prehistoric Copper Mining in the Lake Superior Region: A Collection of
> Reference Articles*. Privately published.

Drooker, Penelope B.
> 1992   *Mississippian Village Textiles at Wickliffe*. University of Alabama Press,
> Tuscaloosa, Alabama.
> 1995   Asking Old Museum Collections New Questions: Protohistoric Fort Ancient Social
> Organization and Interregional Interaction. *Museum Anthropology* 19(3):3–16.
> 1997   *The View from Madisonville: Protohistoric Western Fort Ancient Interaction
> Patterns*. Memoirs No. 31. Museum of Anthropology, University of Michigan,
> Ann Arbor, Michigan.

Dun, Walter A.
> 1886   Report on Mound Exploration in Greene County, Ohio. *Journal of the Cincinnati
> Society of Natural History* 8(4):231–233.

Duncan, James R.
> 1967   Suggested Cultural Affiliations of the Vavak Copper Plaque. *Central States
> Archaeological Journal* 14(2):75–78.

Duncan, James R. and Carol Diaz-Granados
> 2000   Of Masks and Myths. *Midcontinental Journal of Archaeology* 25(1):1–26.

Dunlap, Joseph
> 1961   Exploration of the Henry Boose Site. *Ohio Archaeologist* 11(1):4–10.

Dunnell, Robert C.
> 1962   *The Hughes Farm Site (46-Oh-9), Ohio County, West Virginia. Publication Series
> No. 7*. West Virginia Archeological Society, Inc., Moundsville, West Virginia.

Durham, James H. and Michael K. Davis
> 1974   Report on Burials Found at Crenshaw Mound "C", Miller County, Arkansas.
> *Bulletin of the Oklahoma Anthropological Society* 23:1–90.

Dustin, Fred
> 1932   A Summary of the Archaeology of Isle Royale, Michigan. *Papers of the Michigan
> Academy of Science, Arts and Letters* 16:1–16.
> 1957   An Archaeological Reconnaissance of Isle Royale. *Michigan History Magazine*
> 41:1–34.

Dyck, Ian
> 1983   The Prehistory of Southern Saskatchewan. In *Tracking Ancient Hunters: Prehistoric
> Archaeology in Saskatchewan*, edited by Henry T. Epp and Ian Dyck, pp. 63–139.
> Saskatchewan Archaeological Society, Regina, Saskatchewan, Canada.

Bibliography

Dye, David H.
  2006   The Transformation of Mississippian Warfare: Four Case Studies from the Mid-South. In *The Archaeology of Warfare: Prehistories of Raiding and Conquest*, edited by Elizabeth N. Arkush and Mark W. Allen, pp. 101–147. University Press of Florida, Gainesville, Florida.
  2007   Ritual, Medicine, and the War Trophy Iconographic Theme in the Mississippian Southeast. In *Ancient Objects and Sacred Realms: Interpretations of Mississippian Iconography*, edited by F. Kent Reilly III and James F. Garber, pp. 152–173. University of Texas Press, Austin, Texas.

Eagle Harbor & Waterbury Mining Companies
  1857   *Report of Directors of the Eagle Harbor & Waterbury Mining Companies*. Daily Free Press Book and Job Office, Detroit.

Eames, Frank
  1924   Regional Notes on Specimens of Primitive Copper Craft. In *Thirty-Fourth Annual Archaeological Report 1923. Being Part of Appendix to the Report of the Minister of Education Ontario*, pp. 108–113.

Early, Ann M., Barbara A. Burnett and Daniel Wolfman
  1988   Standridge: Caddoan Settlement in a Mountain Environment. *Arkansas Archeological Survey Research Series* No. 29.

Easterla, David
  2007   A Copper Adz from Northwest Missouri. *Central States Archaeological Journal* 54(1):17.

Eastman, Jane M.
  2001   Life Courses and Gender among Late Prehistoric Siouan Communities. In *Archaeological Studies of Gender in the Southeastern United States*, edited by Jane M. Eastman and Christopher B. Rodning, pp. 57–76. University Press of Florida, Gainesville, Florida.
  2002   Mortuary Analysis and Gender: The Response of Siouan Peoples to European Contact. In *The Archaeology of Native North Carolina: Papers in Honor of H. Trawick Ward*, edited by Jane M. Eastman, Christopher B. Rodning, and Edmond A. Boudreaux III, pp. 46–56. Southeastern Archaeological Conference Special Publication 7, Biloxi, Mississippi.

Eberle, Don
  1997   A Putnam County Copper Adze. *Ohio Archaeologist* 47(4):23.

Egleston, Thomas
  1879   Copper Mining on Lake Superior. *Transactions of the American Institute of Mining Engineers* 6:275–312.

Ehrhardt, Kathleen L.
  2002   *European Materials in Native American Contexts*. PhD dissertation, Department of Anthropology, New York University, New York.
  2005   *European Metals in Native Hands: Rethinking Technological Change, 1640–1683*. University of Alabama Press, Tuscaloosa, Alabama.
  2009   Copper Working Technologies, Contexts of Use, and Social Complexity in the Eastern Woodlands of Native North America. *Journal of World Prehistory* 22(3):213–235.

2013   "Style" in Crafting Hybrid Material on the Fringes of Empire: An Example from the Native North American Midcontinent. In *The Archaeology of Hybrid Material Culture*, edited by Jeb J. Card, pp. 364–396. Occasional Paper No. 39, Center for Archaeological Investigations, Southern Illinois University, Carbondale, Illinois.

Ehrhardt, Kathleen L., Samuel K. Nash, and Charles P. Swann
2000   Metal-forming Practices among the Seventeenth Century Illinois, 1640–1682. *Materials Characterization* 45:275–288.

Ellis, Chris J., Ian T. Kenyon, and Michael W. Spence
1990   The Archaic. In *The Archaeology of Southern Ontario to A. D. 1650*, edited by Chris J. Ellis and Michael W. Spence, pp. 65–124. Publication No. 5. Occasional Publications of the London Chapter, Ontario Archaeological Society Inc., London, Ontario, Canada.

Ellis, Christopher, Peter A. Timmins, and Holly Martelle
2009   At the Crossroads and Periphery: The Archaic Archaeological Record of Southern Ontario. In *Archaic Societies: Diversity and Complexity across the Midcontinent*, edited by Thomas E. Emerson, Dale L. McElrath, and Andrew C. Fortier, pp. 787–837. State University Press of New York, Albany, New York.

Elrod, Moses N.
1884   Geology of Rush County. In *Thirteenth Annual Report*, Indiana Department of Geology and Natural History, John Collett, State Geologist, pp. 86–115.

Emerson, George D.
1857a   Mining Matters. *Lake Superior Miner* [Ontonagon, Michigan], January 3, p. 2.
1857b   Mining Matters. *Lake Superior Miner* [Ontonagon, Michigan], February 14, p. 2.
1857c   Mining Matters. *Lake Superior Miner* [Ontonagon, Michigan], April 4, p. 2.
1857d   Mining Matters. *Lake Superior Miner* [Ontonagon, Michigan], May 23, p. 2.
1857e   Mining Matters. *Lake Superior Miner* [Ontonagon, Michigan], May 30, p. 2.
1857f   Mining Matters. *Lake Superior Miner* [Ontonagon, Michigan], June 6, p. 2.
1857g   Mining Matters. *Lake Superior Miner* [Ontonagon, Michigan], July 4, p. 2.
1857h   Mining Matters. *Lake Superior Miner* [Ontonagon, Michigan], July 18, p. 2.
1857i   Mining Matters. *Lake Superior Miner* [Ontonagon, Michigan], August 15, pp. 4–5.
1857j   Mining Matters. *Lake Superior Miner* [Ontonagon, Michigan], October 17, p. 44.
1857k   Mining Matters. *Lake Superior Miner* [Ontonagon, Michigan], November 6, p. 68.
1857l   Mining Matters. *Lake Superior Miner* [Ontonagon, Michigan], December 19, p. 117.
1857m   Mining Matters. *Lake Superior Miner* [Ontonagon, Michigan], December 26, p. 125.
1858a   By Whom and When Were the Copper Mines of Lake Superior First Worked? *Lake Superior Miner* [Ontonagon, Michigan], May 29, p. 299.
1858b   Illinois Copper. *Lake Superior Miner* [Ontonagon, Michigan], October 2, p. 28.
1858c   Mining Matters. *Lake Superior Miner* [Ontonagon, Michigan], June 19, 1858, p. 323.
1862a   Mining Intelligence. *Portage Lake Mining Gazette* [Houghton, Michigan], August 16, p. 4.
1862b   Mining Intelligence. *Portage Lake Mining Gazette* [Houghton, Michigan], August 30, p. 4.
1862c   Mining Intelligence. *Portage Lake Mining Gazette* [Houghton, Michigan], October 11, p. 4.

Emerson, George D. and H. McKenzie
1862   Mining Intelligence. *Portage Lake Mining Gazette* [Houghton, Michigan], December 13, p. 2.

Bibliography

1863 Mining Intelligence. *Portage Lake Mining Gazette* [Houghton, Michigan], March 21, p. 2.

Emerson, J. Norman
  1955 The Kant Site: A Point Peninsula Manifestation in Renfrew County, Ontario. *Transactions of the Royal Canadian Institute* 31(1):21–66.

Emerson, Thomas E.
  1997 *Cahokia and the Archaeology of Power.* University of Alabama Press, Tuscaloosa, Alabama.

Emerson, Thomas E. and James A. Brown
  1992 The Late Prehistory and Protohistory of Illinois. In *Calumet & Fleur-de-Lys: Archaeology of Indian and French Contact in the Midcontinent*, edited by John A. Walthall and Thomas E. Emerson, pp. 77–128. Smithsonian Institution Press, Washington, DC.

Emerson, Thomas E. and Eve Hargrave
  2000 Strangers in Paradise? Recognizing Ethnic Mortuary Diversity on the Fringes of Cahokia. *Southeastern Archaeology* 19(1):1–23.

Emerson, Thomas E., Kristin M. Hedman, Robert E. Warren, and Mary L. Simon
  2010 Langford Mortuary Patterns as Reflected in the Material Service Quarry in the Upper Illinois River Valley. *The Wisconsin Archeologist* 91(1):1–78.

Emerson, Thomas E. and Randall E. Hughes
  2001 De-Mything the Cahokia Catlinite Trade. *Plains Anthropologist* 46(175):149–161.

Emerson, Thomas E. and Timothy R. Pauketat
  2002 Embodying Power and Resistance at Cahokia. In *The Dynamics of Power*, edited by Maria O'Donovan, pp. 105–125. Occasional Paper No. 30. Center for Archaeological Investigations, Southern Illinois University, Carbondale, Illinois.

Esary, Duane
  1986 Red Ochre Mound Building and Marion Phase Associations: A Fulton County, Illinois Perspective. In *Early Woodland Archeology*, edited by Kenneth B. Farnsworth and Thomas E. Emerson, pp. 231–243. Kampsville Seminars in Archeology, Vol. 2. Center for American Archeology, Center for American Archeology Press, Kampsville, Illinois.

Esch, Jeanne
  1984 A Suggested Order among the Wulfing Plates. *The Chesopiean* 22(3):2–16.

Espy, Josiah
  1871 Memorandums of a Tour Made by Josiah Espy in the State of Ohio and Kentucky and Indian Territory in 1805. *Ohio Valley Historical Series, Miscellanies, No. 1*, pp. 1–28. Robert Clarke & Co., Cincinnati, Ohio.

Essenpreis, Patricia S. and Michael E. Moseley
  1984 Fort Ancient: Citadel or Coliseum?: Past and Present Field Museum Explorations of a Major American Monument. *Field Museum of Natural History Bulletin* 55(6):5–10, 20–26.

Evans, David R.
1977 Special Burials from the Lilbourn Site—1971. *The Missouri Archaeologist* 38:111–122.

Evergreen Bluff Mining Company
1854 *By-Laws and Articles of the Evergreen Bluff Mining Company, Organized September 1853.* Detroit.
1857 *Report of the Directors of the Evergreen Bluff Mining Company.* Free Press Book and Job Printing Establishment, Detroit.
1858 Report of the Evergreen Bluff Mining Company. *Lake Superior Miner* [Ontonagon, Michigan], pp. 210–211

Everman, Franklin
1971 Figure 101. *The Redskin* 6(4):138.

Everts & Co., L. H.
1877 Samuel W. Hill. In *History of Calhoun County, Michigan.* L. H. Everts & Co., Philadelphia.

Fahlstrom, Paul Gerin
1994 *The Great Copper Boulder of Ontonagon.* Pamphlet. Tracys Landing, Maryland.

Fairbanks, Charles H.
1952 Creek and Pre-Creek. In *Archeology of Eastern United States,* edited by James B. Griffin, pp. 285–300. University of Chicago Press, Chicago.
1956 *Archeology of the Funeral Mound, Ocmulgee National Monument, Georgia.* Archeological Research Series No. 3. National Park Service, U. S. Department of the Interior, Washington, DC.

Fairbanks, Charles H., Arthur R. Kelly, Gordon R. Willey, and Pat Wofford, Jr.
1946 The Leake Mounds, Bartow County, Georgia. *American Antiquity* 12(3):126–127.

Farnsworth, Kenneth B.
2004 *Early Hopewell Mound Explorations: The First Fifty Years in the Illinois River Valley.* Studies in Archaeology No. 3. Illinois Transportation Archaeological Research Program, University of Illinois, Urbana, Illinois

Farnsworth, Kenneth B. and David L. Asch
1986 Early Woodland Chronology, Artifact Styles, and Settlement Distribution in the Lower Illinois Valley Region. In *Early Woodland Archeology,* edited by Kenneth B. Farnsworth and Thomas E. Emerson, pp. 326–457. Kampsville Seminars in Archeology, Vol. 2. Center for American Archeology, Center for American Archeology Press, Kampsville, Illinois.

Farnsworth, Kenneth B. and Karen A. Atwell
2015 *Excavations at the Blue Island and Naples-Russell Mounds and Related Hopewellian Sites in the Lower Illinois Valley.* Research Report 34. Illinois State Archaeological Survey Reports, University of Illinois at Urbana-Champaign, Illinois.

Farquharson, R. J.
1876a Recent Explorations of Mounds near Davenport, Iowa. *Proceedings of the American Association for the Advancement of Science. Twenty-fourth Meeting, Held at Detroit, Michigan, August, 1875,* pp. 297–325.

Bibliography

1876b   Recent Archaeological Discoveries at Davenport, Iowa, of Copper Axes, Cloth, etc., Supposed to Have Come Down to us from a Pre-Historic People, called the Mound-Builders. *Proceedings of the Davenport Academy of Natural Sciences* 1:117–143.

Faulkner, Charles H.
1960   The Red Ochre Culture: An Early Burial Complex in Northern Indiana. *The Wisconsin Archeologist* 41(2):35–49.
1968   *The Old Stone Fort: Exploring an Archaeological Mystery.* University of Tennessee Press, Knoxville, Tennessee.
1972   The Late Prehistoric Occupation of Northwestern Indiana: A Study of the Upper Mississippi Cultures of the Kankakee Valley. *Prehistory Research Series* 5(1):1–222. Indiana Historical Society, Indianapolis.

Feagins, Jim D.
1988   Nebraska Phase Burials in Northwest Missouri: A Study of Three Localities. *The Missouri Archaeologist* 49:41–56.

Featherstonhaugh, Thomas
1899   The Mound-Builders of Central Florida. *Publications of the Southern History Association* 3(1):1–14.

Fecht, William G.
1961   The Snyders Mound Group and Village Site. *Central States Archaeological Journal* 8(3):84–93.
1985   Morse Knives from the Snyders Site, Calhoun County, Illinois. *Central States Archaeological Journal* 32(1):8–10.

Feldser, William
1982   A Wythe County, Virginia, Burial. *Central States Archaeological Journal* 29(4):206–207.

Fenn, Thomas R.
2001   *Geochemical Investigation of Prehistoric Native Copper Artifacts, Northern Wisconsin.* Master's thesis, University of New Orleans, New Orleans, Louisiana.

Ferguson, William P. F.
1923   Michigan's Most Ancient Industry: The Prehistoric Mines and Miners of Isle Royale. *Michigan History Magazine* 7(3–4):155–162.
1924   The Franklin Isle Royale Expedition. *Michigan History Magazine* 8(4):450–468.

Ferone, Troy J.
1999   *Terminal Woodland Copper Procurement Strategies in the Southern Lake Superior Basin.* Master's thesis, Department of Anthropology, Michigan State University, East Lansing, Michigan.

Fetzer, E. W. and William J. Mayer-Oakes
1951   Excavation of an Adena Burial Mound at the Half-Moon Site. *West Virginia Archeologist* 4:1–25.

Finkelstein, J. Joe
1940   The Norman Site Excavations near Wagoner, Oklahoma. *Oklahoma Prehistorian* 3(3):2–15.

Finlayson, William D.
    1977   *The Saugeen Culture: A Middle Woodland Manifestation in Southwestern Ontario.* Paper No. 61. National Museum of Man, Mercury Series, Ottawa, Ontario, Canada.

Finlayson, William D., Mel Brow, Roger Byrne, Jim Esler, Ron Farquhar, Ron Hancock, Larry Pavlish, and Charles Turton
    1998   *Iroquoian Peoples of the Land of Rocks and Water, A.D. 1000–1650: A Study in Settlement Archaeology.* Special Publication 1. London Museum of Archaeology, London, Ontario, Canada.

Finlayson, William D. and Robert H. Pihl
    1980   The Evidence for Culture Contact at the Draper Site, Pickering, Ontario. *Arch Notes* 80(3):5–14.

Finney, Fred A.
    2013   Intrasite and Regional Perspectives on the Fred Edwards Site and the Stirling Horizon in the Upper Mississippi Valley. *The Wisconsin Archeologist* 94(1–2):3–248.

Finney, Fred A., Scott B. Meyer, Kathryn E. Parker, and David L. Omernik
    1992   Phase III Archaeological Investigation of a Middle Archaic Raddatz Occupation at the Bobwhite Site (47Ri185), Richland County, Wisconsin. *Research Papers* 17(2). Office of the State Archaeologist, University of Iowa, Iowa City, Iowa.

Fisher, A. W.
    1895   A Rare Copper Axe. *The Archaeologist* 3(2):75–76.

Fisher, Alton K.
    1932   Discussion and Correspondence: Prehistoric Copper Artifacts. *The Wisconsin Archeologist* 12(1)26–27.

Fisher, Ernest B.
    1918   *Grand Rapids and Kent County, Michigan: Historical Account of Their Progress from First Settlement to the Present Time.* Robert O. Law Company, Chicago.

Fisher, James
    1924   Historical Sketch of the Lake Superior Copper District. In *The Nineteen Twenty Four Keweenawan*, pp. 215–288. Michigan College of Mines, Houghton, Michigan.

Fitting, James E.
    1971   Rediscovering Michigan Archaeology: The Gilman Collections at Harvard. *The Michigan Archaeologist* 16(2):83–114.
    1971   Rediscovering Michigan Archaeology: Notes on the 1885 Converse Mound Collection. *The Michigan Archaeologist* 17(1):33–39.
    1979   Middle Woodland Cultural Development in the Straits of Mackinac Region: Beyond the Hopewell Frontier. In *Hopewell Archaeology: The Chillicothe Conference*, edited by David S. Brose and N'omi Greber, pp. 109–112. MCJA Special Paper No. 3. Kent State University Press, Kent, Ohio.
    1972   ed., *The Schultz Site at Green Point: A Stratified Occupation Area in the Saginaw Valley of Michigan.* Memoirs No. 4. Museum of Anthropology, University of Michigan, Ann Arbor, Michigan.

Bibliography

Fitzgerald, William R.
  1990   *Chronology to Cultural Process: Lower Great Lakes Archaeology, 1500–1650.*
          Department of Anthropology, Faculty of Graduate Studies and Research, McGill
          University, Montreal, Quebec, Canada.

Flanders, Richard E.
  1968   The Spoonville Mound Salvage. *The Coffinberry News Bulletin* 15(1):3–7.
  1969   Hopewell Materials from Crockery Creek. *The Michigan Academician* 1(1–2):147–151.
  1979   New Evidence for Hopewell in Southwestern Michigan. In *Hopewell Archaeology:
          The Chillicothe Conference*, edited by David S. Brose and N'omi Greber, pp. 113–
          114. MCJA Special Paper No. 3. Kent State University Press, Kent, Ohio.
  1986   Evidence for Early Woodland Occupations in the Grand and Muskegon River
          Drainages. In *Early Woodland Archeology*, edited by Kenneth B. Farnsworth and
          Thomas E. Emerson, pp. 78–83. Kampsville Seminars in Archeology, Vol. 2. Center for
          American Archeology, Center for American Archeology Press, Kampsville, Illinois.

Flaskerd, George A.
  1940   A Schedule of Classification, Comparison, and Nomenclature for Copper Artifacts
          in Minnesota. *The Minnesota Archaeologist* 6(2):35–50.
  1943   The A. H. Andersen Site. *The Minnesota Archaeologist* 9(1):4–21.
  1945   Copper Axe. *The Minnesota Archaeologist* 11(2):48.

Fleming, Edward P.
  2004   A Middle Woodland Copper Celt from Meeker County, Minnesota. *The
          Minnesota Archaeologist* 63:25–28.

Flotow, Mark
  2006   The Archaeological, Osteological, and Paleodemographical Analysis of the Ray
          Site: A Biocultural Perspective. *Rediscovery* 5:7–16. Illinois Association for
          Advancement of Archaeology, Lewistown, Illinois.

Fogel, Ira L.
  1963   The Dispersal of Copper Artifacts in the Late Archaic Period of Prehistoric North
          America. *The Wisconsin Archeologist* 44(3):129–180.

Forbes, R. J.
  1950   *Metallurgy in Antiquity: A Notebook for Archaeologists and Technologists.* E. J.
          Brill, Leiden, The Netherlands.

Forbis, Richard G.
  1970   *A Review of Alberta Archaeology to 1964.* Publications in Archaeology No. 1.
          National Museum of Man, Ottawa, Ontario, Canada.

Force, Manning F.
  1879   To What Race Did the Mound Builders Belong? In *Some Early Notices of the Indians
          of Ohio*, by Manning F. Force, pp. 41–75. Robert Clarke & Co., Cincinnati, Ohio.

Ford, Billy
  2010   A Large Hopewell Copper Celt. *Central States Archaeological Journal* 57(3):133–135.

Ford, James A.
  1963   *Hopewell Culture Burial Mounds near Helena, Arkansas.* Anthropological Papers
          Vo. 50(1). American Museum of Natural History, New York.

Ford, James A. and Gordon Willey
> 1940   *Crooks Site, a Marksville Period Burial Mound in La Salle Parish, Louisiana.* Anthropological Study No. 3. Louisiana Geological Survey, Department of Conservation, Baton Rouge, Louisiana.

Forest Mining Company
> 1852a   *Report of the Directors of the Forest Mining Company, Made at the Annual Meeting of the Stockholders, December 13, 1852, and the Charter and By-Laws of the Company.* Press of the Franklin Printing House, Boston.

Forster, John H.
> 1879   Lake Superior Copper Mining Industry. In *First Annual Report of the Commissioner of Mineral Statistics of the State of Michigan, for 1877–8 and Previous Years*, by Charles E. Wright, pp. 125–192. Mining Journal Steam Printing House, Marquette, Michigan.
> 1891   The Explorer and His Work in the Copper Regions of Lake Superior. *The Michigan Engineers' Annual* 12:9–24.
> 1893   Autobiographical Sketch of John H. Forster. *The Michigan Engineers Annual Containing the Proceedings of the Michigan Engineering Society for 1893*, pp. 9–18.

Fortier, Andrew C.
> 2006   The Land between Two Traditions: Middle Woodland Societies of the American Bottom. In *Recreating Hopewell*, edited by Douglas K. Charles and Jane E. Buikstra, pp. 328–338. University Press of Florida, Gainesville, Florida.

Fortier, Andrew C., Thomas O. Maher, Joyce A. Williams, Michael C. Meinkoth, Kathryn E. Parker, and Lucretia S. Kelly
> 1989   *The Holding Site: A Hopewell Community in the American Bottom (11-Ms-118).* Published for the Illinois Department of Transportation by the University of Illinois Press, Urbana and Chicago.

Foster, Gardner
> 1902   A Curious Specimen of Copper—Rare Collection of Indian Relics. *The Michigan Miner* 4(5):11.

Foster, John W.
> 1852   Description of Samples of Ancient Cloth from the Mounds of Ohio. *Proceedings of the American Association for the Advancement of Science, Sixth Meeting, Held at Albany (N. Y.), August 1851*, pp. 375–378.
> 1869a   Descriptions of Certain Stone and Copper Implements Used by the Mound-Builders. *Transactions of the Chicago Academy of Sciences* 1:258–266.
> 1869b   On the Antiquity of Man in North America. *Transactions of the Chicago Academy of Sciences* 1:227–257.
> 1874   *Pre-historic Races of the United States of America.* S. C. Griggs and Company, Chicago.

Foster, J. W. and J. D. Whitney
> 1850   *Report on the Geology and Topography of a Portion of the Lake Superior Land District in the State of Michigan: Part 1. Copper Lands.* Executive Document No. 69. House of Representatives, 31st Congress, Washington, DC.

Fowke, Gerard
> 1888   How a Mound Was Built. *Science* 12(288):65.

Bibliography

1892   Some Interesting Mounds. *American Anthropologist* 5(1):73–82.

1896   Archaeological Work in Ohio. In *Proceedings of the Academy of Natural Sciences of Philadelphia* 47:506–515.

1900   Stone Graves in Brown County, Ohio. *Ohio Archaeological and Historical Quarterly* 9:193–204.

1902   Archaeological History of Ohio: The Mound Builders and Later Indians. Ohio State Archaeological and Historical Society, Press of Fred. J. Heer, Columbus, Ohio.

1910   The Copper Plates from Malden, Dunklin County. In *Antiquities of Central and Southeastern Missouri*, p. 98, Plates 15–19. Bulletin 37. Bureau of American Ethnology, Smithsonian Institution, Washington, DC.

1928   Archaeological Investigations—II. In *Forty-fourth Annual Report of the Bureau of American Ethnology to the Secretary of the Smithsonian Institution, 1926–27*, pp. 399–540. U. S. Government Printing Office, Washington, DC.

Fowke, Gerard and Warren K. Moorehead

1895   Recent Mound Exploration in Ohio. *Proceedings of the Academy of Natural Sciences of Philadelphia, 1894*, pp. 308–321.

Fowler, Melvin L.

1952a   The Clear Lake Site: Hopewellian Occupation. In *Hopewellian Communities in Illinois*, edited by Thorne Deuel, pp. 131–174. Scientific Papers 5(4). Illinois State Museum, Springfield, Illinois.

1952b   The Robinson Reserve Site. *Journal of the Illinois State Archaeological Society* 2(2–3):50–62.

1957   Rutherford Mound, Hardin County, Illinois. *Scientific Papers* 7(1). Illinois State Museum. Springfield, Illinois.

1959a   Modoc Rock Shelter: An Early Archaic Site in Southern Illinois. *American Antiquity* 24(3):257–318.

1959b   Summary Report of Modoc Rock Shelter: 1952, 1953, 1955, 1956. *Report of Investigations* No. 8. Illinois State Museum, Springfield, Illinois.

Fowler, Melvin L., Jerome Rose, Barbara Vander Leest, and Steven R. Ahler

1999   The Mound 72 Area: Dedicated and Sacred Space. *Report of Investigations No. 54*. Illinois State Museum, Springfield, Illinois.

Fox, George R.

1911   The Ancient Copper Workings on Isle Royale. *The Wisconsin Archeologist* 10(2):73–100.

1929   Isle Royale Expedition. *Michigan History Magazine* 13(2):308–323

1952   Historic Mines of Isle Royale. *Inland Seas* 8:234–241.

Fox, William A.

1991   The Serpent's Copper Scales. *Wanikan: Newsletter of the Thunder Bay Chapter, Ontario Archaeological Society* 91-03:3–15.

2004   The North-South Copper Axis. *Southeastern Archaeology* 23(1):85–97.

Fox, William A., R. G. V. Hancock, and L. A. Pavlish

1995   Where East Met West: The New Copper Culture. *The Wisconsin Archeologist* 76 (3–4):269–293.

Franklin, U. M., E. Badone, R. Gotthardt, and B. Yorga

1981   *An Examination of Prehistoric Copper Technology and Copper Sources in*

Western Arctic and Subarctic North America. Archaeological Survey of Canada, Paper No. 101. National Museums of Canada, Ottawa, Ontario, Canada.

Fraser, Ray
    2013   Some Middle Mississippian Artifacts from Cass County, Illinois. *Central States Archaeological Journal* 60(2):88–90.

Fregni, Giovanna
    2009   A Study of the Manufacture of Copper Spearheads in the Old Copper Complex. *The Minnesota Archaeologist* 68:121–130.

Fryklund, P. O.
    1941   A Catalog of Copper from Roseau County, Minnesota. *The Minnesota Archaeologist* 7(3):4–16.

Fugle, Eugene
    1957   *Introduction to the Artifact Assemblages of the Big Sioux and Little Sioux Foci.* Master's thesis, Department of Sociology and Anthropology, University of Iowa, Iowa City, Iowa.

Fulton Mining Company
    1853   The Fulton Mining Company. William H. Arthur & Co., Stationers, New York.

Fundaburk, Emma Lila and Mary Douglas Foreman
    1957   *Sun Circles and Human Hands: The Southeastern Indians – Art and Industries.* Privately published by Emma Lila Fundaburk, Luverne, Alabama.

Funkhouser, W. D. and W. S. Webb
    1935   The Ricketts Site in Montgomery County, Kentucky. *Reports in Archaeology and Anthropology* 3(3):70–100. Department of Anthropology and Archaeology, University of Kentucky, Lexington, Kentucky.
    1937   The Chilton Site in Henry County, Kentucky. *Reports in Archaeology and Anthropology* 3(5):169–206. Department of Anthropology and Archaeology, University of Kentucky, Lexington, Kentucky.

Futato, Eugene M.
    1983   *Archaeological Investigations in the Cedar Creek and Upper Bear Creek Reservoirs.* Report of Investigations No. 13, Office of Archaeological Research, University of Alabama, Tuscaloosa, Alabama and Publications in Anthropology 32, Tennessee Valley Authority, Knoxville, Tennessee.
    1988   Continuity and Change in the Middle Woodland Occupation of the Northwest Alabama Uplands. In *Middle Woodland Settlement and Ceremonialism in the Mid-South and Lower Mississippi Valley: Proceedings of the 1984 Mid-South Archaeological Conference, Pinson Mounds, Tennessee – June, 1984*, edited by Robert C. Mainfort, pp. 31–48. Archaeological Report No. 22. Mississippi Department of Archives and History, Jackson, Mississippi.

Gadus, E. F.
    1979   The Harness Copper Plate. *Ohio Archaeologist* 29(3):27–28.

Galloway, Patricia, ed.
    1989   *The Southeastern Ceremonial Complex: Artifacts and Analysis.* University of Nebraska Press, Lincoln, Nebraska.

Bibliography

Galm, Jerry R.
  1981   *Prehistoric Cultural Adaptations in the Wister Valley, East-Central Oklahoma.*
           PhD dissertation, Washington State University, Pullman, Washington.
  1984   Arkansas Valley Caddoan Formative: The Wister and Fourche Maline Phases. In
           Prehistory of Oklahoma, edited Robert E. Bell, pp. 199–219. Academic Press,
           Inc., New York.

Gant, Robert
  1963   Evidence of Old Copper Culture in South Dakota. *The Wisconsin Archeologist* 44(2):97.

Garden City Mining Company
  1856   Articles of Incorporation and By-Laws of the Garden City Mining Company of
           Chicago. Sterling P. Rounds, Book and Job Printer, Chicago.

Gardner, Joan S.
  1980   The Conservation of Fragile Specimens from the Spiro Mound, Le Flore County,
           Oklahoma. *Contributions from the Stovall Museum, University of Oklahoma* No.
           5. Norman, Oklahoma.

Garland, Elizabeth B.
  1986   Early Woodland Occupations in Michigan: A Lower St. Joseph Valley Perspective.
           In *Early Woodland Archeology*, edited by Kenneth B. Farnsworth and Thomas
           E. Emerson, pp. 47–83. Kampsville Seminars in Archeology, Vol. 2. Center for
           American Archeology, Center for American Archeology Press, Kampsville,
           Illinois.

Garland, Elizabeth B. and Arthur L. DesJardins
  2006   Between Goodall and Norton: Middle Woodland Settlement Patterns and
           Interaction Networks in Southwestern Michigan. In *Recreating Hopewell*, edited
           by Douglas K. Charles and Jane E. Buikstra, pp. 227–260. University Press of
           Florida, Gainesville, Florida.

General Electric Company
  1997   *Hopewell in Mt. Vernon: A Study of the Mt. Vernon Site (12-PO-885).*

Genuine Indian Relic Society, Inc.
  1978   Copper Relics of the North American Indian. *The Redskin* 13(4):4–59.

Gerend, Alphonse
  1902   The Archaeological Features of Sheboygan County. *The Wisconsin Archeologist*
           1(3):12–21.

Gibbon, Guy
  1973   *The Sheffield Site: An Oneota Site on the St. Croix River.* Minnesota Prehistoric
           Archaeology Series No. 10. Minnesota Historical Society, St. Paul, Minnesota.
  1998   Old Copper in Minnesota: A Review. *Plains Anthropologist* 43(163):27–50.
  2012   *Archaeology of Minnesota: The Prehistory of the Upper Mississippi River Region.*
           University of Minnesota Press, Minneapolis.

Gibson, Jon L. and J. Richard Shenkel
  1988   Louisiana Earthworks: Middle Woodland and Predecessors. In *Middle Woodland
           Settlement and Ceremonialism in the Mid-South and Lower Mississippi Valley:*

Proceedings of the 1984 Mid-South Archaeological Conference, Pinson Mounds, Tennessee – June, 1984, edited by Robert C. Mainfort, pp. 7–18. Archaeological Report No. 22. Mississippi Department of Archives and History, Jackson, Mississippi.

Gillman, Henry

    1873   Ancient Works at Isle Royale, Michigan. *Appleton's Journal of Literature, Science, and Art* 10(229):173–175.

    1874   The Mound-Builders and Platycnemism in Michigan. *Annual Report of the Board of Regents of the Smithsonian Institution, 1873*, pp. 364–390.

    1876   The Ancient Men of the Great Lakes. *Proceedings of the American Association for the Advancement of Science, Twenty-fourth Meeting, Held at Detroit, Michigan, August, 1875*, pp. 315–331.

Goad, Sharon I.

    1978   *Exchange Networks in the Prehistoric Southeastern United States*. PhD dissertation, University of Georgia, Athens, Georgia.

    1979   Middle Woodland Exchange in the Prehistoric Southeastern United States. In *Hopewell Archaeology: The Chillicothe Conference*, edited by David S. Brose and N'omi Greber, pp. 239–246. MCJA Special Paper No. 3. Kent State University Press, Kent, Ohio.

Goad, Sharon I. and John Noakes

    1978   Prehistoric Copper Artifacts in the Eastern United States. In *Archaeological Chemistry—II*, pp. 335–346. Advances in Chemistry Series 171. American Chemical Society, Washington, DC.

Goggin, John M.

    1947   Manifestations of a South Florida Cult in Northwestern Florida. *American Antiquity* 12(4):273–276.

    1949   A Southern Cult Specimen from Florida. *The Florida Anthropologist* 2(1–2):36–37.

    1952   *Space and Time Perspective in Northern St. Johns Archeology, Florida*. Yale University Publications in Anthropology No. 47. Yale University Press, New Haven, CO.

    1954   Historic Metal Plummet Pendants. *The Florida Anthropologist* 7(1):27.

Goodman, Claire Garber

    1984   *Copper Artifacts in Late Eastern Woodlands Prehistory*. Edited by Ann-Marie Cantwell. Center for American Archaeology at Northwestern University, Evanston, Illinois.

Great Western Copper Mining Company

    1852   *Description of the Lands and Mines of the Great Western Copper Mining Co. of Lake Superior*. Pudney & Russell, Printers, New York.

    1863   *Prospectus for the Formation of a Mining Company on the South-East Quarter Sec. 30, Town. 51 N. of R. 37 W. and the South-West Quarter Sec. 29, Town. 51 N. of R. 37 W. 320 Acres Mineral Land, in the County of Ontonagon, Michigan*. W. S. Haven, Pittsburgh, Pennsylvania.

Greber, N'omi B.

    1979   A Comparative Study of Site Morphology and Burial Patterns at Edwin Harness Mound and Seip Mounds 1 and 2. In *Hopewell Archaeology: The Chillicothe Conference*, edited by David S. Brose and N'omi Greber, pp. 27–38. MCJA Special Paper No. 3. Kent State University Press, Kent, Ohio.

    2005   Adena and Hopewell in the Middle Ohio Valley: To Be or Not to Be? In *Woodland*

Bibliography

Systematics in the Middle Ohio Valley, edited by Darlene Applegate and Robert C. Mainfort, pp. 19–39. University of Alabama Press, Tuscaloosa, Alabama.

2006   Enclosures and Communities in Ohio Hopewell: An Essay. In *Recreating Hopewell*, edited by Douglas K. Charles and Jane E. Buikstra, pp. 74–105. University Press of Florida, Gainesville, Florida.

2010   Coda: Still Seeking "Hopewell". In *Hopewell Settlement Patterns, Subsistence, and Symbolic Landscapes*, edited by A. Martin Byers and DeeAnne Wymer, pp. 335–348. University Press of Florida, Gainesville, Florida.

Greber, N'omi B. and Katherine C. Ruhl

1989   *The Hopewell Site: A Contemporary Analysis Based on the Work of Charles C. Willoughby*. Westview Press, Boulder, CO.

Greeley, Horace

1847   Life on Lake Superior. *New-York Weekly Tribune* [New York]. July 17, 1847, p. 1.

Greenman, Emerson F.

1931   Department of Archaeology. *Museum Echoes* 4(8):55.

1932   Excavation of the Coon Mound and an Analysis of the Adena Culture. *Ohio Archaeological and Historical Quarterly* 41:366–523.

Greer, E. S., Jr.

1966   A Tukabahchee Plate from the Coosa River. *Journal of Alabama Archaeology* 12(2):156–158.

Greusel, Joseph

1903   A Mysterious Race. *The Gateway* 1(1):5–7.

Griffin, James B.

1941a   *Additional Hopewell Material from Illinois*. Prehistory Research Series Vol. 2(3). Indiana Historical Society, Indianapolis.

1941b   Reel-shaped Gorgets. *American Antiquity* 6(3):265.

1952a   Prehistoric Cultures of the Central Mississippi Valley. In *Archeology of Eastern United States*, edited by James B. Griffin, pp. 226–238. University of Chicago Press, Chicago.

1952b   Culture Periods in Eastern United States Archeology. In *Archeology of Eastern United States*, edited by James B. Griffin, pp. 352–364. University of Chicago Press, Chicago.

1961a   Early American Mining in the Upper Peninsula of Michigan and the First Recognition of Prehistoric Mining Activities. In *Lake Superior Copper and the Indians: Miscellaneous Studies of Great Lakes Prehistory*, edited by James B. Griffin, pp. 46–76. Anthropological Papers No. 17. Museum of Anthropology, University of Michigan, Ann Arbor, Michigan.

1961b   Prehistoric Finds Made by Early American Mining Activities on Isle Royale, 1870–1880. In *Lake Superior Copper and the Indians: Miscellaneous Studies of Great Lakes Prehistory*, edited by James B. Griffin, pp. 16–24. Anthropological Papers No. 17. Museum of Anthropology, University of Michigan, Ann Arbor, Michigan.

1961c   Post-Glacial Ecology and Culture Changes in the Great Lakes Area of North America. In *Proceedings: Fourth Conference on Great Lakes Research*, pp. 147–155. Publication No. 7, Great Lakes Research Division, Institute of Science and Technology, University of Michigan, Ann Arbor, Michigan.

1961d ed., *Lake Superior Copper and the Indians: Miscellaneous Studies of Great Lakes Prehistory*. Anthropological Papers No. 17. Museum of Anthropology, University of Michigan, Ann Arbor, Michigan.

1964 The Northeast Woodlands Area. In *Prehistoric Man in the New World*, edited by Jesse D. Jennings and Edward Norbeck, pp. 223–258, University of Chicago Press, Chicago.

1966a The Calumet Ancient Pit. *The Michigan Archaeologist* 12(3):130–133.

1966b *The Fort Ancient Aspect: Its Cultural and Chronological Position in Mississippi Valley Archaeology*. Anthropological Papers No. 28. Museum of Anthropology, University of Michigan, Ann Arbor, Michigan.

1966c A Non-Neolithic Copper Industry in North America. In *XXXVI Congreso Internacional de Americanistas, Sevilla, España, Actas y Memorias* Vol. 1, pp. 281–285.

1972 An Old Copper Point from Chippewa County, Michigan. *The Michigan Archaeologist* 18(1):35–36.

1984 A Short Talk about a Small Hopewell Site in Ohio. Paper presented at the Midwest Archaeological Conference, Chicago.

Griffin, James B., Richard E. Flanders, and Paul F. Titterington
1970 *The Burial Complexes of the Knight and Norton Mounds in Illinois and Michigan*. Memoirs No. 2. Museum of Anthropology, University of Michigan, Ann Arbor, Michigan.

Griffin, James B. and Dan F. Morse
1961 The Short-nosed God from the Emmons Site. *American Antiquity* 26(4):560–563.

Griffin, James B. and George I. Quimby
1961 The McCollum Site, Nipigon District, Ontario. In *Lake Superior Copper and the Indians: Miscellaneous Studies of Great Lakes Prehistory*, edited by James B. Griffin, pp. 91–102, Plates XIX–XXII. Anthropological Papers No. 17. Museum of Anthropology, University of Michigan, Ann Arbor, Michigan.

Griffin, John W.
1952 Prehistoric Florida: A Review. In *Archeology of Eastern United States*, edited by James B. Griffin, pp. 322–334. University of Chicago Press, Chicago.

Groce, Nora
1980 Ornaments of Metal: Rings, Medallions, Combs, Beads and Pendants. In *Burr's Hill: a 17th Century Wampanoag Burial Ground in Warren, Rhode Island*, edited by Susan B. Gibson, pp. 108–117. Studies in Anthropology Vol. 2, Haffenreffer Museum of Anthropology, Brown University, Providence, Rhode Island.

Hagar, Albert D.
1865 Ancient Mining on the Shores of Lake Superior. *Atlantic Monthly* 15:308–315.

Hall, Robert L.
1962 *The Archaeology of Carcajou Point (With an Interpretation of the Development of Oneota Culture in Wisconsin)*. University of Wisconsin Press, Madison, Wisconsin.

1979 In Search of the Ideology of the Adena-Hopewell Climax. In *Hopewell Archaeology: The Chillicothe Conference*, edited by David S. Brose and N'omi Greber, pp. 258–265. MCJA Special Paper No. 3. Kent State University Press, Kent, Ohio.

Bibliography

1989   The Cultural Background of Mississippian Symbolism. In *The Southeastern Ceremonial Complex: Artifacts and Analysis*, edited by Patricia Galloway, pp. 239–278. University of Nebraska Press, Lincoln, Nebraska.

2006   The Enigmatic Copper Cutout from Bedford Mound 8. In *Recreating Hopewell*, edited by Douglas K. Charles and Jane E. Buikstra, pp. 464–474. University Press of Florida, Gainesville, Florida.

Halsey, John R.

1966   Radiocarbon Dates from Archaeological Sites of Old Copper and Related Cultures in the Great Lakes Area. *Arti-Facts* (Newsletter of the Clinton Valley Chapter, Michigan Archaeological Society) 4(4):7–11.

1972   The Molash Creek Red Ocher Burial. *The Wisconsin Archeologist* 53(1):1–15.

1981   The Wayne Mortuary Complex: A New Chapter in Michigan's Prehistoric Past. *Michigan History Magazine* 65(5):17–23.

1996   Without Forge or Crucible: Aboriginal Native American Use of Metals and Metallic Ores in the Eastern Woodlands. *The Michigan Archaeologist* 42(1):1–58.

1998   Jackson's Quiet Promoter. *Michigan History Magazine* 82(6):19–21.

2008   "Ancient Diggings": A Review of Nineteenth-Century Observations in the Prehistoric Copper Mining Pits of the Lake Superior Basin. Paper presented at the Annual Meeting of the Midwest Archaeological Conference, Milwaukee, Wisconsin.

2009   Ancient Pits of the Copper Country. *Michigan History Magazine* 93(3):40–47.

2010 [2015] Copper from the Drift. *The Michigan Archaeologist* 56:1–76.

Halsey, John R. and Janet G. Brashler

2013   More Than Grave Lots?: The Jack's Reef Horizon in Michigan. *Archaeology of Eastern North America* 41:145–192.

Haltiner, Robert E.

2002   *Stories the Red People Have Told....and....More.* Privately published by the author.

Hamilton, Henry W.

1952   The Spiro Mound. *The Missouri Archaeologist* 14:1–276.

Hamilton, Henry W., Jean Tyree Hamilton, Eleanor F. Chapman

1974   *Spiro Mound Copper.* Memoir No. 11. Missouri Archaeological Society, Columbia, Missouri.

Harrington, Mark R.

1920   *Certain Caddo Sites in Arkansas.* Indian Notes and Monographs No. 10. Museum of the American Indian, Heye Foundation, New York.

Harrison, William Henry

1883   *A Discourse on the Aborigines of the Ohio Valley in Which the Opinions of Its Conquest in the Seventeenth Century by the Iroquois or Six Nations Supported by Cadwallader Colden of New York, Gov. Thomas Pownall of Massachusetts, Dr. Benjamin Franklin of Pennsylvania, Hon. DeWitt Clinton of New York, and Judge John Haywood of Tennessee are Examined and Contested; To Which are Prefixed Some Remarks on the Study of History Prepared at the Request of the Historical Society of Ohio, by William Henry Harrison of North Bend. Major-General, U. S. A., President of the United States. Etc.* Fergus Printing Company, Chicago.

Harvey, Amy
   1979   *Oneota Culture in Northwestern Iowa*. Report No. 12, Office of the State
             Archaeologist, University of Iowa, Iowa City, Iowa.

Harvey, Arthur
   1890   Broad Outlines of the Geology of the Northwest of Lake Superior. *Proceedings of
             the Canadian Institute* (Third Series) 7(2):218–225.

Hatch, James W.
   1974   *Social Dimensions of Dallas Mortuary Patterns*. Master's thesis, Pennsylvania
             State University, State College, Pennsylvania.
   1976a   The Citico Site (40HA65): A Synthesis. *Tennessee Anthropologist* 1(2):74–103.
   1976b   *Status in Death: Principles of Ranking in Dallas Culture Mortuary Remains*. PhD
             dissertation, Pennsylvania State University, State College, Pennsylvania.

Hatchcock, Roy
   1976   An Unusual Mississippian Burial. *The Redskin* 11(1):26–27.

Hauck, Robert
   2002   Grounds and Events. *Newsletter of the Sterling Hill Mine* 11(1):24–25.

Haymond, Rufus
   1869   Dr. Rufus Haymond's Report of a Geological Survey of Franklin County,
             Indiana, Made during the Summer and Fall of 1869. In *First Annual Report of the
             Geological Survey of Indiana, Made during the Year 1869*, pp. 175–202.

Hays, Christopher T.
   1994   *Adena Mortuary Patterns and Ritual Cycles in the Upper Scioto Valley, Ohio*.
             PhD dissertation, State University of New York at Binghamton, Binghamton, New
             York.

Hedican, Edward J.
   1981   What's the Problem with the Old Copper Culture? *Wanikan: Newsletter of the
             Thunder Bay Chapter, Ontario Archaeological Society* 81(7):7–13.

Hedican, Edward and James McGlade
   1982   The Old Copper Problem. *Anthropological Journal of Canada* 20(1):16–21.
   1993   A Taxometric Analysis of Old Copper Projectile Points. *Man in the Northeast*
             45:21–38.

Heestand, Norman D.
   1973   The Temple Effigy Mound. *Artifacts* 3(1):18–20.

Heintzelman, Andrea J.
   1974   The Significance of the Panpipe in the Hopewellian Culture. *Kansas
             Anthropological Association Newsletter* 19(7–8):4–35.

Hemmings, E. Thomas
   1978   Exploration of an Early Adena Mound at Willow Island, West Virginia. *Report of
             Archaeological Investigations* 7. West Virginia Geological and Economic Survey,
             Morgantown, West Virginia.
   1984   Fairchance Mound and Village: An Early Middle Woodland Settlement in the
             Upper Ohio Valley. *West Virginia Archeologist* 36(1):3–68.

Bibliography

Henderson, John G.
  1872   Notes on Aboriginal Relics Known as "Plummets." *The American Naturalist* 6
            (11):641–650.
  1884   Aboriginal Remains near Naples, Ill. *Annual Report of the Board of Regents of the
            Smithsonian Institution,* 1882, pp. 686–721.

Henriksen, Harry C.
  1965   Utica Hopewell, a Study of the Early Hopewellian Occupation in the Illinois
            River Valley. In *Middle Woodland Sites in Illinois,* pp. 1–67. Bulletin No. 5.
            Illinois Archaeological Survey, Urbana, Illinois.

Henry, Alexander
  1809   *Travels and Adventures in Canada and the Indian Territories between the Years
            1760 and 1776.* I. Riley, New York.

Henry, Joseph
  1851   American Antiquities. *Fifth Annual Report of the Board of Regents of the
            Smithsonian Institution, 1850,* pp. 22–23.

Henshaw, H. W.
  1890   Mound Exploration in Georgia. *The American Anthropologist* 3(1):102–104.

Hermann, Barbara
  1948   An Analysis of Two Adena Sites in Ohio. *Papers of the Michigan Academy of
            Science, Arts, and Letters* 32:321–340.

Herold, Elaine Bluhm
  1970   Hopewell: Burial Mound Builders. *The Palimpsest* 51(12):497–529.
  1971   *The Indian Mounds at Albany, Illinois.* Anthropological Paper No. 1. Davenport
            Museum, Davenport, Iowa.

Herold, Elaine Bluhm, Patricia J. O'Brien, and David J. Wenner, Jr.
  1990   Hoxie Farm and Huber: Two Upper Mississippian Archaeological Sites in Cook
            County, Illinois. In *At the Edge of Prehistory: Huber Phase Archaeology in the
            Chicago Area,* edited by James A. Brown and Patricia J. O'Brien, pp. 1–119.
            Center for American Archeology, Kampsville, Illinois.

Herrera, Katie
  2012   An Explanation for the Current Sex Distribution in the Riverside Cemetery
            (20ME01),a Terminal Archaic Site, and Implications for a Possible Site
            Reinterpretation. *Field Notes: A Journal of Collegiate Anthropology* 3(1):36–48.

Herrick, Ruth
  1957   Report of a Burial in Montcalm County. *The Michigan Archaeologist* 3(1):8–9.

Hess, Dennis
  2014   A Pair of Copper Earspools with Their Original Wooden Backing. *Central States
            Archaeological Journal* 61(2):105–106.

Hesselberth, Charles
  1946   Notes on the Ogden-Fettie Mounds. *Journal of the Illinois State Archaeological
            Society* 4(1):9–11.

Hewes, Gordon W.
> 1949   Burial Mounds in the Baldhill Area, North Dakota. *American Antiquity* 14(4):322–328.

Heye, George G., F. W. Hodge, and George H. Pepper
> 1918   The Nacooche Mound in Georgia. *Contributions from the Museum of the American Indian, Heye Foundation* 4(3):1–103.

Hiestand, Joseph E.
> 1951   *An Archaeological Report on Newton County, Indiana.* Indiana Historical Bureau, Indianapolis.

Hildreth, Samuel P.
> 1837   American Antiquities. *The Family Magazine, or, Monthly Abstract of General Knowledge* 4:105–106.
> 1842   Ancient Mound at Marietta. *American Pioneer* 1(10):340–342.
> 1843   Pyramids at Marietta. *American Pioneer* 2(6):243–248.

Hill, A. T. and Waldo R. Wedel
> 1936   Excavations at the Leary Indian Village and Burial Site, Richardson County, Nebraska. *Nebraska History Magazine* 17(1):2–73.

Hill, C. Gordon
> 1982   Archaeological Survey of the South Geraldton District. In *Studies in West Patricia Archaeology,* Archaeology Research Report 19, edited by W. A. Ross. Archaeology and Heritage Planning Branch, Ontario Ministry of Citizenship and Culture, Toronto, Ontario, Canada.

Hill, George W.
> 1878   Ancient Earthworks of Ashland County. *Annual Report of the Board of Regents of the Smithsonian Institutions,* 1877, pp. 261–267.

Hill, George W.
> 1880   *History of Ashland County, Ohio, with Illustrations and Biographical Sketches.* Williams Bros., Cleveland, Ohio.

Hill and Company, H. H.
> 1883   *History of Winona and Olmsted Counties.* H. H. Hill and Company Publishers, Chicago.

Hill, Mark A.
> 1994   *Ottawa North and Alligator Eye: Two Late Archaic Sites on the Ottawa National Forest.* Cultural Resource Management Series, No. 6. Prepared by staff of the Ottawa National Forest, Forest Service, United States Department of Agriculture, Washington, DC.
> 2006   The Duck Lake Site and Implications for Late Archaic Copper Procurement and Production in the Southern Lake Superior Basin. *Midcontinental Journal of Archaeology* 31(2):213–248.
> 2009   *The Benefit of the Gift: Exchange and Social Interaction in the Late Archaic Western Great Lakes.* PhD dissertation, Department of Anthropology, Washington State University, Pullman, Washington.
> 2011   New Dates for Old Copper: Contemporaneity in the Archaic Western Great Lakes. *The Wisconsin Archeologist* 92(2):75–82.

Bibliography

2012   Tracing Social Interaction: Perspectives on Archaic Copper Exchange from the Upper Great Lakes. *American Antiquity* 77(2):279–292.

Hill, N. N., compiler
1881   *History of Licking County, O: Its Past and Present.* A. A. Graham & Co., Publishers, Newark, Ohio.

Hill, Samuel W.
1849   Underground Works of the North West Mines, Surveyed Oct. 1847 by S. W. Hill, U. S. D. S. In "Report on the Geological and Mineralogical Survey of the Mineral Lands of the United States in Michigan." In *Message from the President of the United States, to the Two Houses of Congress, at the Commencement of the First Session of the Thirty-first Congress. December 24, 1849. Read December 27, 1849. Part III,* pp. 371–801. House Executive Document 5(3). 31st Congress, 1st Session. Also issued as Senate Executive Document 1). 31st Congress, 1st Session. Washington, DC.
1858   Report of S. W. Hill. In *Report of Directors of the Eagle Harbor & Waterbury Mining Companies, Submitting to Stockholders Letter of J. R. Grout, and Report of S. W. Hill, Esq., Relative to Mining Locations of the Companies.* Daily Free Press Book and Job Office, Detroit.

Hill, Walter E. and Robert W. Neuman
1966   Copper Artifacts from Prehistoric Archeological Sites in the Dakotas. *Science* 154 (3753):1171–1173.

Hills, Leslie W.
1898   To the Editor. *The American Archaeologist* 2(10):270.

Hilton Mining Company
1863   *Prospectus for the Formation of the Hilton Mining Co. on the East Half of Sec. 36, Town 51, N, R. 38 West Containing 320 Acres Mineral Land, in the County of Ontonagon, Michigan.* C. C. P. Moody, Boston.

Hinkle, Kathleen A.
1984   *Ohio Hopewell Textiles: A Medium for the Exchange of Social and Stylistic Information.* Master's thesis, University of Arkansas, Fayetteville, Arkansas.

Hinsdale, W. B.
1925   *Primitive Man in Michigan.* Michigan Handbook Series, No. 1. University Museum, University of Michigan. Published by the University.

Hobart, C. W. and C. E. Wright
1883   Ontonagon County. In *History of the Upper Peninsula of Michigan*, pp. 508–546. Western Historical Company, Chicago.

Hodge, Frederick Webb
1912a   Copper. In *Handbook of American Indians North of Mexico*, edited by Frederick Webb Hodge, pp. 343–347. Bulletin 30. Bureau of American Ethnology, Smithsonian Institution. Government Printing Office, Washington, DC.
1912b   Metal-work. In *Handbook of American Indians North of Mexico*, edited by Frederick Webb Hodge, pp. 847–849. Bulletin 30. Bureau of American Ethnology, Smithsonian Institution. Government Printing Office, Washington, DC.

1912c   Mines and Quarries. In *Handbook of American Indians North of Mexico*, edited by Frederick Webb Hodge, pp. 864–867. Bulletin 30. Bureau of American Ethnology, Smithsonian Institution. Government Printing Office, Washington, DC.

1913   Copper. In *Handbook of Indians of Canada*, edited by Frederick W. Hodge, pp. 110–113. Published as an Appendix to the Tenth Report of the Geographic Board of Canada. Reprinted under the Direction of James White, F.R.G.S., Secretary, Commission of Conservation. Printed by C. H. Parmelee, Printer to the Kings Most Excellent Majesty, Ottawa, Ontario, Camada.

Hodge, James T.

1849a   Copper Ores of Lake Superior. *American Railroad Journal* 22(694):498.

1849b   Copper Ores of Lake Superior. *American Railroad Journal* 22(697):544–546.

1849c   Copper Ores of Lake Superior. *American Railroad Journal* 22(710):751–752.

1850a   Ancient Mining Operations on Lake Superior. *The Annual of Scientific Discovery*, p. 360.

1850b   On the Mineral Region of Lake Superior. *Proceedings of the American Association for the Advancement of Science*, 1849, pp. 301–308.

1850c   Lake Superior Copper Region. *American Railroad Journal* 23(743):433–435.

1851a   Copper Mines of Lake Superior. *American Railroad Journal* 24(789):337–338.

1851b   Copper Mines of Lake Superior. *Hunt's Merchants' Magazine and Commercial Review* 25(2):254–255.

1851c   Lake Superior Copper Mines. *American Railroad Journal* 24(790):353–354.

1873   Mining Industry of the United States. In *First Century of National Existence; the United States as They Were and Are*, pp. 17–168. L. Stebbins, Hartford, CN.

Hodges, Mary Ellen N.

1993   The Archaeology of Native American Life in in Virginia in the Context of European Contact: Review of Past Research. In *The Archaeology of 17th-Century Virginia*, edited by Theodore R. Reinhart and Dennis J. Pogue, pp. 1–65. Archaeological Society of Virginia, Publication No. 30.

Hoffman, Michael P.

1977   The Kinkead-Mainard Site, 3PU2: A Late Prehistoric Quapaw Phase Site near Little Rock, Arkansas. *The Arkansas Archeologist* 16, 17, 18 (1975, 1976, 1977):1–41.

Hohol, April S. and Lane A. Beck

1985   Copena Celts: Considerations of Wear and Function of Copper and Greenstone Artifacts. Paper presented at a meeting of the Southeastern Archaeological Conference, Birmingham, Alabama.

Hollingsworth, Sandra

1978   *The Atlantic: Copper and Community South of Portage Lake*. John H. Forster Press, Hancock, Michigan.

Holmes, William H.

1884   Eccentric Figures from Southern Mounds. *Science* 3(62):436–438.

1892   Sacred Pipestone Quarries of Minnesota and Ancient Copper Mines of Lake Superior. In *Proceedings of the American Association for the Advancement of Science for the Forty-first meetings Held at Rochester, N. Y., August, 1892*, pp. 277–279.

1896   Prehistoric Textile Art of Eastern United States. In *Thirteenth Annual Report of the Bureau of Ethnology to the Secretary of the Smithsonian Institution, 1891–92 by J. W. Powell, Director*, pp. 3–46. Government Printing Office, Washington, DC.

Bibliography

    1901   Aboriginal Copper Mines of Isle Royale, Lake Superior. *American Anthropologist*
                3(4):684–696.

Holmquist, Carl E.
    1948   A Delta County Village and Workshop Site. *The Totem Pole* 21(5):1–3.

Holsten, Walter
    1938   A Copper Spearpoint. *The Wisconsin Archeologist* 18(3):73–74.

Holtz, Wendy K.
    1992   *A Look Through Time at Oneota Copper.* Bachelor's thesis, University of
                Wisconsin-La Crosse, La Crosse, Wisconsin.

Holtz-Leith, Wendy K.
    2001   Copper, Bone, Red Ochre, and Miscellaneous Artifacts. In *Archaeological
                Investigations at an Oneota Village in the Heart of La Crosse, Wisconsin: Data
                Recovery at the Seventh Street Interchange, USH 14/61, South Avenue, within the
                Sanford Archaeological District, La Crosse, Wisconsin.* (WISDOT Project 1641-
                02-00/SHSW#96-5002/Lc), pp. 8-1 through 8-5.

Holzapfel, Elaine
    2005   The Original Adena Mound. *Ohio Archaeologist* 55(4):33–35.

Homsher, George W.
    1884   The Glidwell Mound, Franklin County, Indiana. In *Annual report of the Board of
                Regents of the Smithsonian Institution,* 1882, pp. 721–728.

Hooge, Paul E.
    1999   *Discovering the Prehistoric Mound Builders of Licking County, Ohio.* 2nd edition.
                Licking County Archaeology and Landmarks Society, Newark, Ohio.

Hooton, Earnest A. and Charles C. Willoughby
    1920   Indian Village Site and Cemetery near Madisonville, Ohio. *Papers of the Peabody
                Museum of American Archaeology and Ethnology, Harvard University* Vol. 8(1).
                Cambridge, Massachusetts.

Hoppa, Robert D., Laura Allingham, Kevin Brownlee, Linda Larcombe, and Gregory Monks
    2005   An Analysis of Two Late Archaic Burials from Manitoba: The Eriksdale Site
                (EfL1). *Canadian Journal of Archaeology* 29(2):234–266.

Hothem, Lar
    1998   The Red Ocher People. *Indian-artifact Magazine* 19(1):20–21, 77.

Houck, Louis
    1908   *A History of Missouri* (Vol. I). R. R. Donnelley & Sons Company, Chicago.

Houghton, Jacob
    1862   Crossing around Lake Superior. *Portage Lake Mining Gazette* [Houghton,
                Michigan], September 6, pp. 1–2.
    1876a   The Ancient Copper Miners of Lake Superior. In *History and Review of the Copper,
                Iron, Silver, Slate and Other Material Interests of the South Shore of Lake Superior,*
                by A. P. Swineford, pp. 78–89. *The Mining Journal,* Marquette, Michigan.

1876b   The Ancient Copper Miners of Lake Superior. *Iron* 8:168–169, 199.

1879   The Ancient Copper Mines of Lake Superior. *Report and Collections of the State Historical Society of Wisconsin, for the Years 1877, 1878 and 1879.* 8:140–151.

House, John H.

1996   East-Central Arkansas. In *Prehistory of the Central Mississippi Valley*, edited by Charles H. McNutt, pp. 137–154. University of Alabama Press, Tuscaloosa, Alabama.

Howard, James H.

1953   The Southern Cult in the Northern Plains. *American Antiquity* 19(2):130–138.

1968   *The Southeastern Ceremonial Complex and Its Interpretation.* Memoir No. 6. Missouri Archaeological Society, Columbia, Missouri.

Howland, Henry R.

1877   Recent Archaeological Discoveries in the American Bottom. *Bulletin of the Buffalo Society of Natural Sciences* 3(5):204–211.

Hoxie, R. David

1980   An Analysis of the Late Woodland Copper Assemblage from the Sand Point Site, Baraga County, Michigan. *The Michigan Archaeologist* 26(3–4):25–38.

Hranicky, Wm. Jack

2011   The Adena Culture of the Sandy Hill Area, Dorchester County, Maryland. AuthorHouse, Bloomington, Indiana.

Hruska, Robert

1967   The Riverside Site: A Late Archaic Manifestation in Michigan. *The Wisconsin Archeologist* 48(3):145–257.

Hubbard, L. L.

1895   The Relation of the Vein at the Central Mine, Keweenaw Point, to the Kearsarge Conglomerate. *Proceedings of the Lake Superior Mining Institute* 3:74–83.

Hudson Mining Company

1864   *Prospectus of the Hudson Mining Company, 1864.* Francis Hart & Company, New York.

Hudson, Charles M.

1990   Conversations with the High Priest of Coosa. In *Lamar Archaeology: Mississippian Chiefdoms in the Deep South*, edited by Mark Williams and Gary Shapiro, pp. 214–230. University of Alabama Press, Tuscaloosa, Alabama.

Hulbert, Edwin J.

1893   *"Calumet-Conglomerate": An Exploration and Discovery Made by Edwin J. Hulbert, 1854 to 1861.* Ontonagon-Miner Press, Ontonagon, Michigan.

1899   *Calumet-Conglomerate Discovery, 1864: Reply of Edwin J. Hulbert, Mining-Engineer to an Article Published in the Detroit-Evening-News of October 16 1899 Attributing Discovery to Richard Tregaskis of Cornwall. Published for Mr. Frederick Mackenzie, Editor of the "Calumet-Evening-News" Calumet, Michigan.* Tipografia Nazionale di G. Bertero, Rome, Italy.

Bibliography

Hulburt, Milton F.
> 1933   Copper Spearpoints from Reedsburg, Wisconsin, Dells. *The Wisconsin Archeologist* 13(1):15–17 (and frontispiece).

Hunt, Freeman
> 1848   Ancient Mining on Lake Superior. *Hunt's Merchants' Magazine and Commercial Review* 19(6):663.

Hunzicker, David A.
> 2002   Little Rice Lake Site (47VI272): An Old Copper Complex Manufacturing Site in Vilas County, WI. *The Wisconsin Archeologist* 83(1):45–62.

Hurley, William M.
> 1974   *Silver Creek Woodland Sites, Southwestern Wisconsin.* Report 6. Office of State Archaeologist, University of Iowa, Iowa City, Iowa.
> 1975   *An Analysis of Effigy Mound Complexes in Wisconsin.* Anthropological Papers No. 49. Museum of Anthropology, University of Michigan, Ann Arbor, Michigan.

Hurst, Vernon J. and Lewis H. Larson
> 1958   On the Source of Copper at the Etowah Site, Georgia. *American Antiquity* 24(2):177–181.

Ingersoll, John N.
> 1846   Lake Superior News. *Lake Superior News and Miners' Journal* [Copper Harbor, Michigan], August 8, p. 2.
> 1848   Antiquity of Mining upon Lake Superior. *Lake Superior News and Mining Journal* [Sault Ste.Marie, Michigan], August 26, p. 2.

Ireton, Jerald
> 1982   The Bils Site. *Ohio Archaeologist* 32(3):18.

Irwin, W. G.
> 1900   Old Fort Ancient. *Scientific American* 82(7):104.

Jackman, F.
> 1880   Mounds and Earthworks of Rush County, Indiana. *Annual Report of the Board of Regents of the Smithsonian Institution, 1879*, pp. 374–376.

Jackson, Charles T.
> 1849   Report on the Geological and Mineralogical Survey of the Mineral Lands of the United States in Michigan. In *Message from the President of the United States, to the Two Houses of Congress, at the Commencement of the First Session of the Thirty-first Congress. December 24, 1849. Read December 27, 1849. Part III,* pp. 371–801. House Executive Document 5(3). 31st Congress, 1st Session. Also issued as Senate Executive Document 1). 31st Congress, 1st Session. Washington, DC.
> 1850   Remarks on the Geology, Mineralogy and Mines of Lake Superior. *Proceedings of the American Association for the Advancement of Science* (1849) 2:283–287.
> 1855   Catalogue of Rocks, Minerals, and Ores, Collected during the Years 1847 and 1848, on the Geological Survey of the United States Mineral Lands in Michigan, by Dr. C. T. Jackson, United States Geologist, and Deposited in the Smithsonian Institution. *Ninth Annual Report of the Board of Regents of the Smithsonian Institution* (1854), pp. 338–367. Washington, DC.

Jamison, James K.
    1946   Newspapers of Keweenaw County. *Michigan History Magazine* 30(1):73–75.
    1948   *This Ontonagon Country*. Third edition. The Ontonagon Herald Company, Ontonagon, Michigan.
    1950   *The Mining Ventures of This Ontonagon Country*. Privately published by the author, Ontonagon, Michigan.

Janzen, Donald E.
    1968   *The Naomikong Point Site and the Dimensions of Laurel in the Lake Superior Region*. Anthropological Papers No. 36. Museum of Anthropology, University of Michigan, Ann Arbor, Michigan.

Jefferies, Richard W.
    1979   The Tunacunhee Site: Hopewell in Northwest Georgia. In *Hopewell Archaeology: The Chillicothe Conference*, edited by David S. Brose and N'omi Greber, pp. 162–170. MCJA Special Paper No. 3. Kent State University Press, Kent, Ohio.
    1987   *The Archaeology of Carrier Mills: 10,000 Tears in the Saline Valley of Illinois*. Southern Illinois University Press, Carbondale and Edwardsville, Illinois.
    1996   Hunters and Gatherers after the Ice Age. In *Kentucky Archaeology*, edited by R. Barry Lewis, pp. 39–77. University Press of Kentucky, Lexington, Kentucky.
    2004   Regional Cultures, 700 B.C.–A.D. 1000. In *Southeast*, edited by Raymond D. Fogelson, pp. 115–127. Vol. 14, Smithsonian Institution, Washington, DC.
    2006   Death Rituals at the Tunacunnhee Site. In *Recreating Hopewell*, edited by Douglas K. Charles and Jane E. Buikstra, pp. 161–177. University Press of Florida, Gainesville, Florida.
    2008   The Archaic Period. In *The Archaeology of Kentucky: An Update*, edited by David Pollack, pp. 193–246. Kentucky Heritage Council, State Historic Preservation Comprehensive Plan Report No. 3, Frankfort, Kentucky.
    2009   Archaic Cultures of Western Kentucky. In *Archaic Societies: Diversity and Complexity across the Midcontinent*, edited by Thomas E. Emerson, Dale L. McElrath, and Andrew C. Fortier, pp. 635–665. State University Press of New York, Albany, New York.

Jefferies, Richard W. and Brian M. Butler
    1982   *The Carrier Mills Archaeological Project: Human Adaptation in the Saline Valley, Illinois. Vols. 1 and 2*. Research Paper No. 33. Center for Archaeological Investigations, Southern Illinois University, Carbondale, Illinois.

Jenkins, Ned J.
    1979   Miller Hopewell of the Tombigbee Drainage. In *Hopewell Archaeology: The Chillicothe Conference*, edited by David S. Brose and N'omi Greber, pp. 171–180. MCJA Special Paper No. 3. Kent State University Press, Kent, Ohio.

Jenkins, Ned J. and Richard A. Krause
    1986   *The Tombigbee Watershed in Southeastern Prehistory*. University of Alabama Press, Tuscaloosa, Alabama.

Jennings, Jesse D.
    1941   Chickasaw and Earlier Indian Cultures of Northeast Mississippi. *The Journal of Mississippi History* 3(3):155–226.
    1952   Prehistory of the Lower Mississippi Valley. In *Archeology of Eastern United States*, edited by James B. Griffin, pp. 256–271. University of Chicago Press, Chicago.

Bibliography

Jensen, Harald P., Jr.
    1968   Coral Snake Mound. *Bulletin of the Texas Archeological Society* 39:9–44.

Jenson, Peter S.
    1962   The J. F. Norman Collection of Copper Artifacts. *The Wisconsin Archeologist* 43(3):65–69.

Jenson, Peter S. and Jeffrey Birch
    1963   Archeological Survey of Southeastern Minnesota. *Minnesota Archeologist* 25(2):45–85.

Jeske, John A.
    1922   Recent Interesting Finds near Kingston. *The Wisconsin Archeologist* 1(1):24–25, 27, 29, 33.
    1927   The Grand River Mound Group and Camp Site. *Bulletin of the Public Museum of the City of Milwaukee* 3(2):139–214.

Jeske, Robert J.
    1994   Residual Effects of Grave Desecrations. *Public Archaeology Review* 2(2):8–13.
    2006   Hopewell Regional Interactions and Southeastern Wisconsin and Northern Illinois. In *Recreating Hopewell*, edited by Douglas K. Charles and Jane E. Buikstra, pp. 285–309. University Press of Florida, Gainesville, Florida.

Jeter, Marvin D.
    1984   Mound Volumes, Energy Ratios, Exotic Materials, and Contingency Tables : Comments on Some Recent Analyses of Copena Burial Practices. *Midcontinental Journal of Archaeology* 9(1):91–104.

Johanson, Bruce H.
    1985   *This Land, the Ontonagon: A Short History of Ontonagon County, Michigan.* The Ralph W. Secord Press of the Mid-Peninsula Library Cooperative, Iron Mountain, Michigan.

Johnson, Alfred E.
    1979   Kansas City Hopewell. In *Hopewell Archaeology: The Chillicothe Conference,* edited by David S. Brose and N'omi Greber, pp. 86–93. MCJA Special Paper No. 3. Kent State University Press, Kent, Ohio.

Johnson, David
    2012   *The Old Copper Complex: North America's First Metal Miners and Metal Artisans.* http://copperculture.homestead.com/

Johnson, Earl W.
    1994   The Old Copper Culture. *Central States Archaeological Journal* 41(3):119.
    1996   Copper Fishhooks. *Central States Archaeological Journal* 43(3):123.
    1997   One Unidentified Artifact. *Central States Archaeological Journal* 44(3):137.

Johnson, Elden
    1962   The Prehistory of the Red River Valley. *Minnesota History* 38(4):157–165.
    1964   Copper Artifacts and Glacial Lake Agassiz Beaches. *Minnesota Archaeologist* 26(1):4–22.
    1967   An Unusual Copper Knife. *Minnesota Archaeologist* 29(4):104–105.
    1973   *The Arvilla Complex.* Publications of the Minnesota Historical Society. Minnesota Prehistoric Archaeology Series No. 9. St. Paul, Minnesota.

Johnson, F., Jr.
  1882 The Upper Peninsula of Michigan. *Harper's New Monthly Magazine* 64(384): 892–902.

Johnson, George L.
  1980 Trading Goods Copper? *Central States Archaeological Journal* 27(3):128.
  1985 Bulldozed Copper Bead, the Clue. *Central States Archaeological Journal* 32(1):18–20.

Johnson, Glen
  1969 Excavation of the McCarter Mound, Panola County. *Newsletter of the Mississippi Archaeological Association* 4(1):5–6.

Johnson, Thorley
  1975 Wading the Shore. *Central States Archaeological Journal* 22(4):176–177.

Johnston, Richard B.
  1968a *Archaeology of Rice Lake, Ontario.* Paper No. 19. Anthropology Papers, National Museum of Canada, Ottawa, Ontario.
  1968b *The Archaeology of the Serpent Mounds Site.* Occasional Paper 10. Art and Archaeology, Royal Ontario Museum, University of Toronto, Toronto, Ontario.
  1984 *The McIntyre Site: Archaeology, Subsistence and Environment.* Paper No. 126. Archaeological Survey of Canada, National Museum of Man, Mercury Series, Ottawa, Ontario.

Jolly, Fletcher, III
  1969 Evidence of Aboriginal Trade in Late Prehistoric Times. *Journal of Alabama Archaeology* 15(2):41–47.
  1975 Pierced Ears and Mississippi Valley Headpots. *Central States Archaeological Journal* 22(1):10–15.
  1976 40Kn37: An Early Woodland Habitation Site in Knox County, Tennessee. *Tennessee Archaeologist* 32(1–2):51–64.

Jones, B. Calvin
  1982 Southern Cult Manifestations at the Lake Jackson Site, Leon County, Florida: Salvage Excavations of Mound 3. *Midcontinental Journal of Archaeology* 7(1):3–44.
  1994 The Lake Jackson Mound Complex (8LE1): Stability and Change in Fort Walton Culture. *The Florida Anthropologist* 47(2):120–146.

Jones, Charles C., Jr.
  1873 *Antiquities of the Southern Indians, Particularly of the Georgia Tribes.* D. Appleton and Company, New York.

Jones, Joseph
  1880 Explorations of the Aboriginal Remains of Tennessee. *Smithsonian Contributions to Knowledge* 22(2). Smithsonian Institution, Washington, DC.

Jones, Monty
  1980 Copper Celt Found in Lower Wabash Region. *Central States Archaeological Journal* 27(3):129.

Jones, Walter B. and David L. DeJarnette
  n.d. *Moundville Culture and Burial Museum.* Museum Paper 13. Alabama Museum of Natural History, Tuscaloosa, Alabama.

Bibliography

Jury, W. Wilfrid
    1965   Copper Artifacts from Western Ontario. *The Wisconsin Archeologist* 46(4):223–246.
    1973   Copper Cache at Penetanguishene, Ontario, Canada. *The Wisconsin Archeologist* 54(2):84–106.
    1978   A Red Ochre Burial from Port Franks, Ontario. *Museum Notes 1*. Museum of Indian Archaeology at the University of Western Ontario, London, Ontario, Canada.

Justice, Noel D.
    2001   *Field Guide to Projectile Points of the Midwest.* Indiana University Press, Bloomington.

    2006   *Looking at Prehistory: Indiana's Hoosier National Forest Region, 12,000 B. C. to 1650.* Forest Service, United States Department of Agriculture, Washington, DC.

Justin, Michael
    1995a   *Archaeological Data Recovery at the Roosevelt Lake Narrows Site, Cass County, Minnesota.* MnDOT No. 93-2510, License No. 94-14. Prepared for Minnesota Department of Transportation, St. Paul, Minnesota, MnDOT Agreement No. 70973. W.O. #4. Final Report, Volume I of II.
    1995b   *Archaeological Data Recovery at the Roosevelt Lake Narrows Site, Cass County, Minnesota.* MnDOT No. 93-2510, License No. 94-14. Prepared for Minnesota Department of Transportation, St. Paul, Minnesota, MnDOT Agreement No. 70973. W.O. #4. Final Report, Volume II of II.

"K." (Knapp, Samuel O.?)
    1849   From the Copper Country—Lake Superior. *Lake Superior News and Mining Journal*, May 31, p. 3.

Kalm, Pehr (aka Peter)
    1772   *Travels into North America; Containing Its Natural History, and a Circumstantial Account of Its Plantations and Agriculture in General, with the Civil, Ecclesiastical and Commercial State of the Country, the Manners of the Inhabitants, and Several Curious and Important Remarks on Various Subjects.* Vol. I. (Second Edition). Printed for T. Lowndes, No. 77, in Fleet Street, London, England.

Kammerer, John J.
    1964   Heavy Copper Artifacts. *Minnesota Archaeologist* 26(1):23–30.

Kannenberg, Arthur P. and Gerald C. Stowe
    1938   Butte des Morts Explorations, 1935–1936. *The Wisconsin Archeologist* 18(2):42–52.

Kayner, D. P.
    1883a   Houghton County. In *History of the Upper Peninsula of Michigan*, pp. 250–320. Western Historical Company, Chicago.
    1883b   Keweenaw County. In *History of the Upper Peninsula of Michigan*, pp. 323–345. Western Historical Company, Chicago.

Keel, Bennie C.
    1976   *Cherokee Archaeology: A Study of the Appalachian Summit.* University of Tennessee Press, Knoxville, Tennessee.

Kellar, James H.
1960   The C. L. Lewis Stone Mound and the Stone Mound Problem. *Prehistory
         Research Series* 3(4):355–481. Indiana Historical Society, Indianapolis.
1979   The Mann Site and "Hopewell" in the Lower Wabash-Ohio Valley. In *Hopewell
         Archaeology: The Chillicothe Conference*, edited by David S. Brose and N'omi Greber,
         pp. 100–107. MCJA Special Paper No. 3. Kent State University Press, Kent, Ohio.
1993   *An Introduction to the Prehistory of Indiana*. Indiana Historical Society,
         Indianapolis.

Kellar, James H., A. R. Kelly, and Edward V. McMichael
1962a   *Final Report on Archaeological Explorations at the Mandeville Site, 9CLA1,
         Clay County, Georgia, Seasons 1959, 1960, 1961*. Report No. 8. Laboratory of
         Archaeology Series, University of Georgia, Athens, Georgia.
1962b   The Mandeville Site in Southwest Georgia. *American Antiquity* 27(3):336–355.

Kellar, James H. and B. K. Swartz, Jr.
1971   Adena: The Western Periphery. In *Adena: The Seeking of an Identity*, edited by B.
         K. Swartz, Jr., pp. 122–131. Ball State University, Muncie, Indiana.

Kellogg, Louise P.
1917   *Early Narratives of the Northwest, 1634–1699*. Charles Scribner's Sons, New York.

Kelly, Arthur R.
1970   *Explorations at Bell Field Mound and Village: Seasons 1965, 1966, 1967, 1968*.
         Report submitted to National Park Service. On file at the University of Georgia
         Laboratory of Archaeology as Manuscript No. 289, Athens, Georgia.

Kelly, Arthur R. and Lewis H. Larson, Jr.
1956   Explorations at Etowah, Georgia 1954–1956. *Archaeology* 10(1):39–48.

Kelly, Arthur R. and R. S. Neitzel
1961   *Chauga Mound and Village Site (38 Oc1) in Oconee County, South Carolina*.
         Report No. 3. Laboratory of Archaeology, Department of Sociology and
         Anthropology, University of Georgia, Athens, Georgia.

Kelly, John E.
1991a   Cahokia and Its Role as a Gateway Center in Interregional Exchange. In *Cahokia
         and the Hinterlands*, edited by Thomas E. Emerson and R. Barry Lewis, pp.
         61–80. University of Illinois Press, Urbana and Chicago.
1991b   The Evidence for Prehistoric Exchange and Its Implications for the Development
         of Cahokia. In *New Perspectives on Cahokia: Views from the Periphery*, edited
         by James B. Stoltman, pp. 65–92. Monographs in World Archaeology No. 2.
         Prehistory Press, Madison, Wisconsin.
1994   The Archaeology of the East St. Louis Mound Center: Past and Present. *Illinois
         Archaeology* 6(1–2):1–57.
2004   The Mitchell Mound Center: Then and Now. In *Aboriginal Ritual and Economy in
         the Eastern Woodlands: Essays in Memory of Howard Dalton Winters*, edited by
         Anne-Marie Cantwell, Lawrence A. Conrad, and Jonathan E. Reyman, pp. 269–284.
         Illinois State Museum Scientific Papers Vol. 30, Springfield, Illinois. Kampsville
         Studies in Archeology and History Vol. 5. Center for American Archeology,
         Kampsville, Illinois.

Bibliography

Kelly, John E., James A. Brown, Jenna M. Hamlin, Lucretia S. Kelly, Laura Kozuch, Kathryn Parker, and Julieann Van Nest
    2007   The Context for the Early Evidence of the Southeastern Ceremonial Complex at Cahokia. In *Southeastern Ceremonial Complex: Chronology, Context*, edited by Adam King, pp. 57–87. University of Alabama Press, Tuscaloosa, Alabama.

Kelly, John E., James A. Brown, and Lucretia S. Kelly
    2008   The Context of Religion at Cahokia: The Mound 34 Case. In *Religion, Archaeology, and the Material World*, edited by Lars Fogelin, pp. 297–318. Occasional Paper No. 36. Center for Archaeological Investigations, Southern Illinois University, Carbondale, Illinois.

Kelly, John E., Steven J. Ozuk, and Joyce A. Williams
    1990   *The Range Site 2: The Emergent Mississippian Dohack and Range Phase Occupations (11-S-47)*. Published for the Illinois Department of Transportation by the University of Illinois Press, Urbana and Chicago.

Kennedy, Clyde C.
    1962   Archaic Hunters in the Ottawa Valley. *Ontario History* 54(2):122–128.
    1967   Preliminary Report on the Morrison's Island-6 Site. *Contributions to Anthropology* V, Bulletin 206, pp. 100–125. National Museum of Canada, Ottawa, Canada.
    1970   *The Upper Ottawa Valley*. Renfrew County Council, Pembroke, Ontario, Canada.

Kent, Barry and Ronald Vanderwal
    1962

Kent, Barry and Ronald Vanderwal
    1962   The Copper Industry at the Juntunen Site. Unpublished manuscript, Museum of Anthropology, University of Michigan, Ann Arbor, Michigan.

Kent, Timothy
    1991   Discovery of Ancient Copper Site Yields Unique Pottery Repair Band of Copper. *Central States Archaeological Journal* 38(3):124–126.

Kenyon, Walter A.
    1959a   *The Inverhuron Site, Bruce County, Ontario*. University of Toronto Press, Toronto, Ontario, Canada.
    1959b   The Mound at Pithers Point. *Ontario History* 51(1):64–66.
    1986   *Mounds of Sacred Earth: Burial Mounds of Ontario*. Archaeology Monograph 9. Royal Ontario Museum, Toronto, Ontario, Canada.

Kett, H. F. & Co.
    1878   *The History of Jo Daviess County, Illinois*. H. F. Kett & Co., Times Building, Chicago.

Kidd, Kenneth E.
    1952   Sixty Years of Ontario Archeology. In *Archaeology of Eastern United States*, edited by James B. Griffin, pp. 71–82. University of Chicago Press, Chicago.

    1953   The Excavation and Historical Identification of a Huron Ossuary. *American Antiquity* 18(4):359–379.

Kidder, Tristram R.
2004   Prehistory of the Lower Mississippi Valley After 800 B.C. In *Southeast*, edited by Raymond D. Fogelson, pp. 545–559. Vol. 14, Smithsonian Institution, Washington, DC.

Kidder, Tristram R. and Kenneth E. Sassaman
2009   The View from the Southeast. In *Archaic Societies: Diversity and Complexity across the Midcontinent*, edited by Thomas E. Emerson, Dale L. McElrath, and Andrew C. Fortier, pp. 667–694. State University Press of New York, Albany, New York.

Kime, Julie
1985   Fragment of the Past: The Ater Mound Blanket. *Timeline* 2(4):48–51.

King, Adam
2003   *Etowah: The Political History of a Chiefdom Capital.* University of Alabama Press, Tuscaloosa, Alabama.
2004   Deciphering Etowah's Mound C: The Construction History and Mortuary Record of a Mississippian Burial Mound. *Southeastern Archaeology* 23(2):153–165.
2007a   Mound C and the Southeastern Ceremonial Complex in the History of the Etowah Site. In *Southeastern Ceremonial Complex: Chronology, Content, Context*, edited by Adam King, pp. 107–133. University of Alabama Press, Tuscaloosa, Alabama.
2007b   Whither SECC? In *Southeastern Ceremonial Complex: Chronology, Content, Context*, edited by Adam King, pp. 251–258. University of Alabama Press, Tuscaloosa, Alabama.

King, Blanche Busey
n.d   Research Data Notes—Copper Find. In *Kentucky's Ancient Buried City, Wickliffe, Kentucky*, p. 12–14.
1937   Recent Excavations at the King Mounds, Wickcliffe, Kentucky. *Transactions of the Illinois State Academy of Science* 30(2):83–90.
1939   *Under Your Feet: The Story of the American Mound Builders.* Dodd, Mead & Company, New York.

King, D. R.
1961   The Bracken Cairn. *The Blue Jay* 19(1):45–53.

King, Fain White
n.d.   Kentucky's Ancient Copper Hoard. In *Kentucky's Ancient Buried City, Wickliffe, Kentucky*, p. 10.

King, Mary Elizabeth
1968   Textile Fragments from the Riverside Site, Menominee, Michigan. *Verhandlungen des XXXVIII Internationalen Amerikanisten-Kongresses* 1 (Stuttgart-München), pp. 117–123.

Kirby, Matthew E., Henry T. Mullins, William P. Patterson and Andrew W. Burnett
2002   Late Glacial-Holocene Atmospheric Circulation and Precipitation in the Northeast United States Inferred from Modern Calibrated Stable Oxygen and Carbon Isotopes. *Geological Society of America Bulletin* 14(10):1326–1340.

Bibliography

Kneberg, Madeline
    1952   The Tennessee Area. In *Archeology of Eastern United States*, edited by James B. Griffin, pp. 190–198. University of Chicago Press, Chicago.

Knight, Miletus
    1895   Letter to the Editor. *The Archaeologist* 3(3):105.

Knight, Vernon J., Jr.
    1982   A Repoussé Copper Plate from Northeast Alabama. *Journal of Alabama Archaeology* 28(2)79–82.
    2004   Ceremonialism until 1500. In *Southeast*, edited by Raymond D. Fogelson, pp. 734–741. Vol. 14, Smithsonian Institution, Washington, DC.

Knight, Vernon J., Jr and Vincas P. Steponaitis
    1998   A New History of Moundville. In *Archaeology of the Moundville Chiefdom*, edited by Vernon J. Knight, Jr, and Vincas P. Steponaitis, pp. 1–25. Smithsonian Institution Press, Washington and London.
    2011   A Redefinition of the Hemphill Style in Mississippian Art. In *Visualizing the Sacred: Cosmic Visions, Regionalism, and the Art of the Mississippian World*, edited by George E. Lankford, F. Kent Reilly, and James F. Garber, pp. 201–239. University of Texas Press, Austin, Texas.

Knoblock, Byron W.
    1939   *Bannerstones of the North American*. Published by the author. LaGrange, Illinois.

Knowlton Mining Company
    1863   *Knowlton Mining Company: Report and Prospectus of the Mine up to January 1, 1863*. Cleveland.

Knudsen, J. P. and R. D. Radford
    1957   An Archaeological Survey of the Rockhouse Ledge Area – Part I: The General Area and Rockhouse Spring Cave. *Journal of Alabama Archaeology* 3(2):1–7. (See also Carstens and Knudsen 1958.)

Kocik, Cynthia
    2012   Regional Variations in Hopewell Copper Use. *University of Wisconsin-La Crosse Journal of Undergraduate Research* 15:1–29.

Kohl, J. G.
    1860   *Kitchi-Gami: Wanderings Round Lake Superior*. Chapman and Hall, 193, Piccadilly, London, England.

Koldehoff, Brad and Larry Kinsella
    1995   The Chapel Hill Copper Plate. *Illinois Antiquity* 30(4):4–7.

Korp, Maureen
    1990   *The Sacred Geography of the American Mound Builders*. Native American Studies Vol. 2, The Edwin Mellen Press. Lewiston, New York.

Kowalewski, Stephen A.
    1996   Clout, Corn, Copper, Core-Periphery, Culture Area. In *Pre-Columbian World Systems*, edited by Peter N. Peregrine and Gary M. Feinman, pp. 27–37. Monographs in World Archaeology No. 26. Prehistory Press, Madison, Wisconsin.

Krakker, James J.
1997   Biface Caches, Exchange, and Regulatory Systems in the Prehistoric Great Lakes Region. *Midcontinental Journal of Archaeology* 22(1):1–41.
2012   The Myers Site, Putnam County, Ohio and Middle Woodland Long Distance Interaction. *Archaeology of Eastern North America* 40:131:144.

Krause, David J.
1992   *The Making of a Mining District: Keweenaw Native Copper, 1500–1870.* Wayne State University Press, Detroit.

Krieger, Alex D.
1945   An Inquiry into Supposed Mexican Influence on the Prehistoric "Cult" in the Southern United States. *American Anthropologist* 47(4):483–515.
1946   The Sanders Site, Lamar County, Texas. In *Culture Complexes and Chronology in Northern Texas*, pp.171–218. University of Texas Publication No. 4640. Austin, Texas.

Kruse, Harvey and Harvey Soulen
1940   A Recent Find in Minnesota. *Minnesota Archaeologist* 6(2):51–52.

Kuhm, Herbert W.
1928   *Wisconsin Indian Fishing—Primitive and Modern. The Wisconsin Archeologist* 7(2):61–114.
1935   Isle Royale Antiquities. *The Milwaukee Journal*, May 29, p. 6.

La Fayette Mining Company
1863   *Prospectus of the La Fayette Mining Company, 1863.* Francis Hart & Co., New York.

Ladassor, Gray
1950   Cahokia Ornaments. In *Cahokia Brought to Life*, edited by R. E. Grimm, pp. 32–41. Greater St. Louis Archaeological Society, St. Louis.

Laidlaw, George E.
1899   Some Copper Implements from the Midland District, Ontario. *The American Antiquarian* 21(1):83–90.
1915   Archaeological Notes on Victoria County, Ont. *The Archaeological Bulletin* 6(4):71–77.

Lake Superior Journal
1850   Copper Mines of Lake Superior. *American Railroad Journal* 23(756):645–646.

Lake Superior News and Mining Journal [Sault Ste. Marie, Michigan]
1848   The Copper Region—Singular Discovery. August 18, p. 2.

Lambert, Peter J. B.
1982   The Archaeological Survey of West Lac Seul. In *Studies in West Patricia Archaeology No. 3:1980–1981, Archaeology Research Report 19*, pp. 133–404. Archaeology and Heritage Branch, Ontario Ministry of Citizenship and Culture, Toronto, Ontario, Canada.

Landon, R. H.
1940   The Mining and Fabrication of Copper by the Aborigines of the Lake Superior Region. *Minnesota Archaeologist* 6(2):27–34.

Bibliography

Lane, Alfred C.
    1898   Geological Report on Isle Royale, Michigan. *Geological Survey of Michigan* 6(1).
        R. Smith Printing Co., State Printers, Lansing, Michigan.

Langevin, Érik, Moira T. McCaffrey, Jean-François Moreau and Ron G. V. Hancock
    1995   Le cuivre natif dans le Nord-Est Québécois: contribution d'un site du lac Saint-
        Jean (Québec central). *Recherches amérindiennes au Québec, Paléo-Québec* No.
        23. Montréal, QB, Canada.

Langford, George
    1927   The Fisher Mound Group, Successive Aboriginal Occupations near the Mouth of
        the Illinois River. *American Anthropologist* 29(3):152–205.

Lankford, George E.
    2007a   The Great Serpent in Eastern North America. In *Ancient Objects and Sacred
        Realms: Interpretations of Mississippian Iconography*, edited by F. Kent Reilly III
        and James F. Garber, pp. 107–135. University of Texas Press, Austin, Texas.
    2007b   Some Cosmological Motifs in the Southeastern Ceremonial Complex. In *Ancient
        Objects and Sacred Realms: Interpretations of Mississippian Iconography*, edited by F.
        Kent Reilly III and James F. Garber, pp. 8–38. University of Texas Press, Austin, Texas.

Lankton, Larry
    1997a   *Beyond the Boundaries: Life and Landscape at the Lake Superior Copper Mines,
        1840–1875.* Oxford University Press, New York.
    1997b   *Keweenaw Copper: Mines, Mills, Smelters, and Communities.* Guide presented to
        attendees of the 26th Annual Conference of the Society for Industrial Archeology,
        Houghton, Michigan.

Lapham, Increase A.
    1855   The Antiquities of Wisconsin, as Surveyed and Described. *Smithsonian
        Contributions to Knowledge* 7(4). Washington, DC.

LaRonge, Michael
    2001   An Experimental Analysis of Great Lakes Archaic Copper Smithing. *North
        American Archaeologist* 22(4):371–385.

Larsen, Clark S. and Patricia J. O'Brien
    1973   The Cochran Mound, 23PL86, Platte County, Missouri. *Newsletter of the
        Mississippi Archaeological Society* 267:1–5.

Larson, Lewis H., Jr.
    1954   Georgia Historical Commission Excavations at the Etowah Site. *Early Georgia*
        1(3):18–22.
    1957   An Unusual Wooden Rattle from the Etowah Site. *The Missouri Archaeologist*
        19(4):6–11.
    1958a   Cultural Relationships between the Northern St. Johns Area and the Georgia
        Coast. *Florida Anthropologist* 11(1):10–21.
    1958b   Southern Cult Manifestations on the Georgia Coast. *American Antiquity* 23(4):426–430.
    1959   *A Mississippian Headdress from Etowah, Georgia. American Antiquity* 25(1):109–112.
    1971   Archaeological Implications of Social Stratification at the Etowah Site, Georgia.
        In *Approaches to the Social Dimensions of Mortuary Practices*, organized and
        edited by James A. Brown, pp. 58–67. Memoirs of the Society for American
        Archaeology No. 25.

1989   The Etowah Site. In *The Southeastern Ceremonial Complex: Artifacts and Analysis*, edited by Patricia Galloway, pp. 133–141. University of Nebraska Press, Lincoln, Nebraska.
1993   An Examination of the Significance of a Tortoise-Shell Pin from the Etowah Site. In *Archaeology of Eastern North America: Papers in Honor of Stephen Williams*, edited by James B. Stoltman, pp. 169–185. Archaeological Report No. 25. Mississippi Department of Archives and History, Jackson, Mississippi.

Latchford, Carl
1985   The Edwards Plate: A Late Mississippian Effigy. *Central States Archaeological Journal* 32(1):4–7.

Lathrop, J. H.
1901a   Mound-Builders in the Copper Country of Michigan. *The Northwest Magazine* 19(2):3–7.
1901b   Prehistoric Mines of Lake Superior. *American Antiquarian and Oriental Journal* 23(4):248–258.
1961   Prehistoric Mines of Lake Superior. In *Prehistoric Copper Mining in the Lake Superior Region: A Collection of Reference Articles*, edited by Roy W. Drier and Octave J. Du Temple, pp. 119–126. Privately published. Calumet, Michigan and Hinsdale, Illinois.

Lattanzi, Gregory D.
2007   The Provenance of Pre-Contact Copper Artifacts : Social Complexity and Trade in the Delaware Valley. *Archaeology of Eastern North America* 35:125–137.
2008   Elucidating the Origin of Middle Atlantic Pre-Contact Copper Artifacts Using Laser Ablation ICP-MS. *North American Archaeologist* 29(3–4):297–326.
2013   *The Value of Reciprocity: Copper Exchange and Social Interaction in the Middle Atlantic Region of the Eastern Woodlands of North America.* PhD dissertation, Temple University, Philadelphia.

Laval, Otto
1923   Indian Relics. In *Proceedings of the Southwestern Indiana Historical Society, Fourth Annual Meeting*, pp. 135–139.

Lawshe, Fred E.
1947   The Mero Site-Diamond Bluff, Pierce County, Wisconsin. *Minnesota Archaeologist* 13(4):75–95.

Lawson, Publius V.
1902   Copper Age in the United States. *The American Antiquarian* 24(6):459–474.
1906   *Story of the Rocks and Minerals of Wisconsin.* The Post Publishing Company, Appleton, Wisconsin.

Lawton, Charles D.
1881   Early History. In *Annual Report of the Commissioner of Mineral Statistics of the State of Michigan, for 1880*, pp. 5–152. W. S. George & Co., Lansing, Michigan.
1883   Copper Mines. In *Annual Report of the Commissioner of Mineral Statistics of the State of Michigan for 1882*, by Charles E. Wright, pp. 49–182. W. S. George & Co., State Printers and Binders, Lansing, Michigan.
1886   Mines and Mineral Interests of Michigan. In *The Semi-Centennial of the Admission of the State of Michigan into the Union*, pp. 53–77. Detroit Free Press Printing Company, Detroit.

Bibliography

Leader, Jonathan M.

1985   *Metal Artifacts from Fort Center: Aboriginal Metal Working in the Southeastern United States.* Master's thesis, University of Florida, Gainesville, Florida.

1988   *Technological Continuities and Specialization in Prehistoric Metalwork in the Eastern United States.* PhD dissertation, University of Florida, Gainesville, Florida.

1991   The South Florida Metal Complex: A Preliminary Discussion of the Effects of the Introduction of an Elite Metal on a Contact Period Native American Society. *MASCA Research Papers in Science and Archaeology* 8(2):19–24.

Leechman, Douglas and Frederica de Laguna

1949   The Parker Site. In *Annual Report of the National Museum for the Fiscal Year 1947–1948*, pp. 29–30. Bulletin No. 113, National Museum of Canada, Ottawa, Ontario, Canada.

Leeson, M. A.

1881   *History of Kent County, Michigan.* Chas. C. Chapman & Co., Chicago.

Lehmer, Donald J.

1954   *Archeological Investigations in the Oahe Dame Area, South Dakota, 1950–51.* Bulletin 158. Bureau of American Ethnology, Smithsonian Institution, Government Printing Office, Washington, DC.

Leidy, Joseph

1867   Untitled. *Proceedings of the Academy of Natural Sciences of Philadelphia* 19:97.

Lemley, Harry J.

1936   Discoveries Indicating a Pre-Caddo Culture on Red River in Arkansas. *Bulletin of the Texas Archeological Society* 8:25–55.

Lemmer, Victor F.

1960   Prehistoric Copper Miners on the Shores of Lake Superior. *Skillings' Mining Review* 48(40):4–5, 15.

Leonard, Joseph A.

1910   *History of Olmsted County, Minnesota.* Goodspeed Historical Association, Chicago.

Leonard, Kevin J. M.

1996   *Mi'kmaq Culture during the Late Woodland and Early Historic Periods.* PhD dissertation, Graduate Department of Anthropology, University of Toronto, Toronto, Ontario, Canada.

Lepper, Bradley T.

1989   An Historical Review of Archaeological Research at the Newark Earthworks. *Journal of the Steward Anthropological Society* 18(1–2):118–140.

1995   *People of the Mounds: Ohio's Hopewell Culture.* Hopewell Culture National Historical Park and Eastern National Park and Monument Association.

2002   *The Newark Earthworks: A Wonder of the Ancient World.* Ohio Historical Society, Columbus, Ohio.

Leskinen, Christine and Lauri Leskinen

1980   *Copper Country History, 3000 B.C.–1980.* F. A. Weber & Sons, Inc., Park Falls, Wisconsin.

Leskinen, Lauri
    1974   *4000 Years of Copper Country History.* Greenlee Printing Co., Calumet, Michigan.

Levine, Mary Ann
    1996   *Native Copper, Hunter-gatherers and Northeastern Prehistory.* Ph.D. dissertation, University of Massachusetts Amherst, Amherst, Massachusetts.
    1999   Native Copper in the Northeast: An Overview of Potential Sources Available to Indigenous Peoples. In *The Archaeological Northeast* edited by Mary Ann Levine, Kenneth E. Sassaman, and Michael S. Nassaney, pp. 183–199. Bergin & Garvey, Westport, Connecticut.
    2007   Determining the Provenance of Native Copper Artifacts from Northeastern North America: Evidence from Instrumental Neutron Activation Analysis. *Journal of Archaeological Science* 34:572–587.
    2007   Overcoming Disciplinary Solitude: The Archaeology and Geology of Native Copper in Eastern North America. *Geoarchaeology: An International Journal* 22(1):49–66.

Lewis, Ann
    2003   *A Comparative Study of Hopewell and Oneota Rolled Copper Beads.* Bachelor's thesis, University of Wisconsin-La Crosse, La Crosse, Wisconsin.

Lewis, Clifford M.
    1970   Artifacts from the Elm Grove Mound (46 Oh 94), Wheeling, West Virginia. *West Virginia Archaeologist* 23:1–10.

Lewis, R. Barry
    1996a   Mississippian Farmers. In *Kentucky Archaeology*, edited by R. Barry Lewis, pp. 127–159. University Press of Kentucky, Lexington, Kentucky.
    1996b   The Western Kentucky Border and the Cairo Lowland. In *Prehistory of the Central Mississippi Valley*, edited by Charles H. McNutt, pp. 47–75. University of Alabama Press, Tuscaloosa, Alabama.

Lewis, Theodore H.
    1886   Mounds on the Red River of the North. *The American Antiquarian and Oriental Journal* 8(6):369–371.
    1889   Copper Mines Worked by the Mound Builders. *The American Antiquarian* 11(5):293–296.
    1890   Description of Some Copper Relics of the Collection of T. H. Lewis in the Macalester Museum of History and Archaeology. *Macalester College Contributions* 6:175–181.
    1894   The "Aztlan" Enclosure Newly Described. *The American Antiquarian and Oriental Journal* 16(4):205–208.
    1895   Ancient Mounds in Northern Minnesota. *The American Antiquarian and Oriental Journal* 17(6):316–320.
    1896a   Pre-Historic Remains at St. Paul, Minnesota. *The American Antiquarian and Oriental Journal* 18(4):207–210.
    1896b   Mounds and Stone Cists at St. Paul, Minnesota. *The American Antiquarian and Oriental Journal* 18(6):314–320.

Lewis, Thomas M. N.
    1950   Copper Ceremonial Axes. *Tennessee Archaeologist* 6(2):20–21.
    1951   A Copper Head Ornament. *Tennessee Archaeologist* 7(2):64.
    1958   Artifacts from Citico. *Tennessee Archaeologist* 14(2):88–93.

Bibliography

Lewis, Thomas M. N. and Madeline Kneberg
 1956 Copper Ornaments from the Talassee Site. *Tennessee Archaeologist* 12(1):30–32.
 1957 The Camp Creek Site. *Tennessee Archaeologist* 13(1):1–48.
 1958 *Tribes that Slumber: Indians of the Tennessee Region.* University of Tennessee Press, Knoxville, Tennessee.
 1959 The Archaic Culture in the Middle South. *American Antiquity* 25(2):161–183.
 1961a *Eva: An Archaic Site.* University of Tennessee Press, Knoxville, Tennessee.
 1961b Tennessee Artifacts in the Museum of the American Indian. *Tennessee Archaeologist* 17(1):38–40.
 1970 *Hiwassee Island: An Archaeological Account of Four Tennessee Indian Peoples.* University of Tennessee Press, Knoxville, Tennessee.

Lewis, Thomas M. N. and Madeline D. Kneberg Lewis
 1995 *The Prehistory of the Chickamauga Basin in Tennessee.* University of Tennessee Press, Knoxville, Tennessee.

Lexington Mining Company
 1864 *Prospectus of the Lexington Mining Company...* Howe & Ferry, Stationers, New York.

Lilly, Eli
 1937a *Prehistoric Antiquities of Indiana.* Indiana Historical Society, Indianapolis.
 1937b The Use of Copper by the American Aborigines. *Proceedings of the Indiana Academy of Science* 46:53–56.
 1942 A Cedar Point "Glacial Kame" Burial. *Proceedings of the Indiana Academy of Science* 51:31–33.

Lindeman, Carla G.
 1967 The Vach Sites (1 & 4E): A Comprehensive Analysis. *Minnesota Archaeologist* 29(4):83–98.

Lindley, C. T.
 1877 Mound Explorations in Jackson County, Iowa. *Proceedings of the Davenport Academy of Natural Sciences* 2:83–84.

Littell's Living Age
 1848 The Copper Region. *Littell's Living Age* 18(223):375.

Little, Keith J. and Cailup B. Curren, Jr.
 1981 Site 1Ce308: A Protohistoric Site on the Upper Coosa River in Alabama. *Journal of Alabama Archaeology* 27(2):116–140.

Lloyd, Timothy C.
 1998 Shedding Light on Small Mounds Lost in the Shadow of the Great Mound at the Hopewell Site. *West Virginia Archeologist* 50(1–2):1–13.

Logan, Brad
 1993 *Quarry Creek: Excavation, Analysis and Prospect of a Kansas City Hopewell Site, Fort Leavenworth, Kansas.* Project Report Series No. 80. Office of Archaeological Research, Museum of Anthropology, University of Kansas, Lawrence, Kansas.
 2006 Kansas City Hopewell: Middle Woodland on the Western Frontier. In *Recreating Hopewell*, edited by Douglas K. Charles and Jane E. Buikstra, pp. 339–358. University Press of Florida, Gainesville, Florida.

Logan, Wilfred D.
        1971   Final Investigation of Mound 33, Effigy Mounds National Monument, Iowa. *Journal of the Iowa Archaeological Society* 18:30–45.
        1976   *Woodland Complexes in Northeastern Iowa.* Publications in Archeology 15. National Park Service, U. S. Department of the Interior, Washington, DC.

Long, Joseph K., III
        1958   The Rice Mound. *Tennessee Archaeologist* 14(1):31–32.

Lord, Thomas H.
        1856   From Lake Superior—Large Yield of Copper. *Detroit Daily Free Press* [Detroit, Michigan], February 15, p. 2.

Loughridge, R. M.
        1855   Antique Muscogee Brass Plates. In *Information Respecting the History, Condition and Prospects of the Indian Tribes of the United States: Collected and Prepared under the Direction of the Bureau of the Interior, Per act of Congress of March 3d, 1847, by Henry R. Schoolcraft, LL.D.*, Part V, p. 660. J. D. Lippincott & Company, Philadelphia.

Lovis, William A.
        1973   *Late Woodland Cultural Dynamics in the Northern Lower Peninsula of Michigan.* PhD dissertation, Department of Anthropology, Michigan State University, East Lansing, Michigan.
        2002   *A Bridge to the Past: The Post Nipissing Archaeology of the Marquette Viaduct Replacement Project Sites 20BY28 and 20BY387.* MSU Museum and Department of Anthropology, Michigan State University, East Lansing, Michigan.
        2009   Hunter-Gatherer Adaptations and Alternative Perspectives on the the Michigan Archaic: Research Problems in Context. In *Archaic Societies: Diversity and Complexity across the Midcontinent*, edited by Thomas E. Emerson, Dale L. McElrath, and Andrew C. Fortier, pp. 725–754. State University Press of New York, Albany, New York.

Low, Charles F.
        1880   Archaeological Explorations near Madisonville, Ohio. *Journal of the Cincinnati Society of Natural History* 3(1):40–68.

Lubbock, John
        1875   *Prehistoric Times, as Illustrated by Ancient Remains, and the Manners and Customs of Modern Savages.* D. Appleton and Company, New York.

Lucas, Stephen
        1973   A Digest of Mound Group Sites along the Upper Kankakee River. *Central States Archaeological Journal* 20(2):90–93.

Ludwickson, John, James N. Gundersen, and Craig Johnson
        1993   Select Exotic Artifacts from Cattkle Oiler (39ST224): A Middle Missouri Tradition Site in Central South Dakota. *Plains Anthropologist* 38(145):151–168.

Luedtke, Vernon G.
        1980   Early Finds in Wisconsin. *Central States Archaeological Journal* 27(1):22–24.

Bibliography

Lueger, Richard
    1977   Prehistoric Occupations at Coteau-du-Lac: A Mixed Assemblage of Archaic and Woodland Artifacts. In *History and Archaeology* 12, pp. 3–100. National Historic Parks and Sites Branch, Parks Canada, Department of Indiana and Northern Affairs, Ottawa, Ontario, Canada.

Lurie, Nancy O.
    1959   Indian Cultural Adjustment to European Civilization. In *Seventeenth-Century America: Essays in Colonial History*, edited by James Morton Smith, pp. 33–60. University of North Carolina Press, Chapel Hill, North Carolina.

Lurie, Rochelle, Douglas Kullen, and Scott J. Demel
    2009   Defining the Archaic in Northern Illinois. In *Archaic Societies: Diversity and Complexity across the Midcontinent*, edited by Thomas E. Emerson, Dale L. McElrath, and Andrew C. Fortier, pp. 755–785. State University Press of New York, Albany, New York.

Maasch, Lloyd P.
    1994   Copper Indian Artifacts from Wisconsin. *Central States Archaeological Journal* 41(3):144–147.

MacCurdy, George Grant
    1917   The Wesleyan University Collection of Antiquities from Tennessee. *Proceedings of the Nineteenth International Congress of Americanists. Held at Washington, December 27–31, 1915*, pp. 75–94. Prepared by the Secretary. Edited by F. W. Hodge. Washington, DC.

MacLean, J. Arthur
    1931   Excavation of Albee Mound, 1926–1927. *Indiana History Bulletin* 8(4):89–176.

MacLean, John P.
    1887   *The Mound Builders*. Robert Clarke & Co., Cincinnati, Ohio.
    1901   *The Archaeological Collection of the Western Reserve Historical Society*. Tract No. 90. Western Reserve Historical Society, Cleveland.
    1903   Ancient Works at Marietta, Ohio. *Ohio Archaeological and Historical Quarterly* 12:37–66.

MacNeish, Richard S.
    1952   The Archeology of the Northeastern United States. In *Archeology of Eastern United States*, edited by James B. Griffin, pp. 46–58. University of Chicago Press, Chicago.

Macpherson, J. C.
    1879   Observations on the Pre-Historic Earthworks of Wayne County, Ind. In *Eighth, Ninth and Tenth Annual Reports of the Geological Survey of Indiana, Made during the Years 1876-77-78*, pp. 219–226.

Magrath, Willis H.
    1940   The Temple of the Effigy. *Scientific American* 163(2):76–78.
    1945   The North Benton Mound: A Hopewell Site in Ohio. *American Antiquity* 11(1):40–47.
    1959   A Hopewell Burial on Flint Ridge. *Ohio Archaeologist* 9(3):92–94.

Mainfort, Robert C., Jr.
  1986   *Pinson Mounds: A Middle Woodland Ceremonial Center.* Research Series No.
          7. Division of Archaeology, Tennessee Department of Conservation, Nashville,
          Tennessee.
  1988a   Middle Woodland Ceremonialism at Pinson Mounds, Tennessee. *American
          Antiquity* 53(1):158–173.
  1988b   Middle Woodland Mortuary Patterning at Helena Crossing, Arkansas. *Tennessee
          Anthropologist* 13(1):35–50.
  1996a   Pinson Mounds and the Middle Woodland Period in the Midsouth and Lower
          Mississippi Valley. In *A View from the Core: A Synthesis of Ohio Hopewell
          Archaeology*, edited by Paul J. Pacheco, pp. 370–391. Ohio Archaeological
          Council, Columbus, Ohio.
  1996b   The Reelfoot Lake Basin, Kentucky and Tennessee. In *Prehistory of the Central
          Mississippi Valley*, edited by Charles H. McNutt, pp. 77–96. University of
          Alabama Press, Tuscaloosa, Alabama.

Mallam, R. Clark
  1978   An Old Copper Complex Artifact from Winneshiek County, Iowa. *Newsletter of
          the Iowa Archaeological Society* 90:2–5.

Mallios, Seth and Shane Emmett
  2004   Demand, Supply, and Elasticity in the Copper Trade at Early Jamestown. *Journal
          of the Jamestown Rediscovery Center 2.*

Mangold, William L. and Robert Nesius
  1983   Blades, Beads and Blizzard: A Turkey Tail cache from Northwestern Indiana.
          Central States Archaeological Journal 30(1):22–27.

Mangold, William L. and Mark R. Schurr
  2006   The Goodall Tradition: Recent Research and New Perspectives. In *Recreating
          Hopewell*, edited by Douglas K. Charles and Jane E. Buikstra, pp. 206–226.
          University Press of Florida, Gainesville, Florida.

Manson, Carl P. and Howard A. MacCord
  1941   An Historic Iroquois Site near Romney, West Virginia. *West Virginia History*
          2(4):290–293.
  1944   Additional Notes on the Herriott Farm Site. *West Virginia History* 5(3):201–211.

Marckel, Tim
  2004   Copper in Ohio. *Ohio Archaeologist* 54(1):45.

Markman, Charles W.
  1988   ed., *Putney Landing: Archaeological Investigations at a Havana-Hopewell
          Settlement on the Mississippi River, West-Central Illinois.* Report of Investigations
          No. 15, Department of Anthropology, Northern Illinois University, DeKalb,
          Illinois.

Marquardt, William H. and Patty Jo Watson
  1983   The Shell Mound Archaic of Western Kentucky. In *Archaic Hunters and
          Gatherers in the American Midwest*, edited by James L. Phillips and James A.
          Brown, pp. 323–339. Academic Press, Inc., New York.

Bibliography

Marsh, O. C.
  1866   Description of an Ancient Sepulchral Mound near Newark, Ohio. *American Journal of Science and Arts* 42(124):1–11.

Marshall, John B.
  1992   The St. Louis Mound Group Historical Accounts and Pictorial Depictions. *The Missouri Archaeologist* 53:43–79.

Marshall, L. H.
  1885   Archaeological Researches in Decatur Co., Ind. *Hoosier Mineralogist and Archaeologist* 1(4).

Martin, Patrick E.
  1988   *Technical Report on Archaeological Survey and Evaluation, Isle Royale National Park, 1986.* Archaeology Laboratory, Michigan Technological University, Houghton, Michigan.

Martin, Patrick E., Susan R. Martin, and Michael Gregory
  1994   *Technical Report: 1987–1988, Isle Royal Archaeology.* Report of Investigations 16. Archaeology Laboratory, Michigan Technological University, Houghton, Michigan.

Martin, Paul S., George I. Quimby, and Donald Collier
  1947   *Indians before Columbus: Twenty Thousand Years of North American History Revealed by Archeology.* University of Chicago Press, Chicago.

Martin, Ramon F. and Sukumar P. Desai
  2015   An Appraisal of the Life of Charles Thomas Jackson as Attention Deficit Hyperactivity Disorder. *Journal of Anesthesia History* 1(2):38–43.

Martin, Susan R.
  1993   ed., 20KE20: Excavations at a Prehistoric Copper Workshop. *The Michigan Archaeologist* 39(3–4):127–193.
  1994   A Possible Beadmaker's Kit from North America's Lake Superior Copper District. *Beads: Journal of the Society of Bead Researchers* 6(6):49–60.
  1995   Michigan Prehistory Facts: The State of Our Knowledge about Ancient Copper Mining in Michigan. *The Michigan Archaeologist* 41(2–3):119–138.
  1999   *Wonderful Power: The Story of Ancient Copper Working in the Lake Superior Basin.* Wayne State University Press, Detroit.
  2004   Evidence for Indigenous Hardrock Mining of Copper in Ancient North America. *Journal of the West* 43(1):8–13.
  2008   Mining: Copper Mining in the Great Lakes (USA). *Encyclopaedia of the History of Science, Technology, and Medicine in Non-Western Cultures* 2:1680–1685.
  2009   Isle Royale, Lake Superior: The Oldest Prehistoric Copper Quarries. In *Archaeology in America: An Encyclopedia,* edited by Francis P. McManamon, pp. 78–82. Greenwood Press, Westport, Connecticut.

Martin, Susan R. and Thomas C. Pleger
  1999   The Complex Formerly Known as a Culture: The Taxonomic Puzzle of "Old Copper." In *Taming the Taxonomy: Toward a New Understanding of Great Lakes Archaeology,* edited by Ronald F. Williamson and Christopher M. Watts, pp. 61–70. eastendbooks, Toronto, Ontario, Canada.

Martoglio, Pamela A., K. A. Jakes and J. E. Katon
    1992   The Use of Infrared Microspectroscopy in the Analysis of Etowah Textiles: Evidence of Dye Use and Pseudomorph Formation. In *Proceedings of the 50ᵗʰ Annual Meeting of the Electron Microscopy Society of America*, pp. 1534–1535.

Mason, Carol I. and Ronald J. Mason
    1961   The Age of the Old Copper Culture. *The Wisconsin Archeologist* 42(4):143–155.
    1967   A Catalogue of Old Copper Artifacts in the Neville Public Museum. *The Wisconsin Archeologist* 48(2):81–128.

Mason, Milton
    1858a   Superintendent's Report. *Lake Superior Miner* [Ontonagon, Michigan], January 2, p. 133.
    1858b   Superintendent's Report. *The Mining and Statistic Magazine* 10(1):85–86.

Mason, Otis T
    1868   Kent Scientific Institute. *The American Naturalist* 2(10):502–503.
    1880   Summary of Correspondence of the Smithsonian Institution Previous to January 1, 1880, in Answer to Circular No. 316. *Annual Report of the Board of Regents of the Smithsonian Institution, 1879*, pp. 428–448.
    1881   Abstracts of the Smithsonian Correspondence Relative to Aboriginal Remains in the United States. *Annual Report of the Board of Regents of the Smithsonian Institution, 1880*, pp. 441–448.

Mason, Richard P.
    1989   The Blair Gravel Pit Site (47-WN-32) in Winnebago County, Wisconsin. *Fox Valley Archeology* 14:18–23.

Mason, Richard P. and Carol L. Mason
    1994a   Copper Plummets (?) from East-Central Wisconsin. *Fox Valley Archeology* 23:55–56.
    1994b   An Unusual Copper Artifact from Winnebago County, Wisconsin. *Fox Valley Archeology* 23:53–54.

Mason, Ronald J.
    1966   *Two Stratified Sites on the Door Peninsula, Wisconsin.* Anthropological Papers No. 26. Museum of Anthropology, University of Michigan, Ann Arbor, Michigan.
    1967   The North Bay Component at the Porte des Morts Site, Door County, Wisconsin. *The Wisconsin Archeologist* 48(4):267–345.
    1981   *Great Lakes Archaeology.* Academic Press, New York.

Mass Mining Company
    1856   *Prospectus for the Organization of a Company for Mining Purposes, in the County of Ontonagon, [Michigan] on Lake Superior.* Free Press Printing Establishment, Detroit.
    1876   *Report of the Mass Mining Company, June 1, 1876.* W. G. Johnston, Pittsburgh, Pennsylvania.

Matheny, Clarence
    1905   Mounds of Athens County. In *The Centennial Atlas of Athens County, Ohio*, pp. 61, 77, 83, 85, 93, 98, 101, 106. The Centennial Atlas Association, Athens, Ohio.

Bibliography

Matson, John S. B.
    1872   Ancient Burial Mound and Its Contents, Hardin Co., Ohio. In *Ancient Rock Inscriptions in Ohio, Ancient Burial Mound, Hardin County, O., and a Notice of Some Rare Polished Stone Ornaments*, pp. 9–16 (With Comments by Charles Whittlesey). Historical and Archaeological Tract No. 11. Western Reserve and Northern Ohio Historical Society, Cleveland.

Matthews, D.
    1908   The Sawtell Avenue Mound. *American Journal of Archaeology* 12:393–394.

Matthias, Robert
    2015   Untitled. *Central States Archaeological Journal* 62(3):143.

Maxwell, Moreau S.
    1952   The Archeology of the Lower Ohio Valley. In *Archeology of Eastern United States*, edited by James B. Griffin, pp. 176–189. University of Chicago Press, Chicago.

Mayer-Oakes, William J.
    1951   Starved Rock Archaic, a Prepottery Horizon from Northern Illinois. *American Antiquity* 16(4):313–324.
    1955   Prehistory of the Upper Ohio Valley: An Introductory Archeological Study. *Annals of Carnegie Museum*. Vol. 34. Pittsburgh, Pennsylvania.

McAdams, William H.
    1881   Ancient Mounds of Illinois. In *Proceedings of the American Association for the Advancement of Science, Twenty-ninth Meeting Held at Boston, Mass.. August, 1880*, pp. 710–718.
    1882   Antiquities. In *History of Madison County, Illinois*, pp. 58–64. W. R. Brink & Co., Edwardsville, Illinois.
    1884   Mounds of the Mississippi Bottom, Illinois. In *Annual Report of the Board of Regents of the Smithsonian Institution, 1882*, pp. 684–686
    1895   Archaeology. In *Report of the Illinois Board of World's Fair Commissioners at the World's Columbian Exposition [Chicago], May 1—October 30, 1893*. H. W. Rokker, Printer and Binder, Springfield, Illinois.

McAllister, J. Gilbert
    1932   The Archaeology of Porter County. *Indiana History Bulletin* 10(1):1–96.

McBeth, Donald
    1944   Untitled. *Ohio Indian Relic Collectors Society Bulletin* 8:3–4.
    1951   Excavation of the Nolan Ross Mound, Number One. *Ohio Archaeologist* 1(3):26–29.
    1954   Ohio Hopewell Culture Material. *Ohio Archaeologist* 4(2):24–25.
    1960   Bourneville Mound, Ross County, Ohio. *Ohio Archaeologist* 10(1):12–14.

McCharles, A.
    1887   The Mound-Builders of Manitoba. *American Journal of Archaeology* 3(1–2):70–74.

McClurkan, Burney B., William T. Field, and J. Ned Woodall
    1966   *Excavations in Toledo Bend Reservoir, 1964–65*. Papers of the Texas Archeological Salvage Project No. 8. Austin, Texas.

McClurkan, Burney B., Edward B. Jelks. And Harald P. Jensen
    1980   Jonas Short and Coral Snake Mounds: A Comparison. *Louisiana Archaeology* 6:173–206.

McConaughy, Mark A.
    1990   Early Woodland Mortuary Practices in Western Pennsylvania. *West Virginia Archeologist* 42(2):1–10.
    2011   Burial Ceremonialism at Sugar Run Mound (36WA359), a Hopewellian Squawkie Hill Phase Site, Warren County, Pennsylvania. *Mid-Continental Journal of Archaeology* 36(1):73–104.

McCracken, S. B.
    1876a   State Representation at Philadelphia. In *Michigan and the Centennial; Being a Memorial Record Appropriate to the Centennial Year*, pp. 461–666. Printed for the Publisher at the Office of the Detroit Free Press, Detroit.
    1876b   *The State of Michigan: Embracing Sketches of Its History, Position, Resources and Industries.* W. S. George & Co., State Printers and Binders, Lansing, Michigan.

McDonald, S. Edgar
    1950   The Crable Site, Fulton County, Illinois. *Journal of the Illinois State Archaeological Society* 7(4):16–18.

McElrath, Dale L., Andrew C. Fortier, Brad Koldehoff and Thomas E. Emerson
    2009   The American Bottom: An Archaic Cultural Crossroads. In *Archaic Societies: Diversity and Complexity across the Midcontinent*, edited by Thomas E. Emerson, Dale L. McElrath, and Andrew C. Fortier, pp. 317–375. State University Press of New York, Albany, NY.

McElrath, Thomas
    1858   Lake Superior Copper Region. *The Mining and Statistic Magazine* 11(2):149–150.

McGee, W J and Cyrus Thomas
    1905   Prehistoric North America. In *The History of North America*. Vol. 19. George Barrie & Sons, Philadelphia.

McGregor, John C.
    1952   The Havana Site. In *Hopewellian Communities in Illinois*, edited by Thorne Deuel, pp. 43–91. Scientific Papers 5(2). Illinois State Museum, Springfield, Illinois.
    1958   *The Pool and Irving Villages: A Study of Hopewell Occupation in the Illinois River Valley.* University of Illinois Press, Urbana, Illinois.

McGuire, Joseph D.
    1899   Pipes and Smoking Customs of the American Aborigines, Based on Material in the U. S. National Museum. In *Annual Report of the Board of Regents of the Smithsonian Institution, 1897*, pp. 351–465.
    1903a   Discussion of "Sheet-Copper from the Mounds is Not Necessarily of European Origin." *American Anthropologist* 5(1):33–36.
    1903b   McGuire's Concluding Remarks Concerning "Sheet-Copper from the Mounds is Not Necessarily of European Origin." *American Anthropologist* 5(1):42–48.

Bibliography

McKendry, John
    1993   A Double-bitted Copper Adze. *Ohio Archaeologist* 43(2):12.

McKenzie, Douglas H.
    1965   The Burial Complex of the Moundville Phase, Alabama. *The Florida Anthropologist* 18(3):161–174.
    1966   A Summary of the Moundville Phase, Part I: Description of the Phase. *Journal of Alabama Archaeology* 12(1):1–58.

McKenzie, H.
    1864a   Ancient Mining. *Portage Lake Mining Gazette* [Houghton, Michigan], September 17, p. 2.
    1864b   Indian Relic. *Portage Lake Mining Gazette* [Houghton, Michigan], October 1, p. 3.
    1864c   South Pewabic Mining Co. *Portage Lake Mining Gazette* [Houghton, Michigan], November 6, p. 2.
    1865a   South Pewabic Transverse Vein. *Portage Lake Mining Gazette* [Houghton, Michigan] February 11, p. 2.
    1865b   The Calumet Discovery. *Portage Lake Mining Gazette* [Houghton, Michigan], November 25, p. 2.

McKern, W. C.
    1927   Archeological Field Work in Green Lake and Marquette Counties. *Yearbook of the Public Museum of the City of Milwaukee, 1925*, 5:39–53.
    1928   The Neale and McClaughry Mound Groups. *Bulletin of the Public Museum of the City of Milwaukee* 3(3):213–416.
    1929a   Ohio Type of Mounds in Wisconsin. *Year Book of the Public Museum of the City of Milwaukee, 1928*, 8(1):7–21.
    1929b   A Hopewell Type of Culture in Wisconsin. *American Anthropologist* 31(2):307–312.
    1931a   A Wisconsin Variant of the Hopewell Culture. *Bulletin of the Public Museum of the City of Milwaukee* 10(2):185–328.
    1931b   An Unusual Type of Copper Knife. *The Wisconsin Archeologist* 10(3):111–113.
    1932   New Excavations in Wisconsin Hopewell Mounds. *Year Book of the Public Museum of the City of Milwaukee, 1930*, 10:9–27.
    1942   The First Settlers of Wisconsin. *Wisconsin Magazine of History* 26(2):153–169.
    1945   Preliminary Report on the Upper Mississippi Phase in Wisconsin. *Bulletin of the Public Museum of the City of Milwaukee* 16(3):109–285.

McKinley, Daniel
    1977   Archaeozoology of the Carolina Parakeet. *Central States Archaeological Journal* 24(1):4–26.

McKinney, Joe J.
    1954   Hopewell Sites in the Big Bend Area of Central Missouri. *The Missouri Archaeologist* 16(1):1–54.

McKnight, Matthew D.
    2007   *The Copper Cache in Early and Middle Woodland North America*. PhD dissertation, The Graduate School, Pennsylvania State University, State College, Pennsylvania.

McKnight, Norman
    1972   Adena Artifacts. *Ohio Archaeologist* 22(4):7.

McKusick, Marshall B.
    1964a   Exploring Turkey River Mounds. *The Palimpsest* 45(12):473–485.
    1964b   *Men of Ancient Iowa*. Iowa State University Press, Ames, Iowa.

McLeod, Mike
    1980   *The Archaeology of Dog Lake Thunder Bay: 9000 Years of Prehistory*. Vol. II, Lakehead University. Thunder Bay, Ontario, Canada.

McLeod, K. David
    1987   *Land Below the Forks: Archaeology, Prehistory and History of the Selkirk and District Planning Area*. Manitoba Culture, Heritage and Recreation, Historic Resources Branch, Province of Manitoba, Winnipeg, Manitoba, Canada.

McManus, James W.
    1952   A Washington County Adena Mound. *Ohio Archaeologist* 2(3):26–27.

McMichael, Edward V.
    1956   An Analysis of McKees Rocks Mound, Allegheny County, Pennsylvania. *Pennsylvania Archaeologist* 26(3–4):128–151.
    1964   Veracruz, the Crystal River Complex, and the Hopewellian Climax. In *Hopewellian Studies*, edited by Joseph R. Caldwell and Robert L. Hall, pp. 123–132. Scientific Papers 12(5). Illinois State Museum, Springfield, Illinois.

McMichael, Edward V. and Oscar L. Mairs
    1969   Excavation of the Murad Mound, Kanawha County, West Virginia and an Analysis of Kanawha Valley Mounds. *Report of Archaeological Investigations* No. 1. West Virginia Geological and Economic Survey, Morgantown, West Virginia.

McNeal, Kenneth
    1956   Adena Mound Finds. *Ohio Archaeologist* 6(2):72–73.

McNutt, Charles H.
    1996a   The Central Mississippi Valley: A Summary. In *Prehistory of the Central Mississippi Valley*, edited by Charles H. McNutt, pp. 187–257. University of Alabama Press, Tuscaloosa, Alabama.
    1996b   The Upper Yazoo Basin in Northwest Mississippi. In *Prehistory of the Central Mississippi Valley*, edited by Charles H. McNutt, pp. 155–185. University of Alabama Press, Tuscaloosa, Alabama.

McPherron, Alan
    1967   *The Juntunen Site and the Late Woodland Prehistory of the Upper Great Lakes Area*. Anthropological Papers No. 30. Museum of Anthropology, University of Michigan, Ann Arbor, Michigan.

McPherson, Harry R.
    1952   The Pence Mound. *Ohio Archaeologist* 2(4):10–16.

Melbye, J. Jerome
    1963   *The Kane Burial Mounds*. Archaeological Salvage Report No. 15. Southern Illinois University Museum, Carbondale, Illinois.

Bibliography

Merriam, C. Hart
    1923   Erroneous Identifications of "Copper Effigies" from the Mound City Group. *American Anthropologist* 25(3):424–425.

Merriam, Larry G. and Christopher J. Merriam
    2004   *The Spiro Mound: A Photo Essay*. Merriam Station Books, Oklahoma City, Oklahoma.
    2005   History of the Spiro Mound – The Ohio Connection. *Ohio Archaeologist* 55(3):16–23.

Mesnard Mining Company
    1860   *First Report of the Directors of the Mesnard Mining Company to the Stockholders. March 27, 1860*. Geo C. Rand & Avery, City Printers, Boston.

Metz, Charles L.
    1911   *A Brief Description of the Turner Group of Prehistoric Earthworks in Anderson Township, Hamilton County, Ohio, Prepared by Dr. Charles L. Metz, upon the Occasion of His Presenting a Model of this Group to the Cincinnati Museum.* Cincinnati Museum, Ohio.

Meyer, David
    1979   A Projectile Point of Native Copper Found in Northern Saskatchewan. *Saskatchewan Archaeology Newsletter* 54(3–4):8–13.
    1983   The Prehistory of Northern Saskatchewan. In *Tracking Ancient Hunters: Prehistoric Archaeology in Saskatchewan*, edited by Henry T. Epp and Ian Dyck, pp. 141–170. Saskatchewan Archaeological Society, Regina, Saskatchewan, Canada.

Meyer, David and Scott Hamilton
    1994   Neighbors to the North: Peoples of the Boreal Forest. In *Plains Indians, A.D. 500–1500: The Archaeological Past of Historic Groups*, edited by Karl GH. Schlesier, pp. 96–127. University of Oklahoma Press, Norman, Oklahoma.

Michigan State Board of Centennial Managers
    1876   Catalogue of Specimens, Lake Superior Copper and Iron, Upper Peninsula, Michigan. *Catalogue of Products of Michigan in the Centennial Exhibition of All Nations at Fair Mount Park, Philadelphia*. W. S.George & Co., State Printers and Binders, Lansing, Michigan.

Milanich, Jerald T.
    1994   *Archaeology of Precolumbian Florida*. University Press of Florida, Gainesville, Florida.
    1999   *Famous Florida Sites: Crystal River and Mount Royal*. University Press of Florida, Gainesville, Florida.
    2004   Prehistory of Florida after 500 B.C. In *Southeast*, edited by Raymond D. Fogelson, pp. 191–203. Vol. 14, Smithsonian Institution, Washington, DC.

Milanich, Jerald T. and Charles H. Fairbanks
    1980   *Florida Archaeology*. Academic Press, New York.

Miles, Suzanne W.
    1951   A Revaluation of the Old Copper Industry. *American Antiquity* 16(3):240–247.

Millar, J. F. V.
  1978   *The Gray Site: An Early Plains Burial Ground.* Parks Canada Manuscript Report
            No. 304.

Miller, Towne L.
  1922   Explorations of Mounds at Kingston. *The Wisconsin Archeologist* 1(1):22–24.

Mills, Truman B.
  1919   The Ulrich Group of Mounds. *Ohio Archaeological and Historical Publications*
            28:162–175.

Mills, William C.
  1899   Report of the Curator. *Fourteenth Annual Report of the Ohio State Archaeological
            and Historical Society for the Year February 24, 1898 to February 1, 1899*, pp.
            288–290.
  1900   Report of Fieldwork. *Ohio Archaeological and Historical Quarterly* 8:309–344.
  1902a   Excavation of the Adena Mound. *Ohio Archaeological and Historical Quarterly*
            10:451–479.
  1902b   Excavation of the Adena Mound. *Records of the Past* 1(4):130–149.
  1907a   The Explorations of the Edwin Harness Mound. *Ohio Archaeological and
            Historical Quarterly* 16:113–193.
  1907b   *Ohio Archaeological Exhibit at the Jamestown Exposition.* Ohio State
            Archaeological and Historical Society, Columbus, Ohio.
  1909a   Explorations of the Seip Mound. *Ohio Archaeological and Historical Quarterly*
            18:268–321.
  1909b   The Seip Mound. In *Putnam Anniversary Volume: Anthropological Essays
            Presented to Frederic Ward Putnam in Honor of His Seventieth Birthday, April
            16, 1909, by His Friends and Associates*, pp. 102–125. G. E. Stechert & Co.,
            Publishers, New York.
  1912   Archaeological Remains of Jackson County. *Ohio Archaeological and Historical
            Quarterly* 21:175–214.
  1916   Exploration of the Tremper Mound. In *Certain Mounds and Village Sites in Ohio*,
            Vol. 2, Part 3, pp. 102–240.
  1917a   The Feurt Mounds and Village Site. *Ohio Archaeological and Historical
            Quarterly* 26:304–449.
  1917b   The Feurt Mounds and Village Sites. In *Certain Mounds and Village Sites in
            Ohio*, Vol. 3, Part 1, pp. 2–149.
  1921   Flint Ridge. *Ohio Archaeological and Historical Quarterly* 30:90–161.
  1922a   Exploration of the Mound City Group. *Ohio Archaeological and Historical
            Quarterly* 31:422–584.
  1922b   Exploration of the Mound City Group, Ross County, Ohio. *American
            Anthropologist* 24(4):397–431.
  1928   Exploration of the Mound City Group. In *Annaes do XX Congresso Internacional
            de Americanistas Realisado no Rio de Janeiro, de 20 a 30 de Agosto de 1922*, Vol
            II, pp. 134–159.

Milner, George R.
  1984   Social and Temporal Implications of Variation among American Bottom
            Mississippian Cemeteries. *American Antiquity* 49(3):468–488.
  1985   Cultures in Transition: The Late Emergent Mississippian and Mississippian
            Periods in the American Bottom, Illinois. In *The Emergent Mississippian:
            Proceedings of the Sixth Mid-South Archaeological Conference, June 6–9, 1985*,

edited by Richard A. Marshall, pp. 194–211. Occasional papers 87-01, Cobb Institute of Archaeology, Mississippi State University, Starkville, Mississippi.
1990   The Late Prehistoric Cahokia Cultural System of the Missuissippi River Valley: Foundations, Florescence, and Fragmentation. *Journal of World Prehistory* 4(1):1–43.
1998   *The Cahokia Chiefdom: The Archaeology of a Mississippian Society.* Smithsonian Institution Press, Washington, DC.

Miner, Kenneth W.
1941   Big Sandy Lake: An Important Indian Site in Minnesota. *Minnesota Archaeologist* 7(4):42–74.

Minesota Mining Company
1858a   *Articles of Association and By-Laws of the Minesota Mining Company, with the General Incorporation Laws of the State of Michigan under Which the Said Company is Organized and Maps and Plans of the Minesota Mine.* Francis Hart & Co. Printers, 63 Cortland Street, New York.
1858b   Report of the Minesota Mining Company for the Year Ending March 1, 1858. *Lake Superior Miner* [Ontonagon, Michigan], May 1, p. 265–266.

Minong Mining Company
1875   *Prospectus of the Minong Mining Company, 1875.* Daily Post Book and Job Printing, Detroit.

"Miskwabikokewin"
1863   Ontonagon Correspondence. *Portage Lake Mining Gazette* [Houghton, Michigan], February 21, p. 2.

Mitchell, Brainerd
1880   Mound in Pike County, Illinois. In *Annual Report of the Board of Regents of the Smithsonian Institution, 1879*, pp. 367–368.

Mitchell, Edward C.
1908   Archaeological Collections Recently Donated to This Society. *Collections of the Minnesota Historical Society* 12:304–318.

Mitchem, Jeffrey M.
1989   *Redefining Safety Harbor: Late Prehistoric/Protohistoric Archaeology in West Peninsular Florida.* PhD dissertation, University of Florida, Gainesville, Florida.
1996   The Old Okahumpa Site (8LA57): Late Prehistoric Iconography and Mississippian Influence in Peninsular Florida. *The Florida Anthropologist* 49(4):225–237.
2008   Mississippian Copper Artifacts from Arkansas. Paper presented at the 65th Annual Meeting of the Southeastern Archaeological Conference, Charlotte, North Carolina.

Mitchem, Jeffrey M. and Dale L. Hutchinson
1987   *Interim Report on Archaeological Research at the Tatham Mound, Citrus County, Florida: Season III.* Miscellaneous Project Report Series No. 30. Florida State Museum, Department of Anthropology, Gainesville, Florida.

Mitchem, Jeffrey M. and Jonathan M. Leader
1988   Early Sixteenth Century Beads from the Tatham Mound, Citrus County, Florida: Data and Interpretations. *The Florida Anthropologist* 41(1):42–60.

Moffat, C. R. and J. M. Speth
    1999   Rainbow Dam: Two Stratified Late Archaic and Woodland Habitations in the Wisconsin River Headwaters. *The Wisconsin Archeologist* 80(2):111–160.

Moffett, S. E.
    1903   Romances of the World's Great Mines, III-Calumet and Hecla. *The Cosmopolitan* 34(6):679–684.

Moll, Harold and Norman G. Moll
    1959   Manito Stones near Midland. *Michigan Archaeologist* 5(4):53–60.

Monahan, Valery
    1990   *Copper Technology in the Maritimes: An Examination of Indigenous Copper-working in the Maritime Provices during the Prehistoric and Protohistoric Periods.* Honours in Anthropology, Saint Mary's University, Halifax, Nova Scotia, Canada.

Montgomery, Henry
    1906   Remains of Prehistoric Man in the Dakotas. *American Anthropologist* 8(4):640–651.
    1908   Prehistoric Man in Manitoba and Saskatchewan. *American Anthropologist* 10(1):33–40.
    1910   "Calf Mountain" Mound in Manitoba. *American Anthropologist* 12(1):49–57.
    1913   Recent Archaeological Investigations in Ontario. *Transactions of the Canadian Institute* 9:1–10.

Moody, J. D.
    1883   Explorations in Mounds in Whitesides and La Salle Counties, Illinois. *Annual Report of the Board of Regents of the Smithsonian Institution, 1881*, pp. 544–548.

Moore, Charles
    1915   *History of Michigan.* Vol. I. The Lewis Publishing Company, Chicago.

Moore, Clarence B.
    1894a   Certain Sand Mounds of the St. John's River, Florida. Part I. *Journal of the Academy of Natural Sciences of Philadelphia* 10:5–128.
    1894b   Certain Sand Mounds of the St. John's River, Florida. Part II. *Journal of the Academy of Natural Sciences of Philadelphia* 10:129–246.
    1895a   Archaeology of the St. John's, Florida. *The Archaeologist* 3(2):35–38.
    1895b   Archaeology of the St. John's, Florida. *The Archaeologist* 3(5):149–155.
    1896a   Certain River Mounds of Duval County, Florida. *Journal of the Academy of Natural Sciences of Philadelphia* 10:448–502.
    1896b   Certain Sand Mounds on the Ocklawaha River, Florida. *Journal of the Academy of Natural Sciences of Philadelphia* 10:518–543.
    1896c   Two Sand Mounds on Murphy Island, Florida. *Journal of the Academy of Natural Sciences of Philadelphia* 10:503–517.
    1897a   Certain Aboriginal Mounds of the Georgia Coast. *Journal of the Academy of Natural Sciences of Philadelphia* 11:1–144.
    1897b   A Copper Gorget. *Journal of the Academy of Natural Sciences of Philadelphia* 11:185.
    1899   Certain Aboriginal Remains of the Alabama River. *Journal of the Academy of Natural Sciences of Philadelphia* 11:288–347.
    1902   Certain Aboriginal Remains of the Northwest Florida Coast. Part II. *Journal of the Academy of Natural Sciences of Philadelphia* 12:127–358.
    1903a   Sheet-Copper from the Mounds is Not Necessarily of European Origin. *American Anthropologist* 5(1):27–33.

Bibliography

1903b   Moore's Reply to Charles H. McGuire Concerning "Sheet-Copper from the Mounds is Not Necessarily of European Origin." *American Anthropologist* 5(1):36–42.

1903c   Moore's Reply to Mr. McGuire's Closing Remarks. ("The following was written after the appearance of the symposium as to aboriginal copper, which appeared in the "*American Anthropologist*," January-March, 1903). I–XVI.

1903d   Certain Aboriginal Mounds of the Florida Central West-Coast. *Journal of the Academy of Natural Sciences of Philadelphia* 12:361–438.

1905a   Archaeological Research in the Southern United States. In *International Congress of Americanists: Thirteenth Session Held in New York in 1902*, pp. 27–40.

1905b   Certain Aboriginal Remains of the Lower Tombigbee River. *Journal of the Academy of Natural Sciences of Philadelphia* 13:245–278.

1905c   Certain Aboriginal Remains of the Black Warrior River. *Journal of the Academy of Natural Sciences of Philadelphia* 13:123–244.

1907   Moundville Revisited. *Journal of the Academy of Natural Sciences of Philadelphia* 13:334–405.

1909   Antiquities of the Ouachita Valley. *Journal of the Academy of Natural Sciences of Philadelphia* 14:1–170.

1910   Antiquities of the St. Francis, White, and Black Rivers, Arkansas. *Journal of the Academy of Natural Sciences of Philadelphia* 14:253–362.

1911   Some Aboriginal Sites on Mississippi River. *Journal of the Academy of Natural Sciences of Philadelphia* 14:365–480.

1912   Some Aboriginal Sites on Red River. *Journal of the Academy of Natural Sciences of Philadelphia* 14:481–644.

1915   Aboriginal Sites on Tennessee River. *Journal of the Academy of Natural Sciences of Philadelphia* 16:169–428.

Moore, G. R.
    1922   Cache of Copper Chisels. *The Wisconsin Archeologist* 1(1):21–22.

Moore, Joseph
    1895   Concerning a Burial Mound Recently Opened in Randolph County. *Proceedings of the Indiana Academy of Science, 1894*, pp. 46–47.

Moore, Winston D.
    1973   In Search of Baraga County's Ancient Past. In *Baraga County Historical Book, 1972–73*, pp. 7–19.

Moorehead, Warren K.
    1887   Graves at Fort Ancient. *The American Antiquarian and Oriental Journal* 9(5):295–300.
    1889a   A Detailed Account of Mound Opening. *Ohio Archaeological and Historical Quarterly* 2(4):503–508.
    1889b   Mound Explorations by W. K. Moorehead. *The American Naturalist* 23(273):834–839.
    1890a   Exploration of the Porter Mound, Frankfort, Ross County, Ohio. *Journal of the Cincinnati Society of Natural History* 12:27–30.
    1890b   *Fort Ancient: The Great Prehistoric Earthwork of Warren County, Ohio*. Robert Clarke & Co., Cincinnati, Ohio.
    1890c   Relics of the Mound-Builders. *The Illustrated American* 2(21):93–94.
    1890d   Ohio Meteorites. *Science* 15(386):388.
    1892a   New Relics of the Mound Builders. *The Illustrated American* 9(102):509–511.
    1892b   *Primitive Man in Ohio*. G. P. Putnam's Sons, New York.
    1892c   Recent Archaeological Discoveries in Ohio. *Scientific American Supplement* 34:13886–13890.

1892d   Recent Discoveries among the Mound Builders. *The Californian* 1(5):470–482.

1892e   Singular Copper Implements and Ornaments from the Hopewell Group, Ross County, Ohio. *Proceedings of the American Association for the Advancement of Science for the Forty-first Meeting Held at Rochester, N. Y. August, 1892*, p. 291.

1892f   A Synopsis of Archaeological Work in Ross County, Ohio. *The National Magazine* 15(4):384–389.

1894   Anthropology at the World's Columbian Exposition. *The Archaeologist* 2(1):15–24.

1895a   An Altar Skeleton. *The Archaeologist* 3:289–291.

1895b   The Ambos Mound. In *Third Annual Report of the Ohio State Academy of Science*, pp. 7–8.

1895c   Field Work in Ross County, Ohio. *The Archaeologist* 3:287–288.

1896   The Hopewell Find. *The American Antiquarian and Oriental Journal* 18(1):58–62.

1897a   The Hopewell Group. *The Antiquarian* 1(5):113–120.

1897b   The Hopewell Group. *The Antiquarian* 1(7):178–184.

1897c   The Hopewell Group. *The Antiquarian* 1(8):208–214.

1897d   The Hopewell Group. *The Antiquarian* 1(9):236–244.

1897e   The Hopewell Group. *The Antiquarian* 1(10):254–264.

1897f   The Hopewell Group. *The Antiquarian* 1(11):291–295.

1897g   The Hopewell Group. *The Antiquarian* 1(12):312–316.

1897h   Report of Field Work Carried on in the Muskingum, Scioto and Ohio Valleys during the Season of 1896, by Warren King Moorehead, in Charge of Explorations. *Ohio Archaeological and Historical Quarterly* 5:165–274.

1898   The Hopewell Group. *The American Archaeologist* 2(1):6–11.

1899   Report of Field Work in Various Portions of Ohio. *Ohio Archaeological and Historical Quarterly* 7:110–203.

1900   *Prehistoric Implements.* The Robert Clarke Co., Publishers, Cincinnati, Ohio.

1903   Are the Hopewell Copper Objects Prehistoric? *American Anthropologist* 5(1):50–54.

1909   A Study of Primitive Culture in Ohio. In *Putnam Anniversary Volume: Anthropological Essays Presented to Frederic Ward Putnam in Honor of His Seventieth Birthday, April 16, 1909, by His Friends and Associates*, pp. 137–150. G. E. Stechert & Co., Publishers, New York.

1910   *The Stone Age in North America.* Houghton Mifflin Company, Boston and New York.

1922   *The Hopewell Mound Group of Ohio.* Field Museum of Natural History, Anthropological Series 6(5), Chicago.

1924   Mr. W. E. Myers Archeological Collection. *Science* 60 (1546):159–160.

1925a   Exploration of the Etowah Mound. *The Phillips Bulletin* 20:24–30.

1925b   Rare Find by Moorehead in Etowah Mound. *The Museum News* 3(10):1.

1932   *Etowah Papers I: Exploration of the Etowah Site in Georgia.* Department of Archaeology, Phillips Academy, Andover, Massachusetts. Published by the Yale University Press.

1939   Southern Mound Cultures in the Light of Recent Explorations. In *So Live the Works of Men. Seventieth Anniversary Volume Honoring Edgar Lee Hewett*, edited by Donald D. Brand and Fred E. Harvey, pp. 273–275. University of New Mexico Press, Albuquerque, NM.

2000   *The Cahokia Mounds.* University of Alabama Press, Tuscaloosa, Alabama.

Morgan, Richard G.

1941a   Dunlap Mound Contains Impressive Structure. *Museum Echoes* 14:80.

1941b   A Hopewell Sculptured Head. *Ohio Archaeological and Historical Quarterly* 50:384–387.

1946   *Mound City: A Prehistoric Indian Shrine.* United States Department of the Interior, Washington, DC.

1952   Outline of Cultures in the Ohio Region. In *Archeology of Eastern United States*, edited by James B. Griffin, pp. 83–98. University of Chicago Press, Chicago.

1965   *Fort Ancient*. Ohio Historical Society, Columbus, Ohio.

Morgan, Richard G. and H. Holmes Ellis

1943   The Fairport Harbor Village Site. *Ohio Archaeological and Historical Quarterly* 52:3–64.

Morris, Gordon K.

2014   An Old Copper Assemblage from Vilas County, Wisconsin. *Central States Archaeological Journal* 61(4):362–364.

Morris, Robert W.

2004   The Manring Mound Site, Clark County, Ohio. *Ohio Archaeologist* 54(1):4–9.

2006   The Manring Mound Site, Clark County, Ohio – Part II. *Ohio Archaeologist* 56(3):39–42.

2008   Hopewell Copper Celts from the Manring Mound. *Ohio Archaeologist* 58(4):18.

Morrison, R. S.

1885   Westcott et al. v. Minnesota Mining Co. et al. (23 Michigan, 145. Supreme Court, 1871). In *The Mining Reports. A Series Containing the Cases on the Law of Mines Found in the American and English Reports, Arranged Alphabetically by Subjects with Notes and References* 6:336–352. Callaghan & Co., Chicago.

Morse, Dan F.

1959   Preliminary Report on a Red Ocher Mound at the Morse Site, Fulton County, Illinois. *Papers of the Michigan Academy of Science, Arts, and Letters* 44:193–207.

1963   The Steuben Village and Mounds: A Multicomponent Late Hopewell Site in Illinois. Anthropological Papers No. 21. Museum of Anthropology, University of Michigan, Ann Arbor, Michigan.

1986   McCarty (3-Po-467): A Tchula Period Site near Marked Tree, Arkansas. In *The Tchula Period in the Mid-South and Lower Mississippi Valley: Proceedings of the 1982 Mid-South Archaeological Conference*, edited by David H. Dye, pp. 70–92. Archaeological Report No. 17. Mississippi Department of Archives and History, Jackson, Mississippi.

Morse, Dan F. and Phyllis A. Morse

1961   The Southern Cult: The Emmons Site, Fulton County, Illinois. *Central States Archaeological Journal* 8(4):124–140.

1962   The Detweiller Golf Course Site, Peoria. Illinois. *Central States Archaeological Journal* 9(4):152–158.

1964   1962 Excavations at the Morse Site: A Red Ocher Cemetery in the Illinois Valley. *The Wisconsin Archeologist* 45(2):79–98.

1965   The Hannah Site, Peoria County, Illinois. In *Middle Woodland Sites in Illinois*, pp. 129–146. Bulletin No. 5. Illinois Archaeological Survey, Urbana, Illinois.

1983   *Archaeology of the Central Mississippi Valley*. Academic Press, Inc., New York.

1989   The Rise of the Southeastern Ceremonial Complex in the Central Mississippi Valley. In *The Southeastern Ceremonial Complex: Artifacts and Analysis*, edited by Patricia Galloway, pp. 41–44. University of Nebraska Press, Lincoln, Nebraska.

1996a   Northeast Arkansas. In *Prehistory of the Central Mississippi Valley*, edited by Charles H. McNutt, pp. 137–154. University of Alabama Press, Tuscaloosa, Alabama.

1996b   Changes in Interpretation in the Archaeology of the Central Mississippi Valley since 1983. *North American Archaeologist* 17(1):1–35.

Morse, Phyllis A.
1981   *Parkin: The 1978–1979 Archeological Investigations of a Cross County, Arkansas Site.* Arkansas Archeological Survey Research Series No. 13.

Mortine, Wayne A. and Doug Randles
1978   The Martin Mound: An Extension of the Hopewell Interaction Sphere into the Walhonding Valley of Eastern Ohio. *Occasional Papers in Muskingum Valley Archaeology* No. 10. The Muskingum Valley Archaeological Survey, Zanesville, Ohio.
1981   Excavation of Two Adena Mounds in Coshocton County, Ohio. *Occasional Papers in Muskingum Valley Archaeology* No. 12. The Muskingum Valley Archaeological Survey, Zanesville, Ohio.

Morton, James and Jeff Carskadden
1983   The Rutledge Mound, Licking County, Ohio. *Ohio Archaeologist* 33(1):4–9.
1987   Test Excavations at an Early Hopewellian Site near Dresden, Ohio. *Ohio Archaeologist* 37(1):8–11.

Moses, Thomas F.
1878   Report of the Antiquities of Mad River Valley. *Proceedings of Central Ohio Scientific Association* 1(1):23–49.

Moxley. Ronald W.
1988   The Orchard Site: A Proto-Historic Fort Ancient Village Site in Mason County, West Virginia. *The West Virginia Archeologist* 40(1):32–41.

Mulholland, Susan C., Stephen L. Mulholland, and Robert C. Donahue
2008   *The Archaeology of the Fish Lake Dam Site: Pre-Contact Occupations on the East Bank.* Occasional Publication No. 12, Minnesota Archaeological Society, St. Paul, Minnesota.

Mulholland, Susan C. and Mary H. Pulford
2007   Trace-Element Analysis of Native Copper: The View from Northern Minnesota, USA. *Geoarchaeology: An International Journal* 22(1):67–84.

Muller, Jon
1986   *Archaeology of the Lower Ohio River Valley.* Academic Press, Inc., Orlando, Florida.

Munson, Cheryl Ann and Ronald G. V. Hancock
2001   Analysis of Cupric Metal Artifacts from the Murphy and Hovey Lake Sites, Posey County, Indiana. In *Archaeological Survey, Test Excavation, and Public Education in Southwestern Indiana,* Appendix V, by Cheryl Ann Munson. Prepared for the Indiana Department of Natural Resources, Division of Historic Preservation and Archaeology and the National Park Service, U. S. Department of the Interior.

Murphy, James L.
1975   *An Archeological History of the Hocking Valley.* Ohio University Press, Athens, Ohio.
1978   William C. Mills' Notes on the Edwin Harness Mound Excavations of 1903. *Ohio Archaeologist* 28(3):8–11.

Bibliography

1984	Hamilton County's Spearhead Mound (33-Ha-24). *Ohio Archaeologist* 34(1):25–27.
1986	Willard H. Davis: Nineteenth Century Archaeologist of the Lower Muskingum Valley. *Muskingum Annals* 2:59–76.
2001	Two Unrecorded Publications by Warren King Moorehead. *Ohio Archaeologist* 51(3):14–15.
2008	Walter C. Metz and "Prehistoric Remains in Licking County, Ohio." *Ohio Archaeologist* 58(3):4–11.

Myers, Richard
1964	Four Burials at the Citico Site. *Tennessee Archaeologist* 20(1):11–13.

Myron, Robert E.
1954	Hopewellian Two-Dimensional Sculpture: Part 2. *Central States Archaeological Journal* 1(1):4–13.

Nansel, Blane
1977	Clayton County Archaeology. In *Archaeology, Geology and Natural Areas: A Preliminary Survey, Vol II, Iowa's Great River Road Cultural and Natural Resources*, pp. 347–383. Office of the State Archaeologist, Iowa City, Iowa.

Nassaney, Michael S.
2000	The Late Woodland Southeast. In *Late Woodland Societies: Tradition and Transformation across the Midcontinent*, edited by Thomas E. Emerson, Dale L. McElrath, and Andrew C. Fortier, pp. 713–730. University of Nebraska Press, Lincoln, Nebraska.

National Mining Company
1857	Section of the National Mine on the Conglomerate Vein. In *Report of the National Mining Company: August 1ˢᵗ, 1857*. R. M. Riddle, Pittsburgh, Pennsylvania.
1859	Section of the National Mine. In *Report of the Directors of the National Mining Company*. W. S. Haven, Pittsburgh, Pennsylvania.
1860	Section of the National Mine. In *Report of the Directors of the National Mining Company*. W. S. Haven, Pittsburgh, Pennsylvania.
1862	Section of the National Mine. In *Report of the Directors of the National Mining Company*. W. S. Haven, Pittsburgh, Pennsylvania.

Nebraska Mining Company
1859	*Report of the Directors of the Nebraska Mining Company to the Stockholders with Accompanying Papers, Etc.* Free Press Book and Job Office, Detroit.

Neiburger, E. J.
1989	Isotope Radiography of the Largest Prehistoric Copper Celt. *North American Archaeologist* 10(1):55–61.
2008	Old Copper Mandrils. *Central States Archaeological Journal* 55(2):80–83.
2013a	A Comparison of the American Old Copper Culture with the Indus Valley Civilization. *Central States Archaeological Journal* 60(1):18–21.
2013b	The Old Copper Culture: Mini-Ingots and Trade. *Central States Archaeological Journal* 60(2):76–79.
2014	American Old Copper: Was There World Trade in the Archaic? *Central States Archaeological Journal* 61(3):130–133.
2016a	Four Caches of Old Copper Artifacts from Vilas County, Wisconsin. *Central States Archaeological Journal* 63(1):22–23.

2016b   Little Prehistoric Copper Knives. *Central States Archaeological Journal* 63(2):88–90.

Neiburger, E. J. and Steve Livernash
2009a   Old Copper Nose Ornaments. *Central States Archaeological Journal* 56(1):16–19
2009b   A Riveted Old Copper Point. *Central States Archaeological Journal* 56(4):203–205.

Neiburger, E. J. and Donald Spohn
2007   Prehistoric Money. *Central States Archaeological Journal* 54(4):188–192.

Neph, John R.
1980   *The Adventure Story: Copper Mining in the "Old Adventure."* (Reprint with Sidelites.) Adventure Copper Mine, Greenland, Michigan.

Neuman, Robert W.
1984   *An Introduction to Louisiana Archaeology.* Louisiana State University Press, Baton Rouge, Louisiana.

Neumann, Georg K. and Melvin L. Fowler
1952   Hopewellian Sites in the Lower Wabash Valley. In *Hopewellian Communities in Illinois,* edited by Thorne Deuel, pp. 175–248. Scientific Papers 5(5). Illinois State Museum, Springfield, Illinois.

New-York Tribune [New York]
1848   The Copper Country. *Detroit Free Press* [Detroit, Michigan], December 14, p. 2.
1852   Copper Mines at Lake Superior. *American Railroad Journal* 25(836):260.

Niehoff, Arthur
1959   Beads from a Red Ochre Burial in Ozaukee County. *The Wisconsin Archeologist* 40(1):25–28.

Noakes, Tom
2007   Artifacts from the Noakes Collection. *Ohio Archaeologist* 57(4):34.

Noble, William C.
1971   The Sopher Celt: An Indicator of Early Protohistoric Trade in Huronia. *Ontario Archaeology* 16:42–47.
2004   The Protohistoric Period Revisited. In *A Passion for the Past: Papers in Honour of James F. Pendergast,* edited by James V. Wright and Jean-Luc Pilon, pp.179–191. Mercury Series Archaeology Paper 164. Canadian Museum of Civilization, Gatineau, Quebec, Canada.

Nolan, David J. and Richard L. Fishel
2009   Archaic Cultural Variation and Lifeways in West-Central Illinois. In *Archaic Societies: Diversity and Complexity across the Midcontinent,* edited by Thomas E. Emerson, Dale L. McElrath, and Andrew C. Fortier, pp. 401–490. State University Press of New York, Albany, New York.

Norona, Delf
1957a   Crucifixes Found in the Upper Ohio Valley. *West Virginia Archeologist* 8:29–32.
1957b   Moundsville's Mammoth Mound. *The West Virginia Archaeologist* 9:1–55.
1962   Comments on Townsend's Account of the 1838 Excavation of the Grave Creek Mound. *West Virginia Archeologist* 14:7–18.

Bibliography

2008 *Moundsville's Mammoth Mound*. West Virginia Archeological Society, Charleston, West Virginia.

Norris, Rae
 1985 Excavation of the Toepfner Mound. *Archaeology of Eastern North America* 13:128–137.

Norris, Rae and Shaune M. Skinner
 1985 Excavation of the Connett Mound 3. *Ohio Archaeologist* 35(1):21–26.

North Cliff Mining Company
 1858 *Articles of Association and By-Laws of the North Cliff Mining Company, together with the Report of Explorations, by Samuel W. Hill, Esq., 1858*. W. S. Haven, Pittsburgh, Pennsylvania.

North West Mining Company
 1854 *A Statement of the Condition and Prospects of the North West Mining Company of Michigan*. Edward Grattan, Philadelphia.

Norwich Mining Company
 1865 *Report of the Norwich Mining Company, for the Year 1865*. "The Stockholder" Print, New York.

Oakley, Carey B. and Eugene M. Futato
 1975 *Archaeological Investigations in the Little Bear Creek Reservoir*. Research Series No. 1, Office of Archaeological Research, University of Alabama, Tuscaloosa, Alabama.

Oberholtzer, Frances W.
 1924 Cabell County Once Rich in Relics of Early Race of Moundbuilders. *Huntington Sunday Advertiser*, April 13.

O'Brien, Michael J.
 1996 *Paradigms of the Past: The Story of Missouri Archaeology*. University of Missouri Press, Columbia, Missouri.

O'Brien, Michael J. and Thomas B. Holland
 1994 Campbell. In *Cat Monsters and Head Pots: The Archaeology of Missouri's Pemiscot Bayou*, by Michael J. O'Brien, pp. 195–260. (With Contributions by Gregory L. Fox, Thomas D. Holland, Richard A. Marshall, and J. Raymond Williams). University of Missouri Press, Columbia, Missouri.

O'Brien, Michael J. and Richard A. Marshall
 1994 Late Mississippian Period Antecedents: Murphy and Kersey. In *Cat Monsters and Head Pots: The Archaeology of Missouri's Pemiscot Bayou*, by Michael J. O'Brien, pp. 141–194. (With Contributions by Gregory L. Fox, Thomas D. Holland, Richard A. Marshall, and J. Raymond Williams). University of Missouri Press, Columbia, Missouri.

O'Brien, Michael J. and J. Raymond Williams
 1994 Other Late Mississippian Period Sites. In *Cat Monsters and Head Pots: The Archaeology of Missouri's Pemiscot Bayou*, by Michael J. O'Brien, pp. 261–305. (With Contributions by Gregory L. Fox, Thomas D. Holland, Richard A. Marshall, and J. Raymond Williams). University of Missouri Press, Columbia, Missouri.

O'Brien, Michael J. and W. Raymond Wood
    1998   *The Prehistory of Missouri*. University of Missouri Press, Columbia, Missouri.

O'Brien, Patricia J.
    1971   Urbanism, Cahokia and Middle Mississippian. *Archaeology* 25(3):188–197.
    1988   Ancient Kansas City Area Borders and Trails. *Missouri Archaeologist* 49:27–39.
    1992   The "World-System" of Cahokia within the Middle Mississippian Tradition. *Review* 15(3):389–417.

O'Connor, Mallory M.
    1995   *Lost Cities of the Ancient Southeast*. University Press of Florida, Gainesville, Florida.

Ochsner, Eugene E.
    1972   Artifacts from the Ochsner Collection. *Ohio Archaeologist* 22(1):20–21.

Oehler, Charles
    1973   *Turpin Indians*. Popular Publication Series, No. 1. Cincinnati Museum of Natural History, Cincinnati, Ohio.

Ogima Mining Company
    1863   *Report of the Ogima Mining Company, for the Year 1862*. L. Darbee & Son, Printers, Brooklyn, New York.

O'Gorman, Jodie A.
    1989   *The OT Site (47-LC-262) 1987 Archaeological Excavation: Preliminary Report*. Archaeological Report 15, Wisconsin Department of Transportation, Madison, Wisconsin.
    1994   The Filler Site (47 Lc-149). In *The Tremaine Site Complex: Oneota Occupation in the La Crosse Locality, Wisconsin*, Vol. 2, State Historical Society of Wisconsin, Madison, Wisconsin.
    1995   The Tremaine Site (47 Lc-95).). In *The Tremaine Site Complex: Oneota Occupation in the La Crosse Locality, Wisconsin*, Vol. 3, State Historical Society of Wisconsin, Madison, Wisconsin.

Ormandy, P. G.
    1968   *An Introduction to Metallurgical Laboratory Techniques*. Pergamon Press, New York.

Orr, Kenneth G.
    1941   The Eufaula Mound: Contributions to the Spiro Focus. *Oklahoma Prehistorian* 4(1):2–15.
    1946   The Archaeological Situation at Spiro, Oklahoma; a Preliminary Report. *American Antiquity* 11(4):228–256.
    1952   Survey of Caddoan Area Archaeology. In *Archeology of Eastern United States*, edited by James B. Griffin, pp. 239–255. University of Chicago Press, Chicago.

Orr, Rowland B.
    1911   Archaeology of the Province of Ontario. *Annual Archaeological Report, 1911, Including 1908-9-10. Being Part of Appendix to the Report of the Minister of Education, Ontario*, pp. 8–11.
    1912   Pre-Columbian Copper. *Annual Archaeological Report, 1912. Being Part of*

Bibliography

Appendix to the Report of the Minister of Education, Ontario, pp. 46–54.

1913   Pre-Columbian Copper in Ontario. In *International Congress of Americanists, Proceedings of the XVIII. Session, London, 1912*, pp. 313–316.

1917   Ontario Indians: Their Fisheries and Fishing Appliances. In *Twenty-ninth Annual Archaeological Report, 1917. Being Part of Appendix to the Report of the Minister of Education, Ontario*, pp. 24–43.

1922   Red Paint Burial in Ontario. In *Thirty-third Annual Archaeological Report, 1921–22. Being Part of Appendix to the Report of the Minister of Education, Ontario*, pp. 38–40.

1928   Hook or Gaff. In *Thirty-sixth Annual Archaeological Report, 1928. Including 1926–1927. Being Part of Appendix to the Report of the Minister of Education, Ontario*, p. 54.

Osborn, C. E.
1917   Archaeological Notes on Ogle County, Illinois. *The Archaeological Bulletin* 8(1):5–6.

Osburn, Mary H.
1946   Prehistoric Musical Instruments in Ohio. *Ohio State Archaeological and Historical Quarterly* 55:12–20.

Ostberg, Neil J.
1956   Additional material from the Reigh Site, Winnebago County. *The Wisconsin Archeologist* 37(1):28–31.

Ottesen, Ann I.
1979   *A Preliminary Study of Acquisition of Exotic Raw Materials by Late Woodland and Mississippian Groups*. PhD dissertation, New York University, New York.

Overstreet, David F.
1976   *The Grand River, Lake Koshkonong, Green Bay and Lake Winnebago Phases: Eight Hundred Years of Oneota Prehistory in Eastern Wisconsin*. PhD dissertation, University of Wisconsin-Milwaukee, Milwaukee, Wisconsin.

Overton, George
1929   An Ancient Village Site in Winnebago County. *The Wisconsin Archeologist* 8(3):94–100.
1931   Old Beach Camp Sites in Winnebago County. *The Wisconsin Archeologist* 10(2):54–60.

Packard, R. L.
1893a   Pre-Columbian Copper Mining in North America. *Annual Report of the Board of Regents of the Smithsonian Institution*, 1892, pp. 175–198.
1893b   Pre-Columbian Copper Mining in North America. *The American Antiquarian* 15(2):67–78.

Page, John E.
1848   Collateral Testimony of the Truth and Divinity of the Book of Mormon.—No. 2. *Gospel Herald* 3(89) (September 7). http://www.mormonbeliefs.com/collateral_testimony.htm <23 December 2008>

Parker, Jack
1975   The First Copper Miners in Michigan. *Compressed Air Magazine* 80(1):6–11.

Pasco, George L. and W. C. McKern
1947   A Unique Copper Specimen. *The Wisconsin Archeologist* 28(4):72–75.

Pasco, George L. and Robert E. Ritzenthaler
    1949   Copper Discs in Wisconsin. *The Wisconsin Archeologist* 30(3):63–64.

Patterson, J. T.
    1937   *Boat-shaped Artifacts of the Gulf Southwest States.* Bureau of Research in the Social Sciences Study No. 24. Anthropological Papers Vol. 1(2).

Patterson, L. W.
    2000   Late Archaic Mortuary Tradition of Southeast Texas. *La Tierra: Journal of the Southern Texas Archaeological Association* 27(2):28–44.
    2004   Prehistoric Long-Distance Trade in Eastern North America. *Ohio Archaeologist* 54(3):32–33.

Pauketat, Timothy R.
    1988   Notes on the William Karoly Collection (Appendix V). In *A Phase II Archaeological Investigation for the Proposed Development of Lake Erie Metropark, Cherry Island, Wayne County, Michigan,* edited by Claire A. McHale. Great Lakes Division, Museum of Anthropology, University of Michigan, Ann Arbor, Michigan.
    1994   *The Ascent of Chiefs: Cahokia and Mississippian Politics in Native North America.* University of Alabama Press, Tuscaloosa, Alabama.
    1997   Specialization, Political Symbols, and the Crafty Elite of Cahokia. *Southeastern Archaeology* 16(1):1–15.
    2004   *Ancient Cahokia and the Mississippians.* Cambridge University Press, Cambridge, UK.

Payant, Felix
    1929   Lessons in Design from the Ancient Mound Builders of Ohio. *Design* 31(2): 22–26, 29.

Payne, Claudine
    1994   *Mississippian Capitals: An Archaeological Investigation of Precolumbian Political Structure.* PhD dissertation, University of Florida, Gainesville, Florida.
    2002   Architectural Reflections of Power and Authority in Mississippian Towns. In *The Dynamics of Power,* edited by Maria O'Donovan, pp.188–213. Occasional paper No. 30. Center for Archaeological Investigations, Southern Illinois University Carbondale, Carbondale, Illinois.

Peebles, Christopher S.
    1974   *Moundville: The Organization of a Prehistoric Community and Culture.* PhD dissertation, University of California-Santa Barbara, Santa Barbara, California.
    1983   Moundville: Late Prehistoric Sociopolitical Organization in the Southeastern United States. In *The Development of Political Organization in Native North America,* edited by Elisabeth Tooker, pp. 183–198. The American Ethnological Society.

Peebles, Christopher S. and Susan M. Kus
    1977   Some Archaeological Correlates of Ranked Societies. *American Antiquity* 42(3):421–448.

Peet, Stephen D.
    1881   Relics of the Mound Builders near Joliet, Ill. *American Antiquarian and Oriental Journal* 3(2):155.
    1884   The Kanawha Mounds. *American Antiquarian and Oriental Journal* 6(2):133.
    1896   The Hopewell Find and Its Mysteries. *American Antiquarian and Oriental Journal* 18(1):62–63.

Bibliography

| 1906a | The Copper Age in America. *American Antiquarian and Oriental Journal* 28(3):149–164. |

1906b | Copper Relics among the Mounds. *American Antiquarian and Oriental Journal* 28(4): 213–228.

1909 | A Find of Copper Relics. *American Antiquarian and Oriental Journal* 31(4):189–201.

Peithman, Irvin
1947 | Recent Hopewell Finds in Southern Illinois. *Journal of the Illinois State Archaeological Society* 5(2): 51–54.

Pelletier, Gérald
1997 | The First Inhabitants of the Outaouais: 6,000 Years of History. In *History of the Outaouais*, directed by Chad Gaffield, pp. 43–66. Les Presses de l'Universite Laval, Quebec, Canada.

Penman, John T.
1977 | The Old Copper Culture: An Analysis of Old Copper Artifacts. *The Wisconsin Archeologist* 58(1):3–23.
1978 | Wisconsin's Old Copper Culture. *Wisconsin Natural Resources* 2(1):26–28.

Penney, David W.
1985a | The Late Archaic Period. In *Ancient Art of the American Woodland Indians*, edited by Andrea P. A. Belloli, pp. 15–41. Harry N. Abrams, Inc., Publishers, New York, in Association with the Detroit Institute of Arts.
1985b | Continuities of Imagery and Symbolism in the Art of the Woodlands. In *Ancient Art of the American Woodland Indians*, edited by Andrea P. A. Belloli, pp. 146–198. Harry N. Abrams, Inc., Publishers, New York, in Association with the Detroit Institute of Arts.
1987 | The Origins of an Indigenous Ontario Arts Tradition: Ontario Art from the Late Archaic through the Woodland Periods. *Journal of Canadian Studies* 21(4):37–55.

Pepin, John
2001 | Ancient Arrowhead Found on Lake Bottom. *The Mining Journal* [Marquette, Michigan], June 3, pp. 1A, 12A.

Pepper, George H.
1917 | The Nacoochee Mound, White County, Georgia. In *Proceedings of the Nineteenth International Congress of Americanists, held at Washington, December 27–31, 1915*, pp. 103.

Perino, Gregory H.
1940 | Bluff Finds in the Cahokia Region. *Quarterly Bulletin of the Illinois State Archaeological Society* 2:9–11.
1958 | The Kraske Village Site and Mound Group. *The Wisconsin Archeologist* 39(3):181–188.
1959 | Recent Information from Cahokia and Its Satellites. *Central States Archaeological Journal* 6(4):130–138.
1961 | Tentative Classification of Plummets in the Lower Illinois River Valley. *Central States Archaeological Journal* 8(2):43–56.
1966a | *The Banks Village Site*. Memoirs of the Missouri Archaeological Society No. 4. Columbia, Missouri.
1966b | Short History of Some Sea Shell Ornaments. *Central States Archaeological Journal* 13(1):4–8.

1967   What is It? Pick, Adze, Gouge, or Chisel? *Central States Archaeological Journal* 14(4):138–150.

1968a   The Pete Klunk Mound Group, Calhoun County, Illinois: The Archaic and Hopewell Occupations. In *Hopewell and Woodland Site Archaeology in Illinois*, edited by James A. Brown, pp. 9–124. Illinois Archaeological Survey, Inc. Bulletin No. 6. University of Illinois, Urbana, Illinois.

1968b   The Shiny "Red Stone" That Wouldn't Break. *Central States Archaeological Journal* 15(3):98–108.

1969   Salvage Excavations to Obtain a Hopewell Population. *Southeastern Archaeological Conference Bulletin* 11:49–53.

1970   Some Outstanding Artifacts Acquired by the Gilcrease Institute in 1969. *Central States Archaeological Journal* 17(1):30–35.

1971a   The Mississippian Component at the Schild Site (No. 4), Greene County, Illinois. In *Mississippian Site Archaeology in Illinois I: Site Reports from the St. Louis and Chicago Areas*, edited by James A. Brown, pp. 1–148. Illinois Archaeological Survey, Inc. Bulletin No. 8. University of Illinois, Urbana, Illinois.

1971b   The Yokem Site, Pike County, Illinois. In *Mississippian Site Archaeology in Illinois I: Site Reports from the St. Louis and Chicago Areas*, edited by James A. Brown, pp. 149–191. Illinois Archaeological Survey, Inc. Bulletin No. 8. University of Illinois, Urbana, Illinois.

1971c   Au Sagaunashke Village: The Upper Mississippian Occupation of the Knoll Spring Site, Cook County, Illinois. In *Mississippian Site Archaeology in Illinois I: Site Reports from the St. Louis and Chicago Areas*, edited by James A. Brown, pp. 192–250. Illinois Archaeological Survey, Inc. Bulletin No. 8. University of Illinois, Urbana, Illinois.

1971d   The Lundy Site, Craig County, Northeast Oklahoma. *Bulletin of the Oklahoma Anthropological Society* 20:83–89.

1972   Woodland Tombs and Mortuary Practices in Illinois. *Eastern States Archeological Federation Bulletin* 31:13–14.

1973   A Hopewell Bone Scepter. *Central States Archaeological Journal* 20(1):24–26.

1976   Artifacts Found in Red River near Idabel, Oklahoma. *Artifacts* 6(1):19–21.

1985   Points and Barbs. *Central States Archaeological Journal* 32(1):47–48.

1986a   Technology at Cahokia Mounds Should be Studied Further. *Central States Archaeological Journal* 33(1):3–7.

1986b   A Rare Short-nosed God Mask. *Central States Archaeological Journal* 33(4):248–249.

1992   Points and Barbs. *Central States Archaeological Journal* 39(3):154.

2002   Points and Barbs. *Central States Archaeological Journal* 49(3):155.

2006   *Illinois Hopewell and Late Woodland Mounds: The Excavations of Gregory Perino 1950–1975*. Assembled and edited by Kenneth D. Farnsworth and Michael D. Wiant. Studies in Archaeology No. 4. Illinois Transportation Archaeological Research Program, University of Illinois, Urbana, Illinois.

Perkins, Edwin L.
1897   Antiquities of Fox River, Wisconsin. *The Antiquarian* 1(11):305.

Perkins, F. S.
1887   Mr. Perkins' Collections of Copper Relics. *The American Antiquarian* 9(3):172.

Perkins, Raymond W.
1965   The Frederick Site. In *Middle Woodland Sites in Illinois*, pp. 68–94. Illinois Archaeological Survey. Bulletin No. 5. University of Illinois, Urbana.

Bibliography

Perrin, T. M.
    1873   Mounds near Anna, Union County, Illinois. *Annual Report of the Board of Regents of the Smithsonian Institution, 1872*, pp. 418–420.

Perttula, Timothy K.
    1996   Caddoan Area Archaeology since 1990. *Journal of Archaeological Research* 4(4):295–348.

Perttula, Timothy K. and James E. Bruseth
    1990   Trade and Exchange in Eastern Texas, 1100 B.C. – A.D. 800. *Louisiana Archaeology* 17:93–121.

Peter, Robert
    1873   Ancient Mound near Lexington, Kentucky. *Annual Report of the Board of Regents of the Smithsonian Institution, 1871*, pp. 420–423.

Peters, Bernard C.
    1989   Wa-bish-kee-pe-nas and the Chippewa Reverence for Copper. *Michigan Historical Review* 15(2):47–60.

Peterson, David H.
    2003a   Red Metal Poundings and the "Neubauer Process": Copper Culture Metallurgical Technology. *Central States Archaeological Journal* 50(2):102–106.
    2003b   Native Copper Characteristics Demonstrated in the "Neubauer Process." *Central States Archaeological Journal* 50(3):168–172.
    2004   The Neubauer Process: 1999–2003 Observations. *Central States Archaeological Journal* 51(1):56–59.

Pettipas, Leo F.
    1983   *Introducing Manitoba Prehistory.* Papers in Manitoba Archaeology, Popular Series No. 4. Manitoba Department of Cultural Affairs and Historical Resources, Winnipeg, Canada.
    1996   *Aboriginal Migrations: A History of Movements in Southern Manitoba.* Manitoba Museum of Man and Culture, Winnipeg, Canada.

Philadelphia Ledger [Philadelphia, Pennsylvania]
    1883   Hammers of the Prehistoric Races. *New York Times*, February 25.

Phillips, George Brinton
    1923   A Prehistoric Copper Mine. *The Wisconsin Archeologist* 2(3):151–154.
    1925a   The Primitive Copper Industry of America. *American Anthropologist* 27(2):284–289.
    1925b   The Primitive Copper Industry of America. *The Journal of the Institute of Metals* 34(2):261–270.
    1926   The Primitive Copper Industry of America. Part II. *The Journal of the Institute of Metals* 36(2):99–106.

Phillips, Philip and James A. Brown
    1978   *Pre-Columbian Shell Engravings from the Craig Mound at Spiro, Oklahoma, Part 1.* Peabody Museum Press, Cambridge, Massachusetts.

Phoenix Copper Company
    1853   *Fourth Annual Report of the Directors of the Phoenix Copper Company, with the*

Report of Simon Mandlebaum, Superintendent of the Mine, and the Charter and By-Laws of the Company. Damrell & Moore, Boston.

Pickett, Albert C.
1851   *History of Alabama, and Incidentally of Georgia and Mississippi, from the Earliest Period.* Walker and James, Charleston, South Carolina.

Pickett, Thomas E.
1875   *The Testimony of the Mounds: Considered with Especial Reference to the Pre-Historic Archaeology of Kentucky and the Adjoining States.* Thomas A. Davis, Excelsior Printing Works, Maysville, Kentucky.

Pinkham, Henry M.
1888   *The Lake Superior Copper Properties: A Guide to Investors and Speculators in Lake Superior Copper Shares.* Printed for the author, Boston.

Piscataqua Mining Company
1850   *Charter and By-Laws of the Piscataqua Mining Co. of Michigan.* Grattan & Mclean, Philadelphia.

Pitezel, John H.
1859   *Lights and Shadows of Missionary Life: Containing Travels, Sketches, Incidents, and Missionary Efforts, During Nine Years Spent in the Region of Lake Superior.* R. P. Thompson, Printer, Western Book Concern, Cincinnati, Ohio.

Pittsburgh and Boston Mining Company
1855   *Report of the President and Directors of the Pittsburgh and Boston Mining Company of Pittsburgh, with Accompanying Statements from the Treasurer. September, 1855.* W. S. Haven, Pittsburgh, Pennsylvania.

Platcek, Eldon P.
1965   A Preliminary Survey of a Fowl Lakes Site. *The Minnesota Archaeologist* 27(2):51–62.

Platt, Edward J.
1984   *Preliminary Identification of Prehistoric Artifacts from Northeastern Woodland Cultures.* New York Institute of Anthropology, Jackson Heights, New York.

Pleger, Thomas C.
1992   A Functional and Temporal Analysis of Copper Implements from the Chautauqua Grounds Site (20-MT-71), a Multi-Component Site near the Mouth of the Menominee River. *The Wisconsin Archeologist* 73(3–4):160–176.
1998   *Social Complexity, Trade, and Subsistence during the Archaic/ Woodland Transition in the Western Great Lakes (4000–400 B.C.): A Diachronic Study of Copper Using Cultures at the Oconto and Riverside Cemeteries.* PhD dissertation, Department of Anthropology, University of Wisconsin-Madison, Madison, Wisconsin.
2000   Old Copper and Red Ocher Social Complexity. *Midcontinental Journal of Archaeology* 25(2):169–190.
2001   New Dates for the Oconto Old Copper Culture Cemetery. *The Wisconsin Archeologist* 82(1–2):87–100.
2002   A Brief Introduction to the Old Copper Complex of the Western Great Lakes. *Proceedings of the Twenty-seventh Annual Meeting of the Forest History Association of Wisconsin,* pp. 10–18.

Bibliography

Pleger, Thomas C. and James B. Stoltman
    2009   The Archaic Tradition in Wisconsin. In *Archaic Societies: Diversity and Complexity across the Midcontinent*, edited by Thomas E. Emerson, Dale L. McElrath, and Andrew C. Fortier, pp. 697–723. State University of New York Press, Albany, New York.

Pluckhahn, Thomas J.
    2003   *Kolomoki: Settlement, Ceremony, and Status in the Deep South, A.D. 350 to 750.* University of Alabama Press, Tuscaloosa, Alabama.

Poehls, R. L.
    1944   Kingston Lake Site Burials. *Journal of the Illinois State Archaeological Society* 1:36–38.

Polhemus, Richard R.
    1987   *The Toqua Site – 40MR6: A Late Mississippian, Dallas Phase Town.* Report of Investigations No. 41. Department of Anthropology, University of Tennessee, Knoxville, Tennessee.

Pollack, David
    1998   *Intraregional and Intersocietal Relationships of the Late Mississippian Caborn-Welborn Phase of the Lower Ohio River Valley.* PhD dissertation, University of Kentucky, Lexington, Kentucky.
    2004   *Caborn-Welborn: Constructing a New Society after the Angel Chiefdom Collapse.* University of Alabama Press, Tuscaloosa, Alabama.

Pollack, David, Eric J. Schlarb, William E. Scharp, and Teresa W. Tune
    2005   Walker-Noe: An Early Middle Woodland Adena Mound in Central Kentucky. In *Woodland Systematics in the Middle Ohio Valley*, edited by Darlene Applegate and Robert C. Mainfort, pp. 64–75. University of Alabama Press, Tuscaloosa, Alabama.

Pompeani, David P., Mark B. Abbott, Byron A. Steinman, and Daniel J. Bain
    2013   Lake Sediments Record Prehistoric Lead Pollution Related to Early Copper Production in North America. *Environmental Science and Technology* 47(11):5545–5552.

Pompeani, David P., Mark B. Abbott, Daniel J. Bain, Seth DePasqual and Matthew S. Finkenbinder
    2015   Copper Mining on Isle Royale 6500–5400 Years Ago Identified from McCargoe Cove, Lake Superior. *The Holocene* 25(2):253–262.

Pontiac Jacksonian [Pontiac, Michigan]
    1855   Evergreen Bluff Mine. *The Mining and Statistic Magazine* 5(3):252–255.

Pope, Graham
    1901a   Annual Address. *The Michigan Miner* 3(5):10–13.
    1901b   Some Early Mining Days at Portage Lake. *Proceedings of the Lake Superior Mining Institute* 7: 17–31.

Popham, Robert E. and J. N. Emerson
    1954   Manifestations of the Old Copper Industry in Ontario. *Pennsylvania Archaeologist* 24(1):3–15.

Porter, Tom and Donald McBeth
    1960   An Additional Note on the Bourneville Mound, Ross County, Ohio. *Ohio Archaeologist* 10(4):112–115.

Potter, Martha A.
    1971   Adena Culture Content and Settlement. In *Adena: The Seeking of an Identity*, edited by B. K. Swartz, Jr., pp. 4–11. Ball State University, Muncie, Indiana.

du Pouget, Jean-François-Albert (Marquis de Nadaillac)
    1890   *Pre-Historic America.* (Translated by N. D'Anvers; edited by W. H. Dall.) G. P. Putnam's Sons, New York.

Power, Susan C.
    2004   *Early Art of the Southeastern Indians: Feathered Serpents & Winged Beings.* University of Georgia Press, Athens, Georgia.

Prahl, Earl J.
    1970   *The Middle Woodland Period of the Lower Muskegon Valley and the Northern Hopewellian Frontier.* PhD dissertation, University of Michigan, Ann Arbor, Michigan.
    1991   The Mounds of the Muskegon. *The Michigan Archaeologist* 37(2):59–125.

Pratt, W. H.
    1876   Report of Explorations of the Ancient Mounds at Toolesboro, Louisa County, Iowa. *Proceedings of the Davenport Academy of Natural Sciences* 1:106–111.
    1877a   Curator's Report. *Proceedings of the Davenport Academy of Natural Sciences* 2:48–55.
    1877b   Shell Money and Other Primitive Currencies. *Proceedings of the Davenport Academy of Natural Sciences* 2:38–46.
    1883a   Exploration of a Mound on the Allen Farm. *Proceedings of the Davenport Academy of Natural Sciences* 3:90–91.
    1883b   The President's Annual Address. *Proceedings of the Davenport Academy of Natural Sciences* 3:151–160.
    1886   Contributions to the Museum. *Proceedings of the Davenport Academy of Natural Sciences* 4:13–25.

Priest, Josiah
    1835   *American Antiquities and Discoveries in the West...* Hoffman and White, Albany, New York.

Prufer, Olaf H.
    1963   Der Hopewell-Komplex der Ostlichen Vereinigten Staaten. *Paideuma* 9(2):122–147.
    1964   The Hopewell Complex of Ohio. In *Hopewellian Studies*, edited by Joseph R. Caldwell and Robert L. Hall, pp. 35–83. Scientific Papers 12(2). Illinois State Museum, Springfield, Illinois.

Pulford, Mary H.
    1999   *Pre-Contact Mortuary Copper Usage: An Interdisciplinary Investigation with Emphasis in Eastern North America.* PhD dissertation, Department of Interdisciplinary Studies, University of Minnesota, Minneapolis.
    2000   Initial Review of Northeastern Minnesota Reservoir Copper. *The Minnesota Archaeologist* 59:64–80.

Bibliography

2003   An Interdisciplinary Look at the Traditions of the Great Lakes Proto-Historic and Early Contact periods. *The Minnesota Archaeologist* 62:61–78.

2009   Copper Types of Northeastern Minnesota. *The Minnesota Archaeologist* 68:93–106.

Purtill, Matthew T.

2009   The Ohio Archaic: A Review. In *Archaic Societies: Diversity and Complexity across the Midcontinent*, edited by Thomas E. Emerson, Dale L. McElrath, and Andrew C. Fortier, pp. 565–606. State University Press of New York, Albany, New York.

Pustmueller, A. E.

1950   The Copper Serpent. In *Cahokia Brought to Life*, edited by R. E. Grimm, pp. 13–14. Greater St. Louis Archaeological Society, St. Louis.

Putnam, F. W.

1876   Additions to the Museum. In *Ninth Annual Report of the Trustees of the Peabody Museum of American Archaeology and Ethnology*, pp. 16–20.

1878   Archaeological Explorations in Tennessee. In *Eleventh Annual Report of the Trustees of the Peabody Museum of American Archaeology and Ethnology*, pp. 305–360.

1881   Were Ancient Copper Implements Hammered or Molded into Shape? *The Kansas City Review of Science and Industry* 5(8):490.

1882   Notes on the Copper Objects from North and South America, Contained in the Collections of the Peabody Museum. *Fifteenth Annual Report of the Trustees of the Peabody Museum of American Archaeology and Ethnology* 3(2):83–148.

1883a   Altar-mounds in Anderson Township, Ohio. *Science* 1(12):348–349.

1883b   Exploration of Altar Mounds in Ohio. *The Kansas City Review of Science and Industry* 7(1):32–35.

1883c   Iron from the Ohio Mounds: A Review of the Statements and Misconceptions of Two Writers of over Sixty Years Ago. *Proceedings of the American Antiquarian Society (New Series)* 2:349–363.

1886a   Explorations in the Ohio Valley. *The American Architect and Building News* 20(572):279–281.

1886b   List of Additions to the Museum and Library for the Year 1885. In *Eighteenth and Nineteenth Annual Reports of the Trustees of the Peabody Museum of American Archaeology and Ethnology*, pp. 503–512.

1886c   The Marriott Mound, No. 1, and Its Contents. In *Eighteenth and Nineteenth Annual Reports of the Trustees of the Peabody Museum of American Archaeology and Ethnology*, pp. 449–466.

1886d   Report of the Curator. In *Eighteenth and Nineteenth Annual Reports of the Trustees of the Peabody Museum of American Archaeology and Ethnology*, pp. 477–501.

1888   An Account of Recent Explorations of Mounds in Ohio. *Proceedings of the Boston Society of Natural History* 23:215–218.

1903   Remarks by F. W. Putnam (Presented in His Absence by Roland B. Dixon) Concerning "Sheet-Copper from the Mounds is Not Necessarily of European Origin." *American Anthropologist* 5(1):49.

Putnam, F. W. and C. C. Willoughby

1896a   Symbolic Carvings of the Mound Builders, as Found at the Hopewell Group, Ross County, O. *Popular Science News* 30(1):13–15.

1896b   Symbolism in Ancient American Art. *Proceedings of the American Association for the Advancement of Science for the Forty-fourth Meeting Held at Springfield, Mass. 1895*, pp. 302–322.

Quimby, George I.
1941   *The Goodall Focus: An Analysis of Ten Hopewellian Components in Michigan and Indiana*. Prehistory Research Series 2(2):65–161. Indiana Historical Society, Indianapolis.
1944   Some New Data on the Goodall Focus. *Papers of the Michigan Academy of Science, Arts, and Letters* 29:419–443.
1952   The Archeology of the Upper Great Lakes Area. In *Archeology of Eastern United States*, edited by James B. Griffin, pp. 99–107. University of Chicago Press, Chicago.
1954   The Old Copper Assemblage and Extinct Animals. *American Antiquity* 20(2):169–170.
1957a   An Old Copper Site? at Port Washington. *The Wisconsin Archeologist* 38(1):1–5.
1957b   An Old Copper Site at Menominee, Michigan. *The Wisconsin Archeologist* 38(2):37–41.
1959a   The Old Copper Indians and Their World. *Chicago Natural History Museum Bulletin* 30(1):4–5.
1959b   Upper Lakes Farmers and Artists, 100 B.C. *Chicago Natural History Museum Bulletin* 30(3):6–7.
1960a   Burial Yields Clews to Red Ocher Culture. *Chicago Natural History Museum Bulletin* 31(2):5.
1960b   *Indian Life in the Upper Great Lakes: 11,000 B.C. to A.D. 1800*. University of Chicago Press, Chicago.
1961   Old Copper Artifacts from Chicago. In *Chicago Area Archaeology*, edited by Elaine A. Bluhm, pp. 34–36. Bulletin No. 3. Illinois Archaeological Survey, University of Illinois, Urbana, Illinois.
1962   The Old Copper Culture and the Copper Eskimos. In *Prehistoric Cultural Relations between the Arctic and Temperate Zones of North America*, edited by John M. Campbell, pp. 76–79. Technical Paper 11, Arctic Institute of North America, Montreal, Québec, Canada.
1963   Late Period Copper Artifacts in the Upper Great Lakes Region. *The Wisconsin Archeologist* 44(4):193–198.
1966   The Dumaw Creek Site: A Seventeenth Century Prehistoric Indian Village and Cemetery in Oceana County, Michigan. *Fieldiana: Anthropology* 56(1):1–91.

Quimby, George I. and Albert C. Spaulding
1957   The Old Copper Culture and the Keweenaw Waterway. *Fieldiana: Anthropology* 36(8):189–201.

R., C.
1853a   Lake Superior Mine. *Lake Superior Journal* [Sault Ste. Marie], July 9, p. 2.
1853b   Untitled. *Lake Superior Journal* [Sault Ste. Marie], July 23, p. 2.

Radin, Paul
1911   Some Aspects of Winnebago Archeology. *American Anthropologist* 13(3):517–538.
1923   The Winnebago Tribe. *Thirty-seventh Annual Report of the Buereau of American Ethnology, 1915–1916*, pp. 35–60. Government Printing Office, Washington, DC.
1926   *Ojibwa and Ottawa Indians, Notes*. American Philosophical Society, Philadelphia.

Rafferty, Sean M.
2005   The Many Messages of Death: Mortuary Practices in Ohio and the Northeast. In *Woodland Systematics in the Middle Ohio Valley*, edited by Darlene Applegate and Robert C. Mainfort, pp. 150–167. University of Alabama Press, Tuscaloosa, Alabama.

Bibliography

Railey, Jimmy A.
    1990   Woodland Period. In *The Archaeology of Kentucky: Past Accomplishments and Future Directions*, edited by David Pollack, pp. 247–342.
    1996   Woodland Cultivators. In *Kentucky Archaeology*, edited by R. Barry Lewis, pp. 79–125. University Press of Kentucky, Lexington, Kentucky.

Rakestraw, Lawrence
    1965   *Historic Mining on Isle Royale*. Isle Royale Natural History Association, Houghton, Michigan.
    1992   Historic Mining. In *Borealis: An Isle Royale Potpourri*, edited by David Harmon, pp. 61–62. Isle Royale Natural History Association, Houghton, Michigan.

Randall, Emilius O.
    1916   The Mound Builders of Ohio. *Journal of American History* 10(2):288–304.

Rapp, George, Jr., James D. Allert, and Gordon R. Peters
    1990   The Origins of Copper in Three Northern Minnesota Sites: Pauly, River Point, and Big Rice. In *The Woodland Tradition in the Western Great Lakes: Papers Presented to Elden Johnson*, edited by Guy E. Gibbon, pp. 233–238. Publications in Anthropology, University of Minnesota, Minneapolis.

Rapp, George, James Allert, Vanda Vitali, Zhichun Jing, and Eiler Henrickson
    2000   *Determining Geological Sources of Artifact Copper: Source Characterization Using Trace Element Patterns*. University Press of America, Lanham, Maryland.

Rau, Charles
    1884   Prehistoric Fishing in Europe and North America. In *Smithsonian Contributions to Knowledge* Vol. 25, Article 1:1–342.
    1886   Report upon the Work in the Department of Archaeology in the U. S. National Museum for the Year Ending June 30, 1886. In *Report of the United States National Museum under the Direction of the Smithsonian Institution for the Year Ending June 30, 1886*, pp. 101–112.

Rayner, J. A.
    1910   Mound Builders in Miami County, Ohio. *The Archaeological Bulletin* 1(1):3–5.
    1914   Notes from the Miami Valley. *The Archaeological Bulletin* 5(2):35–36.
    1917   Unearths Stone Grave of Prehistoric People. *The Archaeological Bulletin* 8(1):1–3.

Read, M. C.
    1875   Some Curious Works in the Northern Part of Summit County, Ohio. *Minutes of the Ohio State Archaeological Convention, Held in Mansfield, O., Sept. 1ˢᵗ & 2d, 1875*, pp. 38–41. Printed for the Society by Paul & Thrall, Columbus, Ohio.
    1892   *Archaeology of Ohio*. Tract 73. Western Reserve Historical Society, Cleveland.

Read, M. C. and Charles Whittlesey
    1877   Antiquities of Ohio. In *Final Report of the Ohio State Board of Centennial Managers to the General Assembly of the State of Ohio*, pp. 81–141. Nevins & Myers, State Printers, Columbus, Ohio.

Reardon, William
    2014   Oldest Carbon-14 Dated Copper Projectile Points from Wisconsin. *The Wisconsin Archeologist* 95(1):86–87.

Redmond, Brian G.
    2007a   Analysis and Description of an Inundated Ohio Hopewell Mortuary Site in Sandusky County, Ohio: The Pumpkin Site Collection. Archaeological Research Report No. 149. Cleveland Museum of Natural History, Cleveland, Ohio.
    2007b   Hopewell on the Sandusky: Analysis and Description of an Inundated Ohio Hopewell Mortuary-Ceremonial Site in North-Central Ohio. *North American Archaeologist* 28(3):189–232.

Redmond, Brian G. and Robert G. McCullough
    2000   The Late Woodland to Late Prehistoric Occupations of Central Indiana. In *Late Woodland Societies: Tradition and Transformation across the Midcontinent*, edited by Thomas E. Emerson, Dale L. McElrath, and Andrew C. Fortier, pp. 643–683. University of Nebraska Press, Lincoln, Nebraska.

Reeb, Elizabeth and Jeff Carskadden
    1982   Artifacts from a Destroyed Mound, Dillon Reservoir Area. *Ohio Archaeologist* 32(1):18–20.

Reed, Harvey
    1876   The Modern Evidence of Prehistoric Man in the Copper Region of Lake Superior. In *Second Report of the District Historical Society Containing All Papers Read at the Last Meeting.* Tract No. 4, pp. 4–9, Akron, Ohio.

Reid, C. S.
    1975   *The Boys Site and the Early Ontario Iroquois Tradition.* National Museum of Man Mercury Series Paper No. 42. Ottawa, Ontario, Canada.

Reid, H. A.
    1881   Prehistoric Man in Lafayette County, Missouri. *Kansas City Review of Science and Industry* 5(7):405–408.

Reifel, August J.
    1915   *History of Franklin County, Indiana.* B. F. Bowen & Company, Inc., Indianapolis.

Reilly, F. Kent, III and James F. Garber
    2007   Introduction. In *Ancient Objects and Sacred Realms: Interpretations of Mississippian Iconography*, edited by F. Kent Reilly III and James F. Garber, pp. 1–7. University of Texas Press, Austin, Texas.

Republic Mining Company
    1864   *Prospectus of the Republic Mining Company, 1864.* Francis Hart & Company, Printers, New York.

Reynolds, Thomas
    1856   Discovery of Copper and Other Indian Relics, near Brockville. *The Canadian Journal* (New Series) 4:329–336.

Richards, Ed and Orrin C. Shane III
    1974   Tuscarawas County's Kline Mound. *Ohio Archaeologist* 24(3):4–8.

Richards, John D., Thomas J. Zych, and Katie Z. Rudolph
    2012   Archaeological Investigations at Aztalan (47JE1) by the 2011 UIW-Milwaukee Archaeological Field School. *The Wisconsin Archeologist* 93(1):95–101.

Bibliography

Richards, T. T.
    1871   Relics from the Great Mound. *The American Naturalist* 4(1):62–63.

Richmond, J. F.
    1907   The Mounds of Florida and Their Builders. *Ohio Archaeological and Historical Quarterly* 16:445–454.

Richmond, Michael D.
    2000   *A Geochemical Analysis of Select Copper Artifacts from the Midcontinental United States.* Master's thesis, Department of Anthropology, Kent State University, Kent, Ohio.

Richmond, Michael D. and Jonathan P. Kerr
    2005   Middle Woodland Ritualism in the Central Bluegrass: Evidence from the Amburgey Site, Montgomery County, Kentucky. In *Woodland Systematics in the Middle Ohio Valley*, edited by Darlene Applegate and Robert C. Mainfort, pp. 76–93. University of Alabama Press, Tuscaloosa, Alabama.

Richner, Jeffrey J.
    1973   *Depositional History and Tool Industries at the Winter Site: A Lake Forest Middle Woodland Cultural Manifestation.* Master's thesis, Department of Anthropology, Western Michigan University, Kalamazoo, Michigan.

Rickard, Thomas A.
    1934   The Use of Native Copper by the Indigenes of North America. *Journal of the Royal Anthropological Institute of Great Britain and Ireland* 64:265–287.

Riddle, David K.
    1980   Archaeological Survey of the Upper Albany River. In *Studies in West Patricia Archaeology, No. 1: 1978–1979*, edited by C. S. "Paddy" Reid, pp. 152–213. West Patricia Heritage Resource Report 1. Historical Planning and Research Branch, Ontario Ministry of Culture and Recreation, Toronto, Ontario, Canada.
    1981   Archaeological Survey of the Albany River; Year 2: Triangular Lake to Washi Lake. In *Studies in West Patricia Archaeology No. 2:1979–1980*, edited by C. S. "Paddy" Reid and W. A. Ross, pp. 207–279. West Patricia Heritage Resource Report 2. Historical Planning and Research Branch, Ontario Ministry of Culture and Recreation, Toronto, Ontario, Canada.
    1982   An Archaeological Survey of Attawapiskat Lake, Ontario. In *Studies in West Patricia Archaeology No. 3:1980–1981*, edited by W. A. Ross, pp. 1–68. West Patricia Heritage Resource Report 3. Historical Planning and Research Branch, Ontario Ministry of Citizenship and Culture, Toronto, Ontario, Canada.

Riddle, Walter
    2005   My Introduction to Copper. *Central States Archaeological Journal* 52 (1):56.

Riley, Thomas J.
    1979   Middle Woodland Copper from the Utica Mounds, La Salle County, Illinois. *The Wisconsin Archeologist* 60(1):26–46.

Ritchie, James S.
    1858   *Wisconsin and Its Resources; with Lake Superior, Its Commerce and Navigation.* Charles Desilver, No. 714 Chestnut Street, Philadelphia.

Ritchie, William A.
    1955   Recent Discoveries Suggesting an Early Woodland Burial Cult. New York State
                Museum and Science Service, Circular 40, New York.

Ritchie, William A. and Don Dragoo
    1960   *The Eastern Dispersal of Adena*. New York State Museum and Science Service,
                Bulletin Number 379, New York.

Ritzenthaler, Robert E.
    1951   Upper Mississippi Copper Pendants. *The Wisconsin Archeologist* 32(3):52–53.
    1952   Similarities between Copper Pendants from Wisconsin and Georgia. *The
                Wisconsin Archeologist* 33(4):225–228.
    1957a   Six Old Copper Implements from Long Lake, Florence County. *The Wisconsin
                Archeologist* 38(1):35.
    1957b   The Osceola Site: An "Old Copper" Site near Potosi, Wisconsin. *The Wisconsin
                Archeologist* 38(4):186–203.
    1958a   Old Copper Specimens from Upper Michigan. *The Wisconsin Archeologist*
                39(2):151–152.
    1958b   Some Carbon 14 Dates for the Wisconsin Old Copper Culture. *The Wisconsin
                Archeologist* 39(3):173–174.
    1960a   An "Old Copper" Crescent from Alberta, Canada. *The Wisconsin Archeologist* 41(2):34.
    1960b   Radiocarbon Dates for Wisconsin. *The Wisconsin Archeologist* 41(3):65–69.
    1963   A Copper Ingot? *The Wisconsin Archeologist* 44(4):215–216.
    1965   A Red Ocher Site in Fond du Lac County. *The Wisconsin Archeologist* 46(2): 143–147.
    1967   An Unusual Old Copper Point. *The Wisconsin Archeologist* 48(1):1–2.
    1969   An Old Copper Point from Southeastern Iowa. *The Wisconsin Archeologist* 50(1):33.
    1970   Another Radiocarbon Date for the Oconto Site. *The Wisconsin Archeologist* 51(2):77.

Ritzenthaler, Robert, Neil Ostberg, Kirk Whaley, Martin Greenwald, Phil Wiegand, Penny Foust, Ernest
Schug, Warren Wittry, Heinz Meyer, and Edward Lundsted
    1957   Reigh Site Report – Number 3. *The Wisconsin Archeologist* 38(4):278–310.

Ritzenthaler, Robert and Arthur Niehoff
    1958   A Red Ochre Burial in Ozaukee County. *The Wisconsin Archeologist* 39(2):115–120.

Ritzentahler, Robert E. and George I. Quimby
    1962   The Red Ocher Culture of the Upper Great Lakes and Adjacent Areas. *Fieldiana
                Anthropology* 36(11):243–275.

Ritzenthaler Robert E. and Warren L. Wittry
    1957   The Oconto Site – An Old Copper Manifestation. *The Wisconsin Archeologist*
                38(4):222–244.

Robb, Matthew H.
    2010   *The Wulfing Plaques*. Spotlight Series, March 2010. Mildred Lane Kemper Art
                Museum, Washington University, St. Louis.

Roberts, E. J.
    1849a   Lake Superior Copper Mines—The Weather, Gossip, &c. *Detroit Free Press*
                [Detroit, Michigan], April 4, p. 2.
    1849b   Lake Superior Country and Mines. *Daily Free Press* [Detroit, Michigan], April 8,
                p. 2.

Bibliography

Robertson, R. S.
   1875a   Antiquities of La Porte County, Indiana. *Annual Report of the Board of Regents of the Smithsonian Institution, 1874*, pp. 377–380.
   1875b   Antiquities of Allen and De Kalb Counties, Indiana. *Annual Report of the Board of Regents of the Smithsonian Institution, 1874*, pp. 380–384.

Rockland Mining Company
   1856   *Report of the Board of Directors of the Rockland Mining Company Presented to the Annual Meeting of Stockholders, Held on the16th Day of April, 1856.* Francis Hart, Printer and Stationer, New York.
   1858   *Report of the Board of Directors of the Rockland Mining Company Presented to the Annual Meeting of Stockholders, Held on the13th Day of April, 1858.* Francis Hart, Printer and Stationer, New York.
   1859   *Report of the Board of Directors of the Rockland Mining Company Presented to the Annual Meeting of Stockholders, Held on the14th Day of April, 1859.* Francis Hart & Co., Printers, New York.

Rodell, Roland L.
   1991   The Diamond Bluff Site Complex and Cahokia Influence in the Red Wing Locality. In *New Perspectives on Cahokia: Views from the Periphery*, edited by James B. Stoltman, pp. 253–280. Monographs in World Archaeology No. 2. Prehistory Press, Madison, Wisconsin.
   1997   *The Diamond Bluff Site Complex: Time and Tradition in the Northern Mississippi Valley.* PhD dissertation, University of Wisconsin-Milwaukee, Milwaukee, Wisconsin.

Roedl, Leo J. and James H. Howard
   1957   Archaeological Investigations at the Renner Site. *The Missouri Archaeologist* 19(1–2):53–90.

Rolingson, Martha A.
   1961   The Kirtley Site: A Mississippian Village in Mclean County, Kentucky. *Transactions of the Kentucky Academy of Science* 22(3–4):41–59.
   1967   *Temporal Perspective on the Archaic Cultures of the Middle Green River Region, Kentucky.* PhD dissertation, University of Michigan, Ann Arbor, Michigan.
   1990   The Toltec Mounds Site: A Ceremonial Center in the Arkansas River Lowland. In *The Mississippian Emergence*, edited by Bruce D. Smith, pp. 27–49. Smithsonian Institution Press, Washington, DC.
   2004   Prehistory of the Central Mississippi Valley and Ozarks after 500 B.C. In *Southeast*, edited by Raymond D. Fogelson, pp. 534–544. Vol. 14, Smithsonian Institution, Washington, DC.

Rolingson, Martha A. and Douglas W. Schwartz
   1966   *Late Paleo-Indian and Early Archaic Manifestations in Western Kentucky.* Studies in Anthropology No. 3. University of Kentucky Press, Lexington, Kentucky.

Romain, William F.
   2000   *Mysteries of the Hopewell: Astronomers, Geometers, and Magicians of the Eastern Woodlands.* University of Akron Press, Akron, Ohio.

Romano, Anthony D. and Stephen L. Mulholland
   2000   The Robert and Debra Neubauer Site (21PN86), Pine County, Minnesota: An

Archaic Habitation and Copper Working Site on Mission Creek. *The Minnesota Archaeologist* 59:120–141.

Ross, William A.
    1980   The Coopman Site: A Whittrey (sic) Type 1C Copper Point from the City of Thunder Bay. *Wanikan* 80(9):4–6.
    1982   A Group 'VD' Spud or Socketed Adze from the Danielson Collection in the Nipigon Museum. *Wanikan* 82(1):9–12.

Rothschild, Nan A.
    1979   Mortuary Behavior and Social Organization at Indian Knoll and Dickson Mounds. *American Antiquity* 44(4):658–675.

Rountree, Helen C.
    1989   *The Powhatan Indians of Virginia: Their Traditional Culture.* University of Oklahoma Press, Norman, Oklahoma.

Rountree, Helen C. and E. Randolph Turner III
    2002   *Before and After Jamestown: Virginia's Powhatans and Their Predecessors.* University Press of Florida, Gainesville, Florida.

Rowe, Chandler W.
    1956   *The Effigy Mound Culture of Wisconsin.* Milwaukee Public Museum Publications in Anthropology No. 3.

Rowe, Roger
    1984   A Copper Axe from Wayne County, Ohio. *Ohio Archaeologist* 34(1):13.

Royer, Jacob S.
    1959   The Island Park Site. *Ohio Archaeologist* 9(2):44–46.
    1961   Excavation of a Small Mound, Montgomery County, Ohio. *Ohio Archaeologist* 11(4):112–116.

Ruby, Bret J.
    1997   *The Mann Phase: Hopewellian Subsistence and Settlement Adaptations in the Wabash Lowlands of Southwestern Indiana.* PhD dissertation, Department of Anthropology, Indiana University, Bloomington, Indiana.
    2006   The Mann Phase: Hopewellian Community Organization in the Wabash Lowland. In *Recreating Hopewell*, edited by Douglas K. Charles and Jane E. Buikstra, pp. 190–205. University Press of Florida, Gainesville, Florida.

Rudolph, James L. and David J. Halley
    1985   *Archaeological Investigations at the Beaverdam Creek Site (9EB85), Elbert County, Georgia.* Prepared for Archaeological Services Branch-Atlanta, National Park Service, U. S. Department of the Interior, by the Department of Anthropology, University of Georgia, Athens, Georgia.

Rueping, Henry J.
    1944   A Fond du Lac Pit Burial. *The Wisconsin Archeologist* 25(1):13–16.

Ruhl, Donna L.
    1981   *An Investigation into the Relationships between Midwestern Hopewell and*

Bibliography

Southeastern Prehistory. Master's thesis, College of Social Science, Florida Atlantic University, Boca Raton, Florida.

Ruhl, Katharine C.
- 1992   Copper Earspools from Ohio Hopewell Sites. *Midcontinental Journal of Archaeology* 17(1):46–79.
- 1996   *Copper Earspools in the Hopewell Interaction Sphere: The Temporal and Social Implications*. Master's thesis, Kent State University, Kent, Ohio.
- 2005   Hopewellian Copper Earspools from Eastern North America: The Social, Ritual, and Symbolic Significance of Their Contexts and Distribution. In *Gathering Hopewell: Society, Ritual, and Ritual Interaction*, edited by Christopher Carr and D. Troy Case, pp. 696–713. Kluwer Academic/Plenum Publishers, New York.

Ruhl, Katharine C. and Mark F. Seeman
- 1998   The Temporal and Social Implications of Ohio Hopewell Copper Ear Spool Design. *American Antiquity* 63(4):651–662.

Rummel, John C.
- 2006   Exotic Hopewell Artifacts. *Ohio Archaeologist* 56(3):32–33.
- 2009a   The Robinson-Hunt Site: A Middle Woodland Mound along the Headwaters of the Scioto River. *Ohio Archaeologist* 59(1):24–26.
- 2009b   A Possible Interpretation of the Engraved Bone Whistle Found in the Bourneville Mound, Ross County, Ohio. *Ohio Archaeologist* 59(4):32–35.

Rummel, John C. and Chris K. Balazs
- 2006   Ohio's Smallest Hopewell Pipes. *Ohio Archaeologist* 56(1):4.

Rusnak, Michael
- 2009   Harvard's Peabody Museum On-Line Catalog Project Reveals Many Extraordinary Ohio Artifacts. *Ohio Archaeologist* 59(1):13–17.
- 2010   The Face of Hopewell. *Ohio Archaeologist* 60(3):23–28.
- 2011   Creatures of the Beneath World: Hopewell Effigies from Turner Mound, Part 2: The Water Beast and Fish/Snake Effigies. *Ohio Archaeologist* 61(1):4–10.

Russell, Daniel E.
- 2008   Crystals at the Crystal Palace: The Mineralogical Display at the 1853 Crystal Palace Exhibit in New York City. http://www.mindat.org/article.php/196/Minerals+at+the+1853+New+York+City+Crystal+Palace+Exhibition. Accessed 3/12/2012.

Rust, H. N.
- 1877   The Mound Builders in Missouri. *Western Review of Science and Industry* 1(9):531–535.

Ryan, Thomas H., ed.
- 1911   *History of Outagamie County, Wisconsin*. Goodspeed Historical Association, Chicago.

"S. S."
- 1852   Scenery at the Mines—Copper Falls, Phoenix, Cliff and North American Mines. *Lake Superior Journal*, February 2, p. 3.

Sackett, Richard R.
1940   Unusual Copper Artifacts in Collection of the Minnesota Historical Society. *Minnesota Archaeologist* 6(2):54–61.

Sagard, Gabriel
1939   *The Long Journey to the Country of the Hurons*. The Champlain Society, Toronto, Canada.

St. Paul Dispatch [St. Paul, Minnesota]
1889   General. November 26, p. 5.

St. Paul Pioneer Press [St. Paul, Minnesota]
1897   Relic of a Prehistoric Race. December 19, p. 2.

Salisbury, Roland D.
1885   Notes on the Dispersion of Drift Copper. *Transactions of the Wisconsin Academy of Sciences, Arts, and Letters* 6:42–50.

Salkin, Philip H.
1986   The Lake Farms Phase: The Early Woodland Stage in South-Central Wisconsin as Seen from the Lake Farms Archeological District. In *Early Woodland Archeology*, edited by Kenneth B. Farnsworth and Thomas E. Emerson, pp. 92–120. Kampsville Seminars in Archeology, Vol. 2. Center for American Archeology, Center for American Archeology Press, Kampsville, Illinois.

Salo, Lawr V.
1969   *Archaeological Investigations in the Tellico Reservoir, Tennessee, 1967–1968: An Interim Report*. Submitted in Accordance with National Park Service Contracts Nos. 14-10-1-910-20 and 14-10-7:911-12. Department of Anthropology, University of Tennessee, Knoxville, Tennessee.

Salzer, Robert J.
1969   *An Introduction to the Archaeology of Northern Wisconsin*. PhD dissertation, Department of Anthropology in the Graduate School, Southern Illinois University, Carbondale, Illinois.
1974   The Wisconsin North Lakes Project: A Preliminary Report. In *Aspects of Upper Great Lakes Anthropology: Papers in Honor of Lloyd A. Wilford*, edited by Elden Johnson, pp. 40–54. Minnesota Prehistoric Archaeology Series No. 11. Minnesota Historical Society, St. Paul, Minnesota.

Sampson, Kelvin W.
1991   The Mystery of the Peoria Falcon. *Illinois Antiquity* 26(4):10–15.

Sampson, Kelvin and Duane Esarey
1993   A Survey of Elaborate Mississippian Copper Artifacts from Illinois. *Illinois Archaeology* 5(1–2):452–480.

Sanford, Albert H.
1913   The Exploration of Mounds in White's Group in Vernon County. *The Wisconsin Archeologist* 12(1):30–36.

Bibliography

Sank, Karen and Kelvin Sampson
    1994   A Falcon from the Depths. *Illinois Antiquity* 29(4):4–8.

Santure, Sharron K., Alan D. Harn, and Duane Esarey
    1990   *Archaeological Investigations at the Morton Village and Norris Farms 36 Cemetery.* Illinois State Museum Reports of Investigations, No. 45. Springfield, Illinois.

Sargent, Winthrop
    1795   American Antiquities. *The Massachusetts Magazine for July, 1795*, pp. 195–198.
    1799   A Letter from Colonel Winthrop Sargent, to Dr. Benjamin Smith Barton, Accompanying Drawings and Some Account of Certain Articles, Which Were taken Out of an Ancient Tumulus, or Grave, in the Western-Country. *Transactions of the American Philosophical Society, Held at Philadelphia, for Promoting Useful Knowledge* 4:177–180.

Savage, Howard
    1976   Human Tissue Preserved 3,000 Years in Ontario. *Arch Notes* 76(7):14–18.

Scarry, C. Margaret
    1998   Domestic Life on the Northwest Riverbank at Moundville. In *Archaeology of the Moundville Chiefdom*, edited by Vernon James Knight, Jr. and Vincas P. Steponaitis, pp. 63–101. Smithsonian Institution Press, Washington, DC.

Scarry, John F.
    1990   The Rise, Transformation, and Fall of Apalachee: A Case of Political Change in a Chiefly Society. In *Lamar Archaeology: Mississippian Chiefdoms in the Deep South*, edited by Mark Williams and Gary Shapiro, pp. 175–186. University of Alabama Press, Tuscaloosa, Alabama.
    1992   Political Offices and Political Structure: Ethnohistoric and Archaeological Perspectives on the Native Lords of Apalachee. In *Lords of the Southeast: Social Inequality and the Native Elites of Southeastern North America, edited by Alex W. Barker and Timothy Pauketat*, pp.163–183. Archeological Papers of the American Anthropological Association No. 3.
    1999   Elite Identities in Apalachee Province: The Construction of Identity and Cultural Change in a Mississippian Polity. In *Material Symbols: Culture and Economy in Prehistory*, edited by John E. Robb, pp. 342–361. Occasional Paper No. 26. Center for Archaeological Investigations, Southern Illinois University, Carbondale, Illinois.
    2007   Connections between the Etowah and Lake Jackson Chiefdoms: Patterns in the Iconographic and Material Evidence. In *Southeastern Ceremonial Complex: Chronology, Content, Context*, edited by Adam King, pp. 134–150. University of Alabama Press, Tuscaloosa, Alabama.

Schambach, Frank F.
    1982   An Outline of Fourche Maline Culture in Southwest Arkansas. In *Arkansas Archeology in Review*, edited by Neal L. Trubowitz and Marvin D. Jeter, pp. 132–197. Arkansas Archeological Survey Research Series No. 15, Fayetteville, Arkansas.
    1997   The Development of the Burial Mound Tradition in the Caddo Area. *Journal of Northeast Texas Archaeology* 9:53–72.

Scharf, J. Thomas
    1883   *History of Saint Louis City and County, from the Earliest Periods to the Present Day: Including Biographical Sketches of Representative Men*. Louis H. Everts & Co., Philadelphia.

Scheffers, Ronald
    2004   The Hoxie Farm Site Revisited. *Central States Archaeological Journal* 51(1):14–18.

Scheidegger, Edward
    1962   The Crib Mound. *Central State Archaeological Journal* 9(1):4–14.

Schlarb, Eric J.
    2005   The Bullock Site: A Forgotten Mound in Woodford County, Kentucky. In *Woodland Systematics in the Middle Ohio Valley*, edited by Darlene Applegate and Robert C. Mainfort, pp. 52–63. University of Alabama Press, Tuscaloosa, Alabama.

Schoenbeck, Ethel
    1941   Cultural Objects of Clear Lake Village Site. *Transactions of the Illinois State Academy of Science* 34(2):65–67.
    1947   A Seven-pound Copper Axe among 1946 Hopewell Discoveries. *Transactions of the llinois State Academy of Science* 40:36–42.
    1949   More Data on Hopewell Sites in Peoria Region. *Transactions of the Illinois State Academy of Science* 42:41–46.

Schoolcraft, Henry R.
    1819   *A View of the Lead Mines of Missouri; Including Some Observations of the Mineralogy, Geology, Geography, Antiquities, Soil, Climate, Population, and Productions of Missouri and Arkansas, and Other Sections of the Western Country*. Charles Wiley & Co., New York.
    1843   Account of the Mound at Grave Creek Flats in Virginia. *Journal of the Royal Geographical Society of London* 12:259–260.
    1845   Observations Respecting the Grave Creek Mound. *Transactions of the American Ethnological Society* 1:267–420.
    1851   Ancient Copper Mining in the Basin of Lake Superior. In *Historical and Statistical Information, Respecting the History, Condition and Prospects of the Indian Tribes of the United States* (Vol. 1), edited by Henry R. Schoolcraft, pp. 95–101. Lippincott, Grambo & Company, Philadelphia.
    1853   Notices of Some Metallic Plates Exhibited in Annual Dances among the Muscogees. In *Information Respecting the History, Condition and Prospects of the Indian Tribes of the United States: Collected and Prepared under the Direction of the Bureau of Indian Affairs, Per Act of Congress of March 3d, 1847*, by Henry R. Schoolcraft. LL. D, Part III, pp. 87–90.
    1855   Some Considerations on the Mound-Period of the Mississippi Valley, and on the General State of Indian Art in the Present Area of the United States, at the Beginning of the 16th Century. In *Information Respecting the History, Condition and Prospects of the Indian Tribes of the United States...* Part V, pp. 85–117. J. B. Lippincott & Company, Philadelphia.

Schreffler, Virginia L.
    1988   *Burial Status Differentiation as Evidenced by Fabrics from Etowah Mound C, Georgia*. PhD dissertation, Ohio State University, Columbus, Ohio.

Bibliography

Schroeder, David L. and Katherine C. Ruhl
    1968   Metallurgical Characteristics of North American Prehistoric Copper Work. *American Antiquity* 33(2):162–169.

Schultz, Christian, Jr.
    1810   *Travels on an Inland Voyage...* Vol. 1. Printed by Isaac Riley, New York.

Schultz, S. A.
    1913   Copper Hatchet, Camp Sites, Etc. *The Archaeological Bulletin* 4(1):24.

Schumacher, J. P.
    1918   Indian Remains in Door County. *The Wisconsin Archeologist* 16(4):125–145.

Schumacher, J. P. and J. H. Glaser
    1913   An Archaeological Reconnaissance in Northeastern Wisconsin. *The Wisconsin Archeologist* 11(4):107–124.

Schumacher, J. P. and W. A. Titu
    1913   Aboriginal Remains of the Upper Wisconsin Valley. *The Wisconsin Archeologist* 12(3):75–86.

Schupp, Philip C.
    1944   The Schupp Collection—One of the Outstanding Exhibits in the Entire Central West. *Journal of the Illinois State Archaeological Society* 1(3):28–234.

Schurr, Mark R.
    1997   The Bellinger Site (12 SJ 6) and the Origin of the Goodall Tradition. *Archaeology of Eastern North America* 25:125–142.

Sciulli, Paul W. and Bruce W. Aument
    1987   Paleodemography of the Duff Site (33LO111), Logan County, Ohio. *Midcontinental Journal of Archaeology* 12(1):117–144.

Sciulli, Paul W. and James M. Heilman
    2004   Terminal Late Archaic Mortuary Practices at Berryhill Cemetery. *Pennsylvania Archaeologist* 74(2):49–61.

Sciulli, Paul W., Leonard Piotrowski, and David Stothers
    1984   The Williams Cemetery: Biological Variation and Affinity with Three Glacial Kame Groups. *North American Archaeologist* 5(2):139–170.

Sciulli, Paul W., Ray Schuck, and Myra J. Geisen
    1993   Terminal Late Archaic Mortuary Practices at Kirian-Treglia. *Pennsylvania Archaeologist* 63(2):53–63.

Scoville, Selden S.
    1878   The Mound Builders on the Little Miami. *Journal of Cincinnati Society of Natural History* 1:128.

Sears, William H.
    1951   *Excavations at Kolomoki: Season II-1950, Mound E*. University of Georgia Series in Anthropology No. 3. University of Georgia Press, Athens, Georgia.

1953a   *Excavations at Kolomoki: Season III and IV, Mound D.* University of Georgia Series in Anthropology No. 4. University of Georgia Press, Athens, Georgia.

1953b   Kolomoki Burial Mounds and the Weeden Island Mortuary Complex. *American Antiquity* 18(3):223–229

1956   *Excavations at Kolomoki: Final Report.* University of Georgia Series in Anthropology No. 5. University of Georgia Press, Athens, Georgia.

1959   *Two Weeden Island Period Burial Mounds, Florida: The W. H. Browne Mound, Duval County, the MacKenzie Mound, Marion County.* Contributions of the Florida State Museum, Social Sciences No. 5. University of Florida, Gainesville, Florida.

1962   Hopewellian Affiliations of Certain Sites on the Gulf Coast of Florida. *American Antiquity* 28(1):5–18.

1964   The Southeastern United States. In *Prehistoric Man in the New World*, edited by Jesse D. Jennings and Edward Norbeck, pp. 259–287, University of Chicago Press, Chicago.

Seeman, Mark F.

1979   The Hopewell Interaction Sphere: The Evidence for Interregional Trade and Structural Complexity. *Prehistory Research Series* 5(2):224–438. Indiana Historical Society, Indianapolis.

1985   Craft Specialization and Tool Kit Structure: A Systemic Perspective on the Midcontinental Flint Knapper. In *Lithic Resource Procurement: Proceedings from the Second Conference on Prehistoric Chert Exploitation*, edited by Susan C. Vehik, pp. 7–36. Center for Archaeological Investigations, Southern Illinois University-Carbondale, Carbondale, Illinois.

1986   Adena Houses and Their Implications for Early Woodland Settlement Models in the Ohio Valley. In *Early Woodland Archeology*, edited by Kenneth B. Farnsworth and Thomas E. Emerson, pp. 564–580. Kampsville Seminars in Archeology, Vol. 2. Center for American Archeology, Center for American Archeology Press, Kampsville, Illinois.

1992a   Report on the Age, Affiliation and Significance of the GE Mound (12PO885). In *Archeological Resources Protection, Federal Prosecution Sourcebook, August 1994 Cumulative Supplement*, pp. 93–127.

1992b   Woodland Traditions in the Midcontinent: A Comparison of Three Regional Sequences. In *Long-Term Subsistence Change in Prehistoric North America*, edited by Dale E. Croes, Rebecca A. Hawkins, and Barry L. Isaac, pp. 3–46. Research in Economic Anthropology, Supplement 6. JAI Press Inc., Greenwich, Connecticut.

1995   When Words Are Not Enough: Hopewell Interregionalism and the Use of Material Symbols at the GE Mound. In *Native American Interactions: Multiscalar Analyses and Interpretations in the Eastern Woodlands*, edited by Michael S, Nassaney and Kenneth E. Sassaman, pp. 122–143. University of Tennessee Press, Knoxville, Tennessee.

Seeman, Mark F. and Ann C. Cramer

1982   The Manring Mounds: A Hopewell Center in the Mad River Drainage, Clark County, Ohio. *Ohio Journal of Science* 82:151–160.

Seeman, Mark F. and William S. Dancey

2000   The Late Woodland Period in Southern Ohio: Basic Issues and Prospects. In *Late Woodland Societies: Tradition and Transformation across the Midcontinent*, edited by Thomas E. Emerson, Dale L. McElrath, and Andrew C. Fortier, pp. 583–611. University of Nebraska Press, Lincoln, Nebraska.

Seeman, Mark F. and Frank Soday
    1980   The Russel Brown Mounds: Three Hopewell Mounds in Ross County, Ohio. *Midcontinental Journal of Archaeology* 5(1):73–116.

Setzler, Frank M.
    1930   The Archaeology of the Whitewater Valley. *Indiana History Bulletin* 7(12):351–549.
    1931   The Archaeology of Randolph County and the Fudge Mound. *Indiana History Bulletin* 9(1):1–5.

Shaeffer, James B.
    1957   The Nagle Site, Ok-4. *Bulletin of the Oklahoma Anthropological Society* 5:93–99.

Shallenberger, S. M.
    1879   War Paint—Copper Beads, Etc. *The American Antiquarian and Oriental Journal* 2(2):152–154.

Sharp, Frank C.
    1954   Artifacts from the Frank Sharp Collection, Kingston, Ohio. *Ohio Archaeologist* 4(1):22–23.

Sharp, Merle R.
    1952   Artifacts from Sharps. *Ohio Archaeologist* 2(3):1–2.

Sharp, William E.
    1990   Fort Ancient Period. In *The Archaeology of Kentucky: Past Accomplishments and Future Directions,* Vol. 2. State Historic Preservation Comprehensive Plan Report No. 1, edited by David Pollack, pp. 467–557.

Shaw, Cornelius G.
    1847   Diary of Cornelius G. Shaw, Written on Isle Royale, June 7, 1847 to November 21, 1847 and April 24–29, 1848. Bentley Historical Library, 852251 Aa 1, University of Michigan, Ann Arbor, Michigan.

Shepherd, Forrest
    1864   *Report by Professor Forrest Shepherd of New Haven, on Certain Mineral Lands in the Ontonagon District, Selected and Located by the Late Algernon Merryweather, U. S. Surveyor and Mining Engineer*. George C. Rand & Avery, Boston.

Shepherd, Henry A.
    1887   *Antiquities of the State of Ohio*. John C. Yorston & Co., Cincinnati, Ohio.

Shetrone, Henry C.
    1924   Exploration of the Wright Group of Pre-Historic Earthworks. *Ohio Archaeological and Historical Quarterly* 33:341–358.
    1926   Exploration of the Hopewell Group of Prehistoric Earthworks. *Ohio Archaeological and Historical Quarterly* 35:1–227.
    1930   *The Mound Builders*. D. Appleton and Company, New York.
    1938   *Primer of Ohio Archaeology: The Mound-builders and the Indians* (Third Edition). Ohio State Archaeological and Historical Society, Columbus, Ohio.

Shetrone, Henry C. and Emerson F. Greenman
    1931   Explorations of the Seip Group of Prehistoric Earthworks. *Ohio Archaeological and Historical Quarterly* 40:348–509.

Shippee, J. M.
1956   Concerning the Occurance of Mississippian Artifacts in Western Missouri. *Missouri Archaeological Society News Letter* 131:3–10.
1967   *Archaeological Remains in the Area of Kansas City: The Woodland Period, Early, Middle, and Late*. Missouri Archaeological Society Research Series No. 5. Columbia, Missouri.
1972   *Archeological Remains in the Kansas City Area: The Mississippian Occupation*. Missouri Archaeological Society Research Series No. 9. Columbia, Missouri.

Shiras, Virginia
1965   A Burial on White River. *The Arkansas Archeologist* 6(4):82–83.

Short, John T.
1879   *The North Americans of Antiquity: Their Origin, Migrations, and Type of Civilization Considered*. (Third Edition). Harper & Brothers, Publishers, New York.

Shriver, Phillip R.
1982   The Towner Mound of Pippin Lake: A Northern Ohio Hopewell Site. *Ohio Archaeologist* 32(2):40–41.
1983   A Hopewell Copper Axe from the Cuyahoga Valley. *Ohio Archaeologist* 33(1):24–25.
1987   Old Copper Culture Rat-Tailed Spears from the Upper Peninsula. *Ohio Archaeologist* 37(3):26–27.
1988   Rat-Tailed Spears: Another Look. *Ohio Archaeologist* 38(4):8–9.

Sibley, Lucy R. and Kathryn A. Jakes
1986   Characterization of Selected Prehistoric Fabrics of Southeastern North America. In *Historic Textile and Paper Materials: Conservation and Characterization*, edited by Howard L. Needles and S. Haig Zeronian, pp. 253–275. Advances in Chemistry Series 212. American Chemical Society, Washington, DC.
1989   Etowah Textile Remains and Cultural Context: A Model for Inference. *Clothing Textiles Research Journal* 7(2):37–45.
1994   Coloration in Etowah Textiles from Burial No. 57. In *Archaeometry of Pre-Columbian Sites and Artifacts*, edited by David A. Scott and Pieter Meyers, pp. 395–418. Proceedings of a Symposium Organized by the UCLA Institute of Archaeology and the Getty Conservation Institute, March 23–27, 1992, Los Angeles.

Sibley, Lucy R., Kathryn A. Jakes and Lewis H. Larson
1996   Inferring Behavior and Function from an Etowah Fabric Incorporating Feathers. In *A Most Indispensable Art: Native Fiber Industries from Eastern North America*, edited by James B. Petersen, pp. 73–87. University of Tennessee Press, Knoxville, Tennessee.

Sibley, Lucy R., Virginia S. Wimberly, and Kathryn A. Jakes
1986   A Partially Mineralized Fabric from Etowah as Indicator of Cultural Processes. *ACPTC Proceedings: National Meeting, 1986*, p. 93. Association of College Professors of Textiles and Clothing, Inc., Monument, CO.

Sieg, Lauren E.
2005   Valley View: Hopewell Taxonomy in the Middle Ohio Region. In *Woodland Systematics in the Middle Ohio Valley*, edited by Darlene Applegate and Robert C. Mainfort, pp. 178–196. University of Alabama Press, Tuscaloosa, Alabama.

Bibliography

Sieg, Lauren and Jarrod Burks
  2010   The Land between the Mounds: The Role of "Empty" Spaces in the Hopewellian
          Built Environment. In *Hopewell Settlement Patterns, Subsistence, and Symbolic
          Landscapes*, edited by A. Martin Byers and DeeAnne Wymer, pp. 56–71.
          University Press of Florida, Gainesville, Florida.

Silliman, Benjamin, Jr.
  1854   Notice of the Mineralogical Collection in the Crystal Palace. *The Mining
          Magazine* 2(6):593–609.

Silliman, Benjamin, Jr. and Charles R. Goodrich
  1854   Mineral and Mining Products. Section I. Class I, pp. 1–48. In *The World of
          Science, Art, and Industry Illustrated from Examples in the New-York Exhibition,
          1853–54*. G. P. Putnam and Company, New York.

Simpson, Anson M.
  1934   Kingston (Illinois) Focus of the Mississippi Culture. *Transactions of the Illinois
          State Academy of Science* 27(2):55.
  1937   Various Types of Kingston Site Burials. *Transactions of the Illinois State Academy
          of Science* 30(2):95–96.
  1952   The Kingston Village Site. *Journal of the Illinois State Archaeological Society*
          2(2–3):63–79.

Singer, Jerry
  1969   The Great Lakes Incident. *The Duluthian* 5(3):7–10, 22, 29.

Siskowit Mining Company of Michigan
  1850   *Charter and By-laws of the Siskowit Mining Company of Michigan, together with
          the Treasurer's Report, &c. up to January 1, 1850*. John C. Clark, Philadelphia.

Skinner, Alanson
  1920   *A Native Copper Celt from Ontario*. Indian Notes and Monographs No. 1.
          Museum of the American Indian, Heye Foundation, New York.

Skinner, S. Alan, R. King Harris, and Keith M. Anderson
  1969   *Archaeological Investigations at the Sam Kaufman Site, Red River County, Texas*.
          Southern Methodist University Contributions in Anthropology No. 5. Dallas,
          Texas.

Skinner, Shaune M.
  1985   Preliminary Results of the 1983 Excavations at the Connett Mounds #3 and #4,
          the Wolf Plains National Register District, Athens County, Ohio. *Archaeology of
          Eastern North America* 13:138–152.
  1987   A Hopewell Trove: Excavations from Rutledge Mound. *Timeline* 4(2):51–54.

Slafter, Edmund F.
  1879   Pre-historic Copper Implements. In *Report and Collections of the State Historical
          Society of Wisconsin, For the Years 1877, 1878, and 1879*, Vol. 8, pp. 152–164.
          David Atwood, State Printer, Madison, Wisconsin.

Sloane, Howard N. and Lucille L. Sloane
  1970   *A Pictorial History of American Mining*. Crown Publishers, Inc., New York.

Sly, Kyle L.
    1958   Reminiscences of the Horse and Buggy Days. *Central States Archaeological Journal* 4(3):86–95.

Smith, Alison J., Joseph J. Donovan, Emi Ito and Daniel R. Engstrom
    1997   Ground-water Processes Controlling a Prairie Lake's Response to Middle Holocene Drought. *Geology* 25(5):391–394.

Smith, Alison J., Joseph J. Donovan, Emi Ito, Daniel R. Engstrom and Valerie A. Panek
    2002   Climate-driven Hydrologic Transients in Lake Sediment Records: Multiproxy Record of Mid-Holocene Drought. *Quaternary Science Reviews* 21(4–6):625–646.

Smith, Arthur G.
    1955   One on the Old Sarge. *Ohio Archaeologist* 5(2):44–45.

Smith, Betty A.
    1979   The Hopewell Connection in Southwest Georgia. In *Hopewell Archaeology: The Chillicothe Conference*, edited by David S. Brose and N'omi Greber, pp. 181–187. MCJA Special Paper No. 3. Kent State University Press, Kent, Ohio.

Smith, Carlyle S. and Roger T. Grange, Jr.
    1958   *The Spain Site (39LM301), a Winter Village in Fort Randall Reservoir, South Dakota*. Bulletin 169, pp. 79–128. Bureau of American Ethnology, Smithsonian Institution, Washington, DC.

Smith, Cyril S.
    1968   Metallographic Study of Early Artifacts Made from Native Copper. In *Actes du XI Congrès International d'Histoire des Sciences VI*, pp. 237–243. Warsaw, Poland.

Smith, Gerald P.
    1996   The Mississippi River Drainage of Western Tennessee. In *Prehistory of the Central Mississippi Valley*, edited by Charles H. McNutt, pp. 97–118. University of Alabama Press, Tuscaloosa, Alabama.

Smith, Hale G.
    1951   *The Crable Site, Fulton County, Illinois*. Anthropological Papers No. 7. Museum of Anthropology, University of Michigan, Ann Arbor, Michigan.
    1956   *The European and the Indian: European-Indian Contacts in Georgia and Florida*. Florida Anthropological Society Publications No. 4. Gainesville, Florida.

Smith, Harlan I.
    1893   Primitive Remains in the Saginaw Valley, Michigan. *The Archaeologist* 1(3):51–53.
    1906   Archaeological Materials from Wisconsin in the American Museum of Natural History. *The Wisconsin Archeologist* 6(1):20–44.

Smith, Kevin E.
    1992   *The Middle Cumberland Region Mississippian Archaeology in North Central Tennessee*. PhD dissertation, Graduate School of Vanderbilt University, Nashville, Tennessee.
    1994   Potash from Pyramids: Reconstructing DeGraffenreid (40Wm4)—A Mississippian Mound-Complex in Williamson County, Tennessee. *Tennessee Anthropologist* 19(2):91–113.

Bibliography

Smith, Kevin E. and Michael C. Moore
    1999   "Through Many Mississippian Hands": Late Prehistoric Exchange in the Middle Cumberland Valley. In *Raw Materials and Exchange in the Mid-South: Proceedings of the 16th Annual Mid-South Archaeological Conference, Jackson, Mississippi, June 3 and 4, 1995*, edited by Evan Peacock and Samuel O. Brookes, pp. 95–115. Mississippi Department of Archives and History, Jackson, Michigan.

Smith, Lincoln
    2002   Osceola Revisited. *Central States Archaeological Journal* 49(1):13–16.

Smith, Marvin T.
    1987   *Archaeology of Aboriginal Culture Change in the Interior Southeast: Depopulation during the Early Historic Period.* University Press of Florida/ Florida Museum of Natural History, Gainesville, Florida.
    1989   Early Historic Period Vestiges of the Southern Cult. In *The Southeastern Ceremonial Complex: Artifacts and Analysis,* edited by Patricia Galloway, pp. 142–146. University of Nebraska Press, Lincoln, Nebraska.
    2000   *Coosa: The Rise and Fall of a Southeastern Mississippian Chiefdom.* University Press of Florida, Gainesville, Florida.

Smith, Robert
    1966   Excavating the Hopewell Burial Mounds at Grand Rapids. *Research News* 16(8):1–12. Office of Research Administration, University of Michigan, Ann Arbor, Michigan.

Smith, Samuel D.
    1971   Excavations at the Hope Mound with an Addendum to the Safford Mound Report. *Florida Anthropologist* 24(3):107–134.

Smith, Samuel L.
    1915   Pre-Historic and Modern Copper Mines of Lake Superior. *Michigan Historical Collections* 39:137–151.

Smith, Willie
    1961   The Southern Illinois Hopewell Peoples. *Central States Archaeological Journal* 8(2):62–76.

Smith, William E.
    1964   *History of Southwestern Ohio: The Miami Valleys.* Lewis Historical Publishing Company, New York.

Smucker, Isaac
    1875   The Mound Builders of the Mississippi Valley. *The Scientific Monthly* 1(3):100–120.
    1881   Mound Builders' Works near Newark, Ohio. *The American Antiquarian* 3(4):261–270.

Snyder, John F.
    1893   Buried Deposits of Hornstone Disks. *The Archaeologist* 1(10):181–186.
    1895a   A Group of Illinois Mounds. *The Archaeologist* 3(3):77–81.
    1895b   A Group of Illinois Mounds. *The Archaeologist* 3(4):109–113.
    1898   A Group of Illinois Mounds. *The American Archaeologist* 2(2):16–23.
    1908   Prehistoric Illinois: The Brown County Ossuary. *Journal of the Illinois State Historical Society* 1(2–3):33–43.

1909a   Prehistoric Illinois: Certain Indian Mounds Technically Considered. Part Second: Sepulchral and Memorial Mounds. *Journal of the Illinois State Historical Society* 2(1):47–65.

1909b   Prehistoric Illinois: Certain Indian Mounds Technically Considered. Part Third: Temple or Domiciliary Mounds. *Journal of the Illinois State Historical Society* 2(2):71–92.

Sohrweide, Anton
1932   The Origin and Distribution of Copper Artifacts. *The Wisconsin Archeologist* 11(4):153–156.

Solecki, Ralph S.
1953   *Exploration of an Adena Mound at Natrium, West Virginia.* Bulletin No. 151. Bureau of American Ethnology, Smithsonian Institution, Washington, DC.

Soper, Stanley
1939   Ornamentation among the American Indians. *Transactions of the Illinois State Academy of Science* 32(2):63–65.

Spalding, William W.
1901   Early Days in Duluth. *Michigan Pioneer and Historical Collections* 29:677–697.

Spaulding, Albert C.
1952   The Origin of the Adena Culture of the Ohio Valley. *Southwestern Journal of Anthropology* 8(3):260–268.
1957   Old Copper Culture. *American Antiquity* 22(4):436–437

Spaulding, Ken
1983   An Assortment of Copper Relics. *Prehistoric Relics* 18(4):120–121.

Speer, William S., editor
1888   Alonzo W. Brockway. In *Sketches of Prominent Tennesseans: Containing Biographies and Records of Many of the Families Who Have Attained Prominence in Tennessee*, pp. 326–329. Albert B. Tavel, Nashville, Tennessee.

Spence, Michael W.
1967   *A Middle Woodland Burial Complex in the St. Lawrence Valley.* Anthropology Papers No. 14. National Museum of Canada. Ottawa, Ontario, Canada.
1983   Silver Trade in Prehistoric Eastern North America. *Museum of Indian Archaeology (London, Ontario) Newsletter* 5(4):2–6.
1986   Band Structure and Interaction in Early Southern Ontario. *Canadian Journal of Anthropology* 5(2):83–95.

Spence, Michael W., William D. Finlayson, and Robert H. Pihl
1979   Hopewellian Influences on Middle Woodland Cultures in Southern Ontario. In *Hopewell Archaeology: The Chillicothe Conference*, edited by David S. Brose and N'omi Greber, pp. 115–121. MCJA Special Paper No. 3. Kent State University Press, Kent, Ohio.

Spence, Michael W. and William A. Fox
1986   The Early Woodland Occupations of Southern Ontario. In *Early Woodland Archeology*, edited by Kenneth B. Farnsworth and Thomas E. Emerson, pp. 4–46.

Bibliography

Kampsville Seminars in Archeology, Vol. 2. Center for American Archeology, Center for American Archeology Press, Kampsville, Illinois.

Spence, Michael W., Robert H. Pihl, and Carl R. Murphy
    1990   Cultural Complexes of the Early and Middle Woodland Periods. In *The Archaeology of Southern Ontario to A.D. 1650*, edited by Chris J. Ellis and Michael W. Spence, pp. 125–169. Publication No. 5. Occasional Publications of the London Chapter, Ontario Archaeological Society Inc., London, Ontario, Canada.

Spence, Michael W., Ronald F. Williamson, and John H. Dawkins
    1978   The Bruce Boyd Site: An Early Woodland Component in Southwestern Ontario. *Ontario Archaeology* 29:33–46.

Spielmann, Katherine A.
    2009   Ohio Hopewell Ritual Craft Production. In *Footprints: In the Footprints of Squier and Davis Archeological Fieldwork in Ross County, Ohio*, pp. 179–188. Special Report No. 5, Midwest Archeological Center, National Park Service, United States Department of the Interior, Lincoln, Nebraska.

Spiss, Pluma B.
    1968   Old Copper Artifacts from North Dakota. *The Wisconsin Archeologist* 49(3):125–126.

Spitzer, Michael G.
    1994   Other Artifactual Materials. In *Final Report: Phase III Data Recovery Investigations of the Cotiga Mound (46(MO1), Mingo County, West Virginia*, by Susan R. Frankenberg and Grace E. Henning, pp. 283–288. GAI Consultants, Inc., 570 Beatty Road, Monroeville, Pennsylvania.

Springer, Samantha
    2007   An Examination of Alterations to Mississippian Period Native Copper Artifacts from the Collection of the National Museum of the American Indian. *Association of North American Graduate Programs in the Conservation of Cultural Property Student Conference Papers.*

Squier, Ephraim G.
    1850   Use of Copper by the American Aborigines. In *Aboriginal Monuments of the State of New-York*, pp. 176–187. Smithsonian Contributions to Knowledge 2 (15). Washington, DC.

Squier, Ephraim G. and Edwin H. Davis
    1848   *Ancient Monuments of the Mississippi Valley: Comprising the Results of Extensive Original Surveys and Explorations.* Smithsonian Contributions to Knowledge 1. Washington, DC.

Stack, Guy
    1946   Some Interesting Metallic Artifacts from Tennessee. *Tennessee Archaeologist* 3(1):10–14.

Stafford, Barbara D. and Mark S. Sant
    1985   *Smiling Dan: Structure and Function at a Middle Woodland Settlement in the Lower Illinois Valley.* Vol. 2, Kampsville Archeological Center, Research Series, Center for American Archeology, Kampsville, Illinois.

Starr, Frederick
　　1887　　Mounds and Lodge Circles in Iowa. *The American Antiquarian* 9(6):361–363.
　　1888　　Preservation by Copper Salts. *The American Antiquarian* 10(5):279–282.
　　1893　　Mound Explorations in Northwestern Iowa. *Proceedings of the Davenport Academy of Sciences* 5:110–112.
　　1897　　Summary of the Archaeology of Iowa. *Proceedings of the Davenport Academy of Sciences* 6:53–124.

Starr, S. F.
　　1960　　The Archaeology of Hamilton County, Ohio. *Journal of the Cincinnati Museum of Natural History* 23(1):1–130.

Steinbring, John ("Jack")
　　1966　　Old Copper Culture Artifacts in Manitoba. *American Antiquity* 31(4):567–574.
　　1967　　A Copper 'Gaff Hook' from Ontario. *The Wisconsin Archeologist* 48(4):345–358.
　　1968　　A Copper Blade of Possible Paleo-Indian Type. *Manitoba Archaeological Newsletter* 5(1–2):3–12.
　　1970a　　Evidences of Old Copper in a Northern Transitional Zone. In *Ten Thousand Years: Archaeology in Manitoba*, edited by Walter M. Hlady, pp. 47–75. Manitoba Archaeological Society, D. W. Friesen & Sons Ltd., Altona, Manitoba, Canada.
　　1970b　　1970 Field Season. *Minnesota Archaeological Newsletter* 15:3–5.
　　1971a　　The Littlefork Burial: New Light on Old Copper. *Journal of the Minnesota Academy of Science* 37(1):8–15.
　　1971b　　Preliminary Consideration of Old Copper in Manitoba. *Na'pao* 3(1):1–5. Department of Anthropology and Archaeology, University of Saskatchewan, Saskatoon, Saskatchewan, Canada.
　　1974　　The Preceramic Archaeology of Northern Minnesota. In *Aspects of Upper Great Lakes Anthropology: Papers in Honor of Lloyd A. Wilford*, edited by Eldon Johnson, pp. 64–73. Minnesota Prehistoric Archaeology Series No. 11. Minnesota Historic Society, St. Paul, Minnesota.
　　1975　　*Taxonomic and Associational Considerations of Copper Technology during the Archaic Tradition.* PhD dissertation, University of Minnesota, Minneapolis.
　　1980　　*An Introduction to Archaeology on the Winnipeg River.* Miscellaneous Paper No. 9. Papers in Manitoba Archaeology, Department of Cultural Affairs and Historical Resources, Winnipeg, Manitoba, Canada.
　　1991　　Early Copper Artifacts in Western Manitoba. *Manitoba Archaeological Journal* 1(1):25–62.

Steinbring, Jack and Ron Sanders
　　1996　　Comments on Some Archaic Copper Artifacts from Waupaca County. *The Wisconsin Archeologist* 77(2): 83–86.

Steinbring, Jack and J. P. Whelan
　　1970　　Test Excavations at the Fish Lake Dam Site, Minnesota: A Report on the Minnesota Phase of the 1969 Archaeological Field School. *Minnesota Archaeologist* 31(1):3–40.

Steinen, Karl T.
　　2006　　Kolomoki: Cycling, Settlement Patterns, and Cultural Change in a Late Middle Woodland Society. In *Recreating Hopewell*, edited by Douglas K. Charles and Jane E. Buikstra, pp. 178–189. University Press of Florida, Gainesville, Florida.

Bibliography

Stelle, J. Parish
> 1871 Account of Aboriginal Ruins at Savannah, Tennessee. *Annual Report of the Board of Regents of the Smithsonian Institution, 1870*, pp. 408–414.

Stephenson, M. F.
> 1871 Account of Ancient Mounds in Georgia. *Annual Report of the Board of Regents of the Smithsonian Institution, 1870*, pp. 380–381.

Steponaitis, Vincas P.
> 1986 Prehistoric Archaeology in the Southeastern United States, 1970–1985. In *Annual Review of Anthropology, 1986*, pp. 363–404. Annual Reviews, Inc., Palo Alto, California.

> 1991 Contrasting Patterns of Mississippian Development. In *Chiefdoms: Power, Economy, and Ideology*, edited by Timothy Earle, pp. 193–228. Cambridge University Press, New York, New York.

Steve, John E.
> 1953 Largest Copper Nugget Weighed 420 Tons. *Inside Michigan* 3(7):17.

Stevens, Edward T.
> 1870 *Flint Chips. A Guide to Pre-Historic Archaeology, as Illustrated by the Collection in the Blackmore Museum, Salisbury.* Bell and Daldy, York Street, Covent Garden, London, England,

Stevens, Horace J.
> 1900a Active Copper Mines. In *The Copper Handbook*, pp. 135–314. Horace J. Stevens, Houghton, Michigan.
> 1900b Idle Mines, Dead Corporations and Duplicate Names of Active Mines. In *The Copper Handbook*, pp. 87–131. Horace J. Stevens, Houghton, Michigan.
> 1907 Copper Deposits of the United States. In *The Copper Handbook: A Manual of the Copper Industry of the World* 7: 168–206. Horace J. Stevens, Houghton, Michigan.

Stevens, William H.
> 1854 Agate Harbor Mining Region. *Lake Superior Journal* [Sault Ste. Marie, Michigan] June 17, p. 1.

Stevenson, J. E.
> 1879 The Mound Builders. *The American Antiquarian* 2(2):88–104.

Stirling, M. W.
> 1935 Smithsonian Archeological Projects Conducted under the Federal Emergency Relief Administration, 1933–34. *Annual Report of the Board of Regents of the Smithsonian Institution, 1934*, pp. 371–400.

Stoltman, James B.
> 1973 *The Laurel Culture in Minnesota.* Minnesota Prehistoric Archaeology Series No. 8. Minnesota Historical Society, St. Paul, Minnesota.
> 1978 Temporal Models in Prehistory: An Example from Eastern North America. *Current Anthropology* 19(4):703–746.
> 1979 Middle Woodland Stage Communities of Southwestern Wisconsin. In *Hopewell*

Archaeology: The Chillicothe Conference, edited by David S. Brose and N'omi Greber, pp. 122–139. MCJA Special Paper No. 3. Kent State University Press, Kent, Ohio.

1983   Ancient Peoples of the Upper Mississippi River Valley. In *Historic Lifestyles in the Upper Mississippi River Valley*, edited by John S. Wozniak, pp.187–255. University Press of America, Lanham, New York.

1997   The Archaic Tradition. *The Wisconsin Archeologist* 78(1–2):112–139.

2005   Tilmont (47CR460), a Stratified Prehistoric Site in the Upper Mississippi River Valley. *The Wisconsin Archeologist* 86(2):1–126.

2006   Reconsidering the Context of Hopewell Interaction in Southwestern Wisconsin. In *Recreating Hopewell*, edited by Douglas K. Charles and Jane E. Buikstra, pp. 310–327. University Press of Florida, Gainesville, Florida.

Stoltman, James B. and George W. Christiansen

2000   The Late Woodland Stage in the Driftless Area of the Upper Mississippi Valley. In *Late Woodland Societies: Tradition and Transformation across the Midcontinent*, edited by Thomas E. Emerson, Dale L. McElrath, and Andrew C. Fortier, pp. 497–524. University of Nebraska Press, Lincoln, Nebraska.

Storck, Peter L.

1981   *Ontario Prehistory*. Royal Ontario Museum, Toronto, Ontario, Canada.

Story, Dee Ann, Janice A. Guy, Barbara A. Burnett, Martha Doty Freeman, Jerome C. Rose, D. Gentry Steele, Ben W. Olive, and Karl J. Reinhard

1990   Cultural History of the Native Americans. In *The Archeology and Bioarcheology of the Gulf Coastal Plain: Volume 1*, pp. 163–366. Arkansas Archeological Survey Research Series No. 38.

Stothers, David M.

1974   The East Sugar Island Burial Mound. *Pennsylvania Archaeologist* 44(4):20–25.

Stothers, David M. and Timothy J. Abel

1991   Beads, Brass, and Beaver: Archeological Reflections of Protohistoric "Fire Nation" Trade and Exchange. *Archaeology of Eastern North America* 19:121–134.

1993   Archaeological Reflections of the Late Archaic and Early Woodland Time Periods on the Western Lake Erie Region. *Archaeology of Eastern North America* 21:25–109.

Stothers, David M. and Susan K. Bechtel

2000   The Land between the Lakes: New Perspectives on the Late Woodland (ca. A.D. 500–1300) Time Period in the Region of the St. Clair-Detroit River System. In *Cultures before Contact: The Late Prehistory of Ohio and Surrounding Regions*, edited by Robert A. Genheimer, pp. 2–51. Ohio Archaeological Council, Columbus, Ohio.

Stout, A. B. and H. L. Skavlem

1908   The Archaeology of the Lake Koshkonong Region. *The Wisconsin Archeologist* 7(2):47–102.

Stout, Charles B. and Charles H. Stout

1970   Re-Excavation of the Sackett Mound. *The Redskin* 5(1):27–29.

Stowe, Gerald C.

1938   An Enigmatic Copper Artifact. *The Wisconsin Archeologist* 19(2):37–41.

1942   Archaeological History of Douglas County, Wisconsin. *The Wisconsin Archeologist* 23(4):89–128.

Bibliography

Stowe, Noel R.
   1989   The Pensacola Variant and the Southeastern Ceremonial Complex. In *The Southeastern Ceremonial Complex: Artifacts and Analysis*, edited by Patricia Galloway, pp. 125–132. University of Nebraska Press, Lincoln, Nebraska.

Straley, Wilson
   1948   New Finds in the Hopewell Mounds. *Hobbies* 53(9):156.

Straw, Burton
   1962   Copper Mining Hammerstones from Upper Michigan. *The Wisconsin Archeologist* 43(3):76.

Strezewski, Michael
   2003   *Mississippi Period Mortuary Practices in the Central Illinois River Valley: A Region-Wide Survey and Analysis*. PhD dissertation, Department of Anthropology, Indiana University, Bloomington, Indiana.

Strong, John A.
   1989   The Mississippian Bird-Man Theme in Cross-Cultural Perspective. In *The Southeastern Ceremonial Complex: Artifacts and Analysis*, edited by Patricia Galloway, pp. 211–237. University of Nebraska Press, Lincoln, Nebraska.

Strong, William Duncan
   1935   *An Introduction to Nebraska Archeology*. Smithsonian Miscellaneous Collections 93(10):1–323.

Struever, Stuart
   1964   The Hopewell Interaction Sphere in Riverine-Western Great Lakes Culture History. In *Hopewellian Studies*, edited by Joseph R. Caldwell and Robert L. Hall, pp. 85–106. Scientific Papers 12(3). Illinois State Museum, Springfield, Illinois.
   1968   *A Re-Examination of Hopewell in Eastern North America*. PhD dissertation, University of Chicago, Chicago.

Struever, Stuart and Gail L. Houart
   1972   An Analysis of the Hopewell Interaction Sphere. In *Social Exchange and Interaction*, edited by Edwin N. Wilmsen, pp. 47–79. Anthropological Papers No. 46. Museum of Anthropology, University of Michigan, Ann Arbor, Michigan.

Suhm, Dee Ann and Alex D. Krieger
   1954   *An Introductory Handbook of Texas Archeology*. Bulletin of the Texas Archeological Society 25.

Swanton, John R.
   1911   *Indian Tribes of the Lower Mississippi Valley and Adjacent Coast of the Gulf of Mexico*. Bureau of American Ethnology Bulletin 43. Washington, DC.
   1928a   Aboriginal Culture of the Southeast. In *Forty-second Annual Report of the Bureau of American Ethnology, 1924–1925*, pp. 673–726. Washington, DC.
   1928b   Religious Beliefs and Medical Practices of the Creek Indians. In *Forty-second Annual Report of the Bureau of American Ethnology, 1924–1925*, pp. 473–672. Washington, DC.

Swartz, B. K., Jr.
  1973   Mound Three, White Site Hn-10 (IAS-BSU: The Final Report on a Robbins
          Manifestation in East Central Indiana. *Contributions to Anthropological History
          No. 1.* Ball State University, Muncie, Indiana.

Swineford, A. P.
  1876   *History and Review of the Copper, Iron, Silver, Slate and Other Material Interests
          of the South Shore of Lake Superior.* The Mining Journal, Marquette, Michigan.
  1884   Copper. In *Annual Report of the Commissioner of Mineral Statistics of the State
          of Michigan for 1883*, pp. 5–94. Marquette Mining Journal Publishing House,
          Marquette, Michigan.

Syms, E. Leigh
  1977   Cultural Ecology and Ecological Dynamics of the Ceramic Period in
          Southwestern Manitoba. *Plains Anthropologist* 22(76, Part 2):1–160.
  1979   The Devils Lake-Sourisford Burial Complex on the Northeastern Plains. *Plains
          Anthropologist* 24(86):283–308.

Taché, Karine
  2011   *Structure and Regional Diversity of the Meadowood Interaction Sphere.* Memoirs
          No. 48. Museum of Anthropology, University of Michigan, Ann Arbor, Michigan.

Tankersley, Kenneth B.
  2008   Archaeological Geology of the Turner Site Complex, Hamilton County, Ohio.
          North American Archaeologist 28(4):271–294.

Tankersley, Kenneth B. and Patricia A. Tench
  2009   Riker-Todd: A Salvaged Ohio Hopewell Mound. *North American Archaeologist*
          30(2):195–217.

Tanner, Donald R.
  1977   Two Artifacts. *Saginaw Valley Archaeologist* 9(1):3–5.

Tenney, William J.
  1853a   Lake Superior Copper Mines. *The Mining Magazine* 1(4):413–417.
  1853b   Lake Superior Copper Region. *The Mining Magazine* 1(2):171–174.
  1853c   The Ontonagon Mining Range. *The Mining Magazine* 1(6):632–636.
  1855   Lake Superior Region. *The Mining Magazine* 5(3):252–258.
  1856a   Lake Superior Region. *The Mining Magazine* 7(1–2):112–121.
  1856b   Lake Superior Region.—The Great Conglomerate Lode. *The Mining Magazine*
          7(4):308–310.
  1857a   Ancient Mining Operations at Lake Superior. *The Mining Magazine* 9(4):384–385.
  1857b   Evergreen Bluff Mine. *The Mining Magazine* 9(3):280–281.
  1857c   Journal of Copper Mining Operations. *The Mining Magazine* 9(1):97–99.
  1857d   Lake Superior Mining Region. *The Mining Magazine* 8(4):389–393.
  1857e   Superior Mining Company. *The Mining Magazine* 9(4):382–383.
  1857f   The Phalan Tract—Lake Superior Copper Region. *The Mining Magazine*
          9(5):475–476.
  1857g   Untitled. *The Mining Magazine* 9(6):571–572.

Tenney, William J. and Stephen P. Leeds
  1855a   Lake Superior. *The Mining Magazine* 4(4):288–297.

Bibliography

1855b   Lake Superior Region. *The Mining Magazine* 4(3):187–191.
1855c   Lake Superior Region. *The Mining Magazine* 4(5–6):402–409.
1855d   Lake Superior Region. *The Mining Magazine* 5(2):164–173.
1855e   Lake Superior Region. *The Mining Magazine* 5(6):540–545.

Tenney, William J., Stephen P. Leeds, and August Partz
    1854   Lake Superior Region. *The Mining Magazine* 3(4):425–433.

Tenney , William J. and August Partz
    1854a   Agate Harbor Mining Region. *The Mining Magazine* 2(5):554–557.
    1854b   The Ancient Miners. *The Mining Magazine* 2(1):77.
    1854c   Lake Superior Copper Mines. *The Mining Magazine* 2(1):74–76.
    1854d   Lake Superior Copper Mines. *The Mining Magazine* 2(4):428–434.
    1854e   Lake Superior Copper Region. *The Mining Magazine* 2(6):666–676.

Tharp, James
    2004   Terminal Late Archaic Glacial Kame and Its Meadowood Phase. Edited by John Chapin. *Central States Archaeological Journal* 51(1):41–54.

Thatcher, William H.
    1971   Pre-Historic Copper Mining in the Lake Superior Region. *The Coffinberry News Bulletin* 18(11):83–88.

Thayer, Burton W.
    1940   Narrative of George Hodge. *The Minnesota Archaeologist* 6(1):3–11.
    1944   A Minnesota Copper "Sickle." *The Minnesota Archaeologist* 10(2):73–75.

Theler, James L. and Robert F. Boszhardt
    2003   *Twelve Millennia: Archaeology of the Upper Mississippi River Valley*. University of Iowa Press, Iowa City, Iowa.

Thoburn, Joseph B.
    1926   Oklahoma Archaeological Explorations in 1925–26. *Chronicles of Oklahoma* 4(2):143–148.
    1929   The Prehistoric Cultures of Oklahoma. *Chronicles of Oklahoma* 7(3):211–241.
    1931   The Prehistoric Cultures of Oklahoma. In *Archaeology of the Arkansas River Valley*, by Warren King Mooreherad, pp. 53–150. Published for the Department of Archaeology, Phillips Academy, Andover, Massachusetts by the Yale University Press.

Thomas, Cyrus
    1884a   The Etowah Mounds. *Science* 3(73):779–785.
    1884b   Grave Mounds in North Carolina and East Tennessee. *The American Naturalist* 18(3):232–240.
    1884c   Spool-shaped Ornaments from Mounds. *Science* 3(62):434.
    1885   Excavation of a Mound in Tennessee. *American Journal of Archaeology* 1(2–3):183–184.
    1886   Mound Explorations in 1885, under the Ethnological Bureau. *The American Antiquarian and Oriental Journal* 8(1):35–37.
    1888   Burial Mounds of the Northern Sections of the United States. In *Fifth Annual Report of the Bureau of Ethnology to the Secretary of the Smithsonian Institution, 1883–84, by J. W. Powell, Director*, pp. 3–119. Government Printing Office, Washington, DC.

1894   Report of the Mound Explorations of the Bureau of Ethnology. *Twelfth Annual Report of the Bureau of American Ethnology*, pp. 3–742. Washington, DC.
1898   *Introduction to the Study of North American Archaeology.* The Robert Clarke Company, Cincinnati, Ohio.

Thomas, Prentice M., Jr. and L. Janice Campbell
1991   The Elliott's Point Complex: New Data Regarding the Localized Poverty Point Expression on the Northwest Florida Gulf Coast, 2000 B.C. – 500 B.C. In *The Poverty Point Culture: Local Manifestations, Subsistence Practices, and Trade Networks*, edited by Kathleen M. Byrd, pp. 103–119. Geoscience and Man Vol. 29. Depafrtment of Geography and Anthropology, Louisiana State University, Baton Rouge, Louisiana.

Thompson, Ben
1968   Comments about Copper Artifacts. *Central States Archaeological Journal* 15(3):114–115.

Thomson & Co.
1912[1864] *Prospectus for the Organization of a Copper Mining Company, upon the "Hulburt Tract," in Section 28, Town 58 N. of R. 31 W., Lake Superior, Michigan.* Francis Hart & Company, Printers, New York.

Thornberry-Ehrlich, Trista L.
2010   Geologic Resources Inventory Scoping Summary, Hopewell Culture National Historical Park, Ohio. Geologic Resources Division, National Park Service, US Department of the Interior.

Throop, Addison J.
1928   Stone Burials near Fults, Monroe County. In *Mound Builders of Illinois*, pp. 34–35. Call Printing Company, East St. Louis, Illinois.

Thruston, Gates P.
1888   Ancient Society in Tennessee. *Magazine of American History* 19(5):374–400.
1890   *The Antiquities of Tennessee and the Adjacent States and the State of Aboriginal Society in the Scale of Civilization Represented by Them.* Robert Clarke & Co., Cincinnati, Ohio.
1892   New Discoveries in Tennessee. *The American Antiquarian* 14(2):95–100.
1897   Engraved Shell Gorgets and Flint Ceremonial Implements. *The American Antiquarian* 19(3):96–100.
1904   Tennessee Archeology at St. Louis.—The Thruston Exhibit. *The Wisconsin Archeologist* 3(4):132–148.

Thunen, Robert L. and Keith H. Ashley
1995   Mortuary Behavior along the Lower St. Johns: An Overview. *The Florida Anthropologist* 48(1):3–12.

Thwaites, Reuben Gold, ed.
1959   Lower Canada, Iroquois, Ottawas: 1664–1667. In *The Jesuit Relations and Allied Documents*. Vol. 50. Pageant Book Company, New York.

Tichenor, Barcus
1910   Ancient Mounds and Enclosures in Indiana. *Indiana Magazine of History* 6(1):33–42.

Bibliography

Tidball, Eugene C.
2002 *"No Disgrace to My Country": The Life of John C. Tidball*. Kent State University Press, Kent, Ohio.

Tiffany, A. S.
1876a An Ancient Copper Implement Donated by E. B. Baldwin. *Proceedings of the Davenport Academy of Natural Sciences* 1:58–59.
1876b Mound Explorations in 1875. *Proceedings of the Davenport Academy of Natural Sciences* 1:113–114.

Tiffany, Joseph A.
1991 Models of Mississippian Culture History in the Western Prairie Peninsula: A Perspective from Iowa. In *Cahokia and the Hinterlands: Middle Mississippian Cultures of the Midwest*, edited by Thomas E. Emerson and R. Barry Lewis, pp. 183–192. University of Illinois Press, Urbana, Illinois.

Till, Anton
1977 Louisa County Archaeology. In *Iowa's Great River Road Cultural and Natural Resources, Vol. II, Archaeology, Geology, and Natural Areas: A Preliminary Survey*, prepared by John Hotopp, pp. 187–276. Iowa Department of Transportation Highway Division, Ames, Iowa.

Titterington, Paul F.
1938 *The Cahokia Mound Group and Its Village Materials*. P. F. Titterington, St. Louis.
1940 Outline of Cultural Traits of the Jersey County, Illinois, Bluff Focus. *Journal of the Illinois State Archaeological Society* 3(1):15–22.
1950 Some Non-Pottery Sites in the St. Louis Area. *Journal of the Illinois State Archaeological Society* 1(1):19–31.

Titus, W. A.
1914 The Fond du Lac Cache of Copper Implements. *The Wisconsin Archeologist* 13(2):97–100.
1915 Fond du Lac County Antiquities. *The Wisconsin Archeologist* 14(1):1–27.
1916 A Copper Banner Stone. *The Wisconsin Archeologist* 15(4):198–200, frontispiece.

Tomak, Curtis H.
1983 A Proposed Prehistoric Cultural Sequence for a Section of the Valley of the West Fork of the White River in Southwestern Indiana. *Tennessee Anthropologist* 8(1):67–94.
1994 The Mount Vernon Site: A Remarkable Hopewell Mound in Posey County, Indiana. *Archaeology of Eastern North America* 22:1–46.

Tomak, Curtis H. and Frank N. Burkett
1996 Decorated Leather Objects from the Mount Vernon Site. A Hopewell Site in Posey County, Indiana. In *A View from the Core: A Synthesis of Ohio Hopewell Archaeology*, edited by Paul J. Pacheco, pp. 354–369. Ohio Archaeological Council, Columbus, Ohio.

Tomlinson, A. B.
1843 Mr. Tomlinson's Letter. *American Pioneer* 2(5):197–203.

Torbenson, Michael, Odin Langsjoen, and Arthur Aufderheide
1994 Laurel Culture Human Remains from Smith Mounds Three and Four. *Plains Anthropologist* 39(150):429–444.

1996   Human Remains from McKinstry Mound Two. *Plains Anthropologist* 41(155):71–92.

Toth, Alan (aka Edwin A. Toth)
1979   The Marksville Connection. In *Hopewell Archaeology: The Chillicothe Conference*, edited by David S. Brose and N'omi Greber, pp. 189–199. MCJA Special Paper No. 3. Kent State University Press, Kent, Ohio.
1988   *Early Marksville Phases in the Lower Mississippi Valley: A Study of Culture Contact Dynamics.* Archaeological Report No. 21. Mississippi Department of Archives and History, Jackson, Mississippi.

Townsend, J. B.
1858a   Minesota Mine. *The Mining and Statistic Magazine* 10(4):332–335.
1858b   Mr. Townsend's Report. *Lake Superior Miner*, May 1, pp. 265–266.
1864a   Great Mass of Native Copper. *The Mining and Smelting Magazine* 6:25–26.
1864b   Large Mass of Native Copper. *American Journal of Science and Arts* 87(111):431.

Trevelyan, Amelia M.
1985   Powhatan Copper and the Prehistoric Ceremonial Complexes of the Eastern United States. *Phoebus* 4:62–69.
1987   *Prehistoric Native American Copper Work from the Eastern United States.* PhD dissertation, University of California-Los Angeles, Los Angeles.
2003   *Miskwabik, Metal of Ritual: Metallurgy in Precontact Eastern North America.* University Press of Kentucky, Lexington, Kentucky.

Tribble, Scott
2008   Living in the Past: Charles Whittlesey. *Timeline* 25(2):16–31.

Trigger, Bruce G.
1976   *The Children of Aataentsic: A History of the Huron People to 1660.* McGill-Queen's University Press, Montreal and London.

Trocolli, Ruth
2002   Mississippian Chiefs: Women and Men of Power. In *The Dynamics of Power*, edited by Maria O'Donovan, pp. 168–187. Occasional Paper No. 30. Center for Archaeological Investigations, Southern Illinois University-Carbondale, Carbondale, Illinois.

Trottier, Philippe G.
1973   A Preliminary Report on the Falcon Lake Site (C3-UN-35), Whiteshell Provincial Park, Manitoba. *Manitoba Archaeological Newsletter* 10(3–4):3–68.

Trowbridge, C. A.
1881   The Copper-Mines of Isle Royale. *The Popular Science Monthly* 20:122–123.

Trowbridge, Stephen V. R.
1850a   Report of S. V. R. Trowbridge, Assistant Agent United States Mineral Lands. In *Statistical Report of the Secretary of State of the State of Michigan for the Year 1848–1849*, pp. 53–59. Joint Document No. 4. Joint Documents of the Legislature of the State of Michigan, at the Annual Session of 1850. R. W. Ingals, State Printer, Lansing, Michigan.
1850b   Report of S. V. R. Trowbridge, Assistant Agent United States Mineral Lands. *Detroit Free Press* [Detroit, Michigan], January 16, p. 2.

Bibliography

Troyer, Byron
    1955   The Boudeman Copper Collection. *The Totem Pole* 35(4):1–4.

Turff, Gina M.
    1997   *A Synthesis of Middle Woodland Panpipes in Eastern North America.* Master's thesis, Trent University, Peterborough, Ontario, Canada.

Turff, Gina M. and Christopher Carr
    2005   Hopewellian Panpipes from Eastern North America: Their Social, Ritual, and Symbolic Significance. In *Gathering Hopewell: Society, Ritual, and Ritual Interaction,* edited by Christopher Carr and D. Troy Case, pp. 648–695. Kluwer Academic/Plenum Publishers, New York.

Van Blair, Dale
    1981   Artifacts from a Sumner County, Tennessee, Site. *Central States Archaeological Journal* 28(4):208–211.

Van Ert, Tyler
    2015   Some Things are Best Left Well Enough Alone. *Central State Archaeological Journal* 62(3):158.

Van Gorder, W. B.
    1916   Geology of Greene County. In *Fourteenth Annual Report of Department of Natural Resources, Indiana: 1915,* pp. 240–266.

Van Tassel, Charles S.
    1901   *The Book of Ohio and Its Centennial, or One Hundred Years of the Buckeye State.* C. S. Van Tassel, Publisher, Bowling Green and Toledo, Ohio.

Vansteen, Mick
    2007   A Collection of Michigan Copper. *Ohio Archaeologist* 57(2):18.

Vastokas, Romas
    1970   *Aboriginal Use of Copper in the Great Lakes Area.* PhD dissertation, Columbia University, New York.

Vavak, Floyd
    1967   Finding of a Copper Eagle. *Central States Archaeological Journal* 14(1):24–27.

Veakis, Emil
    1979   *Archaeometric Study of Native Copper in Prehistoric North America.* PhD dissertation, State University of New York at Stony Brook, Stony Brook, New York.

Vecsey, Christopher
    1983   *Traditional Ojibwa Religion and Its Historical Changes.* The American Philosophical Society, Philadelphia.

Vehik, Susan C.
    2002   Conflict, Trade and Political Development on the Southern Plains. *American Antiquity* 67(1):37–64.

Vehik, Susan C. and Timothy G. Baugh
    1994   Prehistoric Plains Trade. In *Prehistoric Exchange Systems in North America*, edited by Timothy G. Baugh and Jonathan E. Ericson, pp. 249–274. Plenum Press, New York.

Vernon, William W.
    1985   New Perspectives on the Archaeometallurgy of the Old Copper Industry. *MASCA Journal* 3(5):154–163.
    1990   New Archaeometallurgical Perspectives on the Old Copper Industry of North America. In *Archaeological Geology of North America*, Centennial Special Volume 4, edited by N. P. Lasca and J. Donahue, pp. 499–512. Geological Society of North America, Boulder, CO.

Vickers, Chris
    1945   Archaeology of the Rock and Pelican Lake Area of South-Central Manitoba. *American Antiquity* 11(2):88–94.

Vickers, Chris and Ralph D. Bird
    1949   A Copper Trade Object from the Headwaters Lakes Aspect in Manitoba. *American Antiquity* 15(2):157–160.

Vickery, Kent D.
    1979   "Reluctant" or "Avant-garde" Hopewell? Suggestions of Middle Woodland Culture Change in East-central Indiana and South-central Ohio. In *Hopewell Archaeology: The Chillicothe Conference*, edited by David S. Brose and N'omi Greber, pp. 59–63. MCJA Special Paper No. 3. Kent State University Press, Kent, Ohio.

Vietzen, Raymond C.
    1965   *Indians of the Lake Erie Basin or Lost Nations*. Ludi Printing Co., Wahoo, Nebraska.
    1974   The Riker Site. Published by the Sugar Creek Valley Chapter, Archaeological Society of Ohio.
    1976   *Shakin' the Bushes: From Paleo to Historic Indians*. White Horse Publishers, Elyria, Ohio.

Wagner, Herbert
    1995   Wisconsin's Ancient Copper Miners. *Wisconsin Outdoor Journal* 9(3):62–64.

Walbrandt, Jacqueline
    1979   A Birthday Gift from the Past. *Central States Archaeological Journal* 26(1):31.

Walley, Dave
    2015   Untitled. *Central States Archaeological Journal* 62(3):154–155.

Walthall, John A.
    1973   *Copena: A Tennessee Valley Middle Woodland Culture*. PhD dissertation, University of North Carolina at Chapel Hill, Chapel Hill, North Carolina.
    1979   Hopewell and the Southern Heartland. In *Hopewell Archaeology: The Chillicothe Conference*, edited by David S. Brose and N'omi Greber, pp. 200–210. MCJA Special Paper No. 3. Kent State University Press, Kent, Ohio.
    1985   Early Hopewellian Ceremonial Encampments in the South Appalachian Highlands. In *Structure and Process in Southeastern Archaeology*, edited by Roy

S. Dickens, Jr. and H. Trawick ward, pp. 243–262. University of Alabama Press, Tuscaloosa, Alabama.

1990   *Prehistoric Indians of the Southeast: Archaeology of Alabama and the Middle South*. University of Alabama Press, Tuscaloosa, Alabama.

Walthall, John A. and David L. DeJarnette
    1974   Copena Burial Caves. *Journal of Alabama Archaeology* 20(1):1–62.

Wang, Lois
    1969   An Old Copper Implement. *Saginaw Valley Archaeologist* 6(2):18–19.

Warburton, George
    1850   *The Conquest of Canada* (Vol. 1). Harper and Brothers, New York.

Wardle, H. Newell
    1906   The Treasures of Prehistoric Moundville. *Harper's Monthly Magazine* 112(668):200–210.

Ware, John M.
    1917   *A Standard History of Waupaca County, Wisconsin*. The Lewis Publishing Company: Chicago and New York.

Waring, A. J., Jr.
    1945   "Hopewellian" Elements in Northern Georgia. *American Antiquity* 11(2):119–120.
    1968a   The Southern Cult and Muskhogean Ceremonial. In *The Waring Papers: The Collected Papers of Antonio J. Waring, Jr.*, edited by Stephen Williams, pp. 30–69. Papers of the Peabody Museum of Archaeology and Ethnology 58. Peabody Museum, Harvard University, Cambridge, Massachusetts and the University of Georgia Press, Athens, Georgia.

Waring, A. J., Jr. and Preston Holder
    1945   A Prehistoric Ceremonial Complex in the Southeastern United States. *American Anthropologist* 47(1):1–34.

Warner, Beers & Co.
    1883   *The History of Hardin County, Ohio...* Warner, Beers & Co. Chicago

Watson, Gordon D.
    1972   A Copper Projectile Point from Lake Winnipeg, Manitoba. *Manitoba Archaeological Newsletter* 9(3):3–6.
    1990   Palaeo-Indian and Archaic Occupations of the Rideau Lakes. *Ontario Archaeology* No. 50, pp. 5–26.

Watson, Virginia D.
    1950   *The Wulfing Plates: Products of Prehistoric Americans*. Social and Philosophical Sciences No. 8. Washington University Studies, New Series, St. Louis.

Wayman, M. L., J. C. H. King, and P. T. Craddock
    1992   Analysis of the Copper Artifacts in the "Squier and Davis" Collection, Museum of Mankind. In *Aspects of Early North American Metallurgy*, pp. 95–144. Occasional Paper 79, Departments of Scientific Research and Ethnography, British Museum, London, England.

Webb, Clarence H.
    1982   The Poverty Point Culture (Second Edition, Revised). *Geoscience and Man* 17:1–73.

Webb, Clarence H. and Monroe Dodd, Jr.
    1939   Further Excavations of the Gahagan Mound; Connections with a Florida Culture. *Texas Archeological and Paleontological Society* 11:92–128.

Webb, Clarence H. and Hiram F. Gregory
    1986   The Caddo Indians of Louisiana. *Anthropological Study No. 2*. Louisiana Archaeological Survey and Antiquity Commission, Department of Culture, Recreation and Tourism, Baton Rouge, Louisiana.

Webb, Clarence H. and Ralph R. McKinney
    1975   Mounds Plantation (16CD12), Caddo Parish, Louisiana. *Louisiana Archaeology* 2:39–127.

Webb, William S.
    1938   *An Archaeological Survey of the Norris Basin in Eastern Tennessee*. Bulletin 118. Bureau of American Ethnology, Smithsonian Institution, Washington, DC.
    1940   The Wright Mounds, Sites 6 and 7, Montgomery County, Kentucky. *Reports in Anthropology* 5(1). Department of Anthropology and Archaeology, University of Kentucky, Lexington, Kentucky.
    1941   Mt. Horeb Earthworks, Site 1 and the Drake Mound, Site 11, Fayette County, Kentucky. *Reports in Anthropology and Archaeology* 5(2). Department of Anthropology and Archaeology, University of Kentucky, Lexington, Kentucky.
    1942   The C. and O. Mounds at Paintsville, Sites Jo 2 and Jo 9, Johnson County, Kentucky. *Reports in Anthropology and Archaeology* 5(4). Department of Anthropology and Archaeology, University of Kentucky, Lexington, Kentucky.
    1943a   The Crigler Mounds, Sites Be 20 and Be 27 and the Hartman Mound, Site Be 32, Boone County, Kentucky. *Reports in Anthropology and Archaeology* 5(6). Department of Anthropology and Archaeology, University of Kentucky, Lexington, Kentucky.
    1943b   The Riley Mound and the Lansing Mound, Site Be 17, Boone County, Kentucky, with Additional Notes on the Mount Horeb Site, Fa 1 and Sites Fa 14 and 15, Fayette County. *Reports in Anthropology and Archaeology* 5(7). Department of Anthropology and Archaeology, University of Kentucky, Lexington, Kentucky.
    1950   The Carlson Annis Mound, Site 5, Butler County, Kentucky. *Reports in Anthropology* 7(4). Department of Anthropology, University of Kentucky, Lexington, Kentucky.
    1951   The Parrish Village Site, Site 45, Hopkins County, Kentucky. *Reports in Anthropology* 7(6). Department of Anthropology, University of Kentucky, Lexington, Kentucky.
    1974   *Indian Knoll*. Introduction by Howard D. Winters. University of Tennessee Press, Knoxville, Tennessee.

Webb, William S. and Raymond S. Baby
    1957   *The Adena People-No. 2*. The Ohio Historical Society, Columbus, Ohio.

Webb, William S. and David L. DeJarnette
    1942   *An Archeological Survey of Pickwick Basin in the Adjacent Portions of the States of Alabama, Mississippi and Tennessee*. Bulletin 129. Bureau of American Ethnology, Smithsonian Institution, Washington, DC.

Bibliography

Webb, William S. and John B. Elliott
    1942   The Robbins Mounds, Sites Be 3 and Be 14, Boone County, Kentucky. *Reports in Anthropology and Archaeology* 5(5). Department of Anthropology, University of Kentucky, Lexington, Kentucky.

Webb, William S. and W. D. Funkhouser
    1940   Ricketts Site Revisited, Site 3, Montgomery County, Kentucky. *Reports in Anthropology and Archaeology* 3(6). Department of Anthropology and Archaeology, University of Kentucky, Lexington, Kentucky.

Webb, William S. and William G. Haag
    1947a   Archaic Sites in McLean County, Kentucky. *Reports in Anthropology and Archaeology* 7(1). Department of Anthropology and Archaeology, University of Kentucky, Lexington, Kentucky.
    1947b   The Fisher Site, Fayette County, Kentucky. *Reports in Anthropology and Archaeology* 7(2). Department of Anthropology, University of Kentucky, Lexington, Kentucky.

Webb, William S. and Charles E. Snow
    1959   *The Dover Mound.* University of Kentucky Press, Lexington, Kentucky.
    1974   *The Adena People.* University of Tennessee Press, Knoxville, Tennessee.

Webb, William S. and Charles G. Wilder
    1951   *An Archaeological Survey of Guntersville Basin on the Tennessee River in Northern Alabama.* University of Kentucky Press, Lexington, Kentucky.

Webster, David L.
    1973   Nonceramic Artifacts from Laurel Culture Sites Excavated Prior to 1961. In *The Laurel Culture in Minnesota*, edited By James B. Stoltman, pp. 94–111. Minnesota Prehistoric Archaeology Series No. 8. Minnesota Historical Society, St. Paul, Minnesota.

Webster Mining Company
    1853   *The Webster Mining Company of Lake Superior.* Sleeper & Rogers, Boston, Massachusetts.

Wedel, Mildred Mott
    1959   Oneota Sites on the Upper Iowa River. *The Missouri Archaeologist* 21(2–4):1–181.

Wedel, Waldo R.
    1938   Hopewellian Remains near Kansas City, Missouri. *Proceedings of the United States National Museum* 86(3045):99–106.
    1940   Culture Sequence in the Central Great Plains. In *Smithsonian Miscellaneous Collections* 100, pp. 291–352.
    1961   *Prehistoric Man on the Great Plains.* University of Oklahoma Press, Norman, Oklahoma.
    1964   The Great Plains. In *Prehistoric Man in the New World*, edited by Jesse D. Jennings and Edward Norbeck, pp. 193–210, University of Chicago Press, Chicago.
    1986   *Central Plains Prehistory: Holocene Environments and Culture Change in the Republican River Basin.* University of Nebraska Press, Lincoln and London.

Weed, Walter H.
　　1918　　*The Mines Handbook: An Enlargement of the Copper Handbook* (Vol. 13). International Edition. W. H. Weed, New York.

Weer, Paul
　　1945　　Indian Curiosities. *American Antiquity* 11(2):121.

Weinland, Walter D.
　　1910　　Exploration of an Indian Mound, Decatur County, Indiana. *The Archaeological Bulletin* 1(3):63–65.

Weisman, Brent R.
　　1995　　Crystal River: A Ceremonial Mound Center on the Florida Gulf Coast. *Florida Archaeology* 8. Florida Bureau of Archaeological Research, Division of Historical Resources, Florida Department of State, Tallahassee, Florida.

Welch, Paul D.
　　1990　　Mississippian Emergence in West-Central Alabama. In *The Mississippian Emergence*, edited by Bruce D. Smith, pp. 197–225. Smithsonian Institution Press, Washington, DC.

Wellman, Howard B.
　　1994　　*The Provenance of Copper Artifacts from the Boucher Site.* Master's thesis, Boston University, Boston.

West, George A.
　　1884　　Copper Relics in the Mounds of Wisconsin. *The American Antiquarian and Oriental Journal* 6(2):107–108.
　　1929　　Copper: Its Mining and Use by the Aborigines of the Lake Superior Region. *Bulletin of the Public Museum of the City of Milwaukee* 10(1):1–182.
　　1931　　Superimposed Aboriginal Implement. *The Wisconsin Archeologist* 10(3):89–90.
　　1932　　Exceptional Prehistoric Copper Implements. Bulletin of the Public Museum of the City of Milwaukee 10(4):375–400.
　　1933　　The Greater Copper Pike. *The Wisconsin Archeologist* 12(2):31–33.
　　1939　　A Copper Banner-Stone. In *Banner-Stones of the North American Indian*, by Byron W. Knoblock, pp. 782–73. Published by the author, La Grange, Illinois

Wheeler, Ryan J.
　　2000　　*Treasure of the Calusa; The Johnson/Willcox Collection from Mound Key, Florida.* Monographs in Florida Archaeology No. 1. Tallahassee, Florida.

White, Andrew A.
　　2003　　Mortuary Assemblages and the Chronology of the Jack's Reef Cluster. *The Michigan Archaeologist* 49(3–4):43–71.

White, Ellanor P.
　　1987　　*Excavating in the Field Museum: A Survey and Analysis of Textiles from the 1891 Hopewell Mound Group Excavation.* Master's thesis, Department of Public History, Sangamon State University, Springfield, Illinois.

Whiteford, Andrew H.
　　1952　　A Frame of Reference for the Archeology of Eastern Tennessee. In *Archeology*

Bibliography

of Eastern United States, edited by James B. Griffin, pp. 207–225. University of Chicago Press, Chicago.

Whitman, Janice K.
1977   Kohl Mound, a Hopewellian Mound in Tuscarawas County. *Ohio Archaeologist* 27(3):4–8.

Whitney, Josiah D.
1849   Field Notes for 1847. In Report on the Geological and Mineralogical Survey of the Mineral Lands of the United States in Michigan, by Charles T, Jackson, pp. 713–758. In *Message from the President of the United States, to the Two Houses of Congress, at the Commencement of the First Session of the Thirty-first Congress. December 24, 1849. Read December 27, 1849. Part III*, pp. 371–801. House Executive Document 5(3). 31ˢᵗ Congress, 1ˢᵗ Session. Also issued as Senate Executive Document 1). 31ˢᵗ Congress, 1ˢᵗ Session. Washington, DC.
1854   *The Metallic Wealth of the United States, Described and Compared with that of Other Countries*. Lippincott, Grambo & Co., Philadelphia.

Whittlesey, Charles
1850   Lake Superior Copper Region. *American Railroad Journal* 23(736):325–326.
1852a   The Ancient Miners of Lake Superior. *Annals of Science* 1(2):15–18, 1(3):27–30.
1852b   The Ancient Miners of Lake Superior. *The Canadian Journal* 1(5):106–109.
1852c   *Fugitive Essays, upon Interesting and Useful Subjects...* Sawyer, Ingersoll and Co., Hudson, Ohio.
1853   The Ancient Miners of Lake Superior. *The Canadian Journal* 1(6):132–135.
1857a   Ancient Mining on Lake Superior. *Lake Superior Miner* [Ontonagon, Michigan], September 12, p. 2.
1857b   Ancient Mining on the Shores of Lake Superior. *Canadian Naturalist and Geologist* 2:292–293.
1863   Ancient Mining on the Shores of Lake Superior. *Smithsonian Contributions to Knowledge* 13(4). Washington, DC.
1868   On the Weapons and Military Character of the Race of the Mounds. *Memoirs of the Boston Society of Natural History* 1(4):473–481.
1871   *Ancient Earth Forts of the Cuyahoga Valley, Ohio*. Tract No. 5. Western Reserve and Northern Ohio Historical Society, Cleveland.

Wiant, Michael D., Kenneth B. Farnsworth, and Edwin R. Hajic
2009   The Archaic Period in the Lower Illinois River Basin. In *Archaic Societies: Diversity and Complexity across the Midcontinent*, edited by Thomas E. Emerson, Dale L. McElrath, and Andrew C. Fortier, pp. 229–285. State University Press of New York, Albany, New York.

Wickersham, James
1885   Mounds of Sangamon County, Illinois. *Annual Report of the Board of Regents of the Smithsonian Institution, 1883*, pp. 825–835.

Wied, Prince Maximilian
1906   *Travels in the Interior of North America, 1832–1834*. The Arthur H. Clark Company, Cleveland.

Wiegand, Phil
1955   A Native Copper Harpoon Point. *The Wisconsin Archeologist* 36(1):22–23.

Wilford, Lloyd A.
    1950a   The Prehistoric Indians of Minnesota: Some Mounds of the Rainy River Aspect. *Minnesota History* 31(3):163–171.
    1950b   The Prehistoric Indians of Minnesota: The McKinstry Mounds of the Rainy River Aspect. *Minnesota History* 31(4):231–237.
    1955   A Revised Classification of the Prehistoric Cultures of Minnesota. *American Antiquity* 21(2):130–142.
    1960   The First Minnesotans. In *Minnesota Heritage: A Panoramic Narrative of the Historical Development of the North Star State*, edited by Lawrence M. Brings, Truly Latchaw, Edward Olderen and Howard Lindberg, pp. 40–79. T. S. Denison & Company, Inc., Minneapolis.

Wilkins, Katherine S.
    2001   Oblong Copper Gorgets from Moundville. *Journal of Alabama Archaeology* 47(1):69–74.

Willey, Gordon R.
    1998   *Archeology of the Florida Gulf Coast*. University Press of Florida, Gainesville, Florida.

Williams Brothers
    1880   *History of Ross and Highland Counties, Ohio, with Illustrations and Biographical Sketches*. W. W. Williams, Printer, Cleveland.

Williams, Ephraim S.
    1888   Remembrances of Early Days in Saginaw in 1833. *Michigan Pioneer and Historical Collections* 10:142–147.

Williams, H. Z. and Bro.
    1881   *History of Preble County, Ohio, with Illustrations and Biographical Sketches*. From the Printing House of W. W. Williams, Cleveland.

Williams, Mark
    1990   Medlin Copper Plate. In *Archaeological Excavations at Shinholser (9BL1), 1985 & 1987*, pp. 232–233. LAMAR Institute, Watkinsville, Georgia.

Williams, Mentor L.
    1946   Men with Hammers. *The Michigan Alumnus Quarterly Review* 52(20):238–254.
    1950   Horace Greeley and Michigan Copper. *Michigan History Magazine* 34(2):120–132.

Williams, Ralph D.
    1907   *The Honorable Peter White: A Biographical Sketch of the Lake Superior Iron Country*. The Penton Publishing Co., Cleveland.

Williams, Ray
    1968   Southeast Missouri Land Leveling Salvage Archaeology: 1967. Conducted for the National Park Service Midwest Region by the University of Missopuri Archaeological Research Division, Department of Anthropology, College of Arts and Science, Columbia, Missouri.

Williams, Stephen
    1980   Armorell: A Very Late Phase in the Lower Mississippi Valley. *Southeastern Archaeological Conference Bulletin* 22:105–110.

Bibliography

Williams, Stephen and Jeffrey P. Brain
    1983   *Excavations at the Lake George Site, Yazoo County, Mississippi: 1958–1960.*
                Vol 74. Papers of the Peabody Museum of Archaeology and Ethnology, Harvard
                University, Cambridge, Massachusetts.

Williams, Stephen and John M. Goggin
    1956   The Long Nosed God Mask in Eastern United States. *The Missouri Archaeologist*
                18(3):3–71.

Williamson, C. W.
    1887   An Ohio Mound. *Science* 9(210):135.

Williamson, Margaret H.
    2003   *Powhatan Lords of Life and Death: Command and Consent in Seventeenth
                Century Virginia.* University of Nebraska Press, Lincoln, Nebraska.

Williamson, Ronald F.
    1980   The Liahn II Site and Early Woodland Mortuary Ceremonialism. *Ontario
                Archaeology* 33:3–11.

Willoughby, Charles C.
    1903   Primitive Metal Working. *American Anthropologist* 5(1):54–57.
    1917   The Art of the Great Earthwork Builders of Ohio. *Annual Report of the Board of
                Regents of the Smithsonian Institution, 1916,* pp. 489–500. Government Printing
                Office, Washington, DC.
    1935   Michabo, the Great Hare: A Patron of the Hopewell Mound Settlement. *American
                Anthropologist* 37(2):280–286.
    1938   Textile Fabrics from the Burial Mounds of the Great Earthwork Builders of Ohio.
                Ohio State Archaeological and Historical Quarterly 47:273–287.

Willoughby, Charles C. and Earnest A. Hooton
    1922   The Turner Group of Earthworks, Hamilton County, Ohio. *Papers of the Peabody
                Museum of American Archaeology and Ethnology, Harvard University* Vol. 8(3).
                Cambridge, Massachusetts.

Wilson, Curtis L. and Melville Sayre
    1935   A Brief Metallographic Study of Primitive Copper Work. *American Antiquity*
                1(2):109–112.

Wilson, Daniel
    1855   Antiquities of the Copper Region of the North American Lakes. *Proceedings of
                the Society of Antiquaries of Scotland* 2:203–212.
    1856   The Ancient Miners of Lake Superior. *The Canadian Journal* 1(3):225–237.
    1876   *Prehistoric Man* (Vol. I). Macmillan and Co., London, England.

Wilson, James G. and John Fiske
    1898   Hodge, James Thatcher. *Appleton's Cyclopaedia of American Biography* (Revised
                Edition) 3:224. D. Appleton and Company, New York.

Wilson, John
    1853   The Michigan Block of Copper. *New York Times*, October 26, 1853.

Wilson, John P. and C. J. Maginel
   1976 A Red Ocher Earthwork Enclosure in Illinois. *Central States Archaeological Journal* 17(1):4–16.

Wilson, Lee Anne
   1981 A Possible Interpretation of the Bird-man Figure Found on Objects Associated with the Southern Cult of the Southeastern United States, A. D. 1200 to 1350. *Phoebus* 3:6–18.

Wilson, Rex L.
   1965 Excavations at the Mayport Mound, Florida. *Contributions of the Florida State Museum, Social Sciences* 13. University of Florida, Gainesville, Florida.

Wilson, Thomas
   1893 Report on the Department of Prehistoric Anthropology in the U. S. National Museum, 1892. *Annual Report of the Board of Regents of the Smithsonian Institution, 1892*, pp. 135–142. Government Printing Office, Washington, DC.
   1896 The Swastika: The Earliest Known Symbol, and Its Migrations; With Observations on the Migration of Certain Industries in Prehistoric Times. In *Annual Report of the Board of Regents of the Smithsonian Institution, 1894*, pp. 757–1011. Government Printing Office, Washington, DC.
   1898 Prehistoric Art; or, The Origin of Art as Manifested in the Works of Prehistoric Man. In *Annual Report of the Board of Regents of the Smithsonian Institution, 1896*, pp. 325–664. Government Printing Office, Washington, DC.
   1899 Arrowpoints, Spearheads, and Knives of Prehistoric Times. In *Annual Report of the Board of Regents of the Smithsonian Institution, Year Ending June 30, 1897*, pp. 811–988. Government Printing Office, Washington, DC.

Wimberly, Steven B. and Harry A. Tourtelot
   1941 *The McQuorquodale Mound: A Manifestation of the Hopewellian Phase in South Alabama*. Museum Paper 19. Alabama Museum of Natural History, Tuscaloosa, Alabama.

Wimberly, Virginia S.
   2004 Preserved Textiles on Hopewell Copper. In *Perishable Material Culture in the Northeast*, edited by Penelope Ballard Drooker, pp. 69–85. New York State Museum Bulletin 500. University of the State of New York, State Education Department, Albany, New York.

Winchell, Alexander
   1867 Statement of Operations in the Museum of the University of Michigan, in the Department of "Geology, Zoölogy and Botany," and the Department of Ethnology and Relics," for the Year Ending September 21st, 1867. *Annual Report on the Museum of the University of Michigan*. Ann Arbor, Michigan.

Winchell, Newton H.
   1881a Ancient Copper-Mines of Isle Royale. *The Popular Science Monthly* 19:601–620.
   1881b Ancient Copper Mines of Isle Royale. *The Engineering and Mining Journal* 32(12):184–186.
   1881c Catalogue of Archaeological Specimens in the General Museum. In *Geological and Natural History Survey of Minnesota, Ninth Annual Report*, 1880, pp. 162–164. J. L. Moore, State Printer, St. Peter, Minnesota.

Bibliography

1882a   Archaeological Additions to the Museum in 1881. *Geological and Natural History Survey of Minnesota, Tenth Annual Report*, 1881, p. 174. J. W. Cunningham, State Printer, St. Paul, Minnesota.

1882b   Preliminary List of Rocks Collected by N. H. Winchell in 1879—Continued from the Last Report. *Geological and Natural History Survey of Minnesota, Tenth Annual Report*, 1881, pp. 9–122. J. W. Cunningham, State Printer, St. Paul, Minnesota.

1911   *The Aborigines of Minnesota: A Report Based on the Collections of Jacob V. Brower, and on the Field Surveys and Notes of Alfred J. Hill and Theodore H. Lewis.* The Pioneer Company, St. Paul. Minnesota.

Winn, Vetal

1924a   The Minocqua Lake Region. *The Wisconsin Archeologist* 3(2):41–51.

1924b   A Cache of Copper Chisels. *The Wisconsin Archeologist* 3(2):51–52.

1942   Ornamented Coppers of the Wisconsin Area. *The Wisconsin Archeologist* 23(3):49–85.

1946   Unusual Varieties of Common Types of Indian Implements. *The Wisconsin Archeologist* 27(4):91–93.

1947a   Unusual Varieties of Common Types of Indian Implements. 3. Copper Knives of Pointed Flat Tang Type. *The Wisconsin Archeologist* 28(2):40–42.

1947b   Unusual Varieties of Common Types of Indian Implements. 5. Copper Chisels, *The Wisconsin Archeologist* 28(4):76–78.

Wintemberg, W. J.

1900   Indian Village Sites in the Counties of Oxford and Waterloo. *Archaeological Report, 1899. Being Part of Appendix to the Report of the Minister of Education, Ontario,* pp. 83–92. Warrick Bro's & Rutter, Printers, Toronto, Ontario, Canada.

1926   Foreign Aboriginal Artifacts from Post European Iroquoian Sites in Canada. *Transactions of the Royal Society of Canada,* Third Series, Vol. 20, Sec. 2, pp. 37–61.

1928   Artifacts from Ancient Graves and Mounds in Ontario. *Transactions of the Royal Society of Canada,* Third Series, Vol. 22, Sec. 2, pp. 175–202.

1931   Distinguishing Characteristics of Algonkian and Iroquoian Cultures. In *Annual Report for 1929,* pp. 65–126. Bulletin No. 67. National Museum of Canada, Ottawa, Ontario, Canada.

1935   Archaeological Evidences of Algonkian Influence on Iroquoian Culture. *Transactions of the Royal Society of Canada,* Third Series, Vol. 29, Sec. 2, pp. 231–242.

1948   The Middleport Prehistoric Village Site. Bulletin No. 109. National Museum of Canada, Ottawa, Ontario, Canada.

Winters, Howard D.

1968   Value Systems and Trade Cycles of the Late Archaic in the Midwest. In *New Perspectives in Archeology,* edited by Sally R. Binford and Lewis R. Binford, pp. 175–221. Aldine Publishing Company, Chicago.

1974   Some Unusual Grave Goods from a Mississippian Burial Mound. *Indian Notes* 10 (2):34–46.

1981   Excavating in Museums: Notes on Mississippian Hoes and Middle Woodland Copper Gouges and Celts. In *The Research Potential of Anthropological Museum Collections,* edited by Anne-Marie Cantwell, James B. Griffin, and Nan A. Rothschild, pp. 17–34. Annals of the New York Academy of Sciences Vol. 376. New York.

Wisconsin Archeologist

1956a   Copper Pendants. Photograph. 37(1):24.

1956b   Copper Harpoon; Copper Bannerstone. Photographs. 37(2):49.

1956c	Copper Crescent. Photograph. 37(4):130.
1957a	Copper Punch. Photograph. 38(1):32.
1957b	Copper Needle; Copper Knife. Photographs. 38(3):182.
1958a	Ornamented Copper Point. Photograph. 39(2):154.
1958b	Decorated Copper Knife. Photograph. 39(3):189.
1958c	Copper Necklace; Copper Harpoon. Photographs. 39(4):274.
1961	Serrate Tanged Copper Spear. Photograph. 42(2):96.
1962a	Unfinished Copper Spear Point. 43(1):26.
1962b	Old Copper Knife. 43(3):85.
1964	Old Copper Spearpoint. 45(4):184.
1968	Copper Gorget. 49(3):148.

Wittry, Warren L.
1954   Evidence of Ancient Man in Wisconsin. *Lore* 4(4):114–117.
1957   A Preliminary Study of the Old Copper Complex. *The Wisconsin Archeologist* 38(4):204–221.
1959   The Wakanda Park Mound Group, Dn1, Menomonie, Wisconsin. *The Wisconsin Archeologist* 40(3):95–115.

Wittry, Warren L. and Robert E. Ritzenthaler
1957   The Old Copper Complex: An Archaic Manifestation in Wisconsin. *The Wisconsin Archeologist* 38(4):312–329.

Wolbach, T. D.
1877   Evidences of Commercial Intercourse among the Mound-Builders. In *First Report, Containing the Constitution and an Account of the Organization of the Society, Together with "Man: His Origin in Geological Time," by Edward Brown, A. M., and Other Interesting Papers.* District Historical Society (Ohio).

Wolf, Elinor
1916   Evidences of Prehistoric Man. *The Archaeological Bulletin* 7(1):9–10.

Wood, Alvinus B.
1907   The Ancient Copper-Mines of Lake Superior. *Transactions of the American Institute of Mining Engineers* 37:288–296.

Wood, E. F.
1936   A Central Basin Manifestation in Eastern Wisconsin. *American Antiquity* 1(3):215–219.

Wood, James A. and John C. Allman
1961   The Irvin Coy Mound, Greene County, Ohio. *Ohio Archaeologist* 11(1):52–56.

Wood, James A. and Jack Berner
1969   An Unusual Mound Builder Trait. *The Redskin* 4(3):100–101.

Wood, Kathryn M.
1992   Map of West Mound. *Ohio Archaeologist* 42(3):9–15,

Wood, W. Raymond
1963   A Preliminary Report on the 1962 Excavations at the Crenshaw Site – 3MI6. In *Arkansas Archeology 1962*, edited by Charles R. McGimsey III, pp. 1–14. Arkansas Archeological Society, Central Office, University Museum, Fayetteville, Arkansas.

Bibliography

1967 The Fristoe Burial Complex of Southwestern Missouri. *The Missouri Archaeologist* 29:1–128.

Wood, W. Raymond and Sharon L. Brock
  1984 The Bolivar Burial Complex of Southwestern Missouri. *The Missouri Archaeologist* 45:1–133.
  2000 Six Late Prehistoric Burial Mounds in Southwestern Missouri. *The Missouri Archaeologist* 61:1–69.

Woods, John
  1855 Antique Copper Implements Discovered in the Valley of the Great Miami. In *Information Respecting the History, Condition and Prospects of the Indian Tribes of the United States: Collected and Prepared under the Direction of the Bureau of Indian Affairs, Department of the Interior, Per Act of Congress of March 3d, 1847. By Henry R. Schoolcraft, LL.D*, Part 5, pp. 665–666. J. B. Lippincott & Company, Philadelphia.

Woodworth, J. B.
  1897 Charles Thomas Jackson. *The American Geologist* 20(2):69–110.

Woolworth, Alan R. and Nancy L. Woolworth
  1963 Thirty Old Copper Culture Artifacts from Michigan. *The Michigan Archaeologist* 9(2):17–19.

Woolworth, Nancy L.
  1968 Some Indiana Artifacts in the Minnesota Historical Society Collection. *Central States Archaeological Journal* 15(3):109–111.

Wormington, H. M.
  1964 *Ancient Man in North America*. (Fourth Edition, Fully Revised). Popular Series No. 4. Denver Museum of Natural History, Denver, CO.

Wormington, Hanna M. and Richard G. Forbis
  1965 *An Introduction to the Archaeology of Alberta, Canada*. Proceedings of the Denver Museum of Natural History No. 11. Denver, CO.

Worthing, Ivy
  1915 Indians' Copper Implements. *The Magazine of History* 20(3–4):160–167.

Wray, Donald E.
  1937 A Red Ochre Mound in Fulton County. *Transactions of the Illinois State Academy of Science* 30(2):82.
  1952 Archeology of the Illinois Valley: 1950. In *Archeology of Eastern United States*, edited by James B. Griffin, pp. 152–164. University of Chicago Press, Chicago.

Wright, Alice P.
  2014 History, Monumentality, and Interaction in the Appalachian Summit Middle Woodland. *American Antiquity* 79(2):277–294.

Wright, Carl M.
  2000 Avocational Archaeology: Old Copper Industry Studies. *Central States Archaeological Journal* 47(4):178.

Wright, G. Frederick
    1885   The Use of Copper Implements by the American Aborigines. *Kansas City Review of Science and Industry* 8(12):705–707.

Wright, G. Frederick and Frederick B. Wright
    1906   The Mitchell Collection. *Records of the Past* 5(5):159–160.

Wright, Henry T.
    1973   *An archeological sequence in the Middle Chesapeake Region, Maryland.* Maryland Geological Survey Archeological Studies, John's Hopkins University. Baltimore: Maryland Geological Survey.

Wright, James V.
    1963   *An Archaeological Survey along the North Shore of Lake Superior.* Anthropology Papers No. 3. National Museum of Canada, Department of Northern Affairs and National Resources, Ottawa, Ontario, Canada.
    1967   The Pic River Site: A Stratified Late Woodland Site on the North Shore of lake Superior. In *Contributions to Anthropology* 5, pp. 54–99. Anthropological Series No. 72, Bulletin No. 206, National Museum of Canada, Ottawa, Ontario, Canada.
    1969   Michipicoten Site. In *Contributions to Anthropology* 6, pp. 3–85. Anthropological Series No. 82, Bulletin No. 224, National Museum of Canada, Ottawa, Ontario, Canada.
    1972a   *The Shield Archaic.* Publications in Archaeology No. 3. National Museum of Man, Ottawa, Ontario, Canada.
    1972b   *Ontario Prehistory: An Eleven-Thousand Year Archaeological Outline.* National Museums of Canada. Ottawa, Ontario, Canada.
    1974   *The Nodwell Site.* National Museum of Man Mercury Series Paper No. 22. National Museum of Canada, Ottawa, Ontario, Canada.
    1981a   The Glen Site: An Historic Cheveux Relevés Campsite on Flowerpot Island, Georgian Bay, Ontario. *Ontario Archaeology* 35:45–59.
    1981b   Prehistory of the Canadian Shield. In *Handbook of North American Indians, Volume 6: Subarctic,* edited by June Helm, pp. 86–96. Smithsonian Institution, Washington, DC.
    1995   *A History of the Native People of Canada, Volume I (10,000–1,000 B.C.).* Paper 152, Mercury Series, Archaeological Survey of Canada. Canadian Museum of Civilization, Hull, Quebec, Canada.

Wyckoff, Don G.
    1974   The Caddoan Cultural Area: An Archaeological Perspective. In *Caddoan Indians I*, pp. 1–250. Garland Publishing Inc., New York.
    1980   *Caddoan Adaptive Strategies in the Arkansas Basin, Eastern Oklahoma.* PhD dissertation, Washington State University, Pullman, Washington.
    2001   Spiro: Native American Trade a Millennium Ago. *Gilcrease Journal* 9 (Winter):48–63.

Wyman, Jeffries
    1869   Report of the Curator. *Second Annual Report of the Trustees of the Peabody Museum of American Archaeology and Ethnology,* pp. 5–20.
    1872   Report of the Curator. *Fifth Annual Report of the Trustees of the Peabody Museum of American Archaeology and Ethnology,* pp. 5–30.

Wymer, DeeAnne
    2004   Organic Preservation on Prehistoric Copper Artifacts of the Ohio Hopwewell. In *Perishable Material Culture in the Northeast,* edited by Penelope Ballard

Bibliography

        Drooker, pp. 45–68. New York State Museum Bulletin 500, University of the State of New York, State Education Department, Albany, New York.

Yarnell, Richard Asa
    1964   *Aboriginal Relationships between Culture and Plant Life in the Upper Great Lakes Region.* Anthropological Papers No. 23. Museum of Anthropology, University of Michigan, Ann Arbor, Michigan.

Young, Bennett H.
    1910   *The Prehistoric Men of Kentucky.* Filson Club Publications No. 25. John P. Morton & Company, Louisville, Kentucky.

Young, Gloria A.
    1970   Reconstruction of an Arkansas Hopewellian Panpipe. *Arkansas Academy of Science Proceedings* 24:28–32.
    1976   A Structural Analysis of Panpipe Burials. *Tennessee Archaeologist* 32(1–2):1–10.

Young, Philip D., David J. Wenner, Jr., and Elaine A. Bluhm
    1961   Two Early Burial Sites in Lake County. In *Chicago Area Archaeology*, edited by Elaine A. Bluhm, pp. 20–28. Bulletin No. 3. Illinois Archaeological Survey, University of Illinois, Urbana, Illinois.

Yu, Zicheng, John H. McAndrews and Ueli Eicher
    1997   Middle Holocene Dry Climate Caused by Change in Atmospheric Circulation Patterns: Evidence from Lake Levels and Stable Isotopes. *Geology* 25(3):251–254.

Zakucia, John A.
    1992   Artifacts from My Collection. *Ohio Archaeologist* 42(3):48.